DIGITAL SYSTEMS
Principles and Applications

Ronald J. Tocci

Monroe Community College

DIGITAL SYSTEMS
Principles and Applications

REVISED AND ENLARGED

Prentice/Hall International, Inc., London

Library of Congress Cataloging in Publication Data

Tocci, Ronald J
   Digital systems.

   Includes index.
   1. Electronic digital computers. 2. Digital
electronics. I. Title.
TK7888.3.T6 1980   621.3819′58   79-22697
ISBN 0-13-212357-6

Editorial/production supervision and interior
   design by Gary Samartino
Manufacturing buyer: Gordon Osbourne

Printed in the United States of America

10 9 8 7 6

Prentice-Hall International, Inc., *London*

Prentice-Hall of Australia Pty. Limited, *Sydney*

Prentice-Hall of Canada, Ltd., *Toronto*

Prentice-Hall of India Private Limited, *New Delhi*

Prentice-Hall of Japan, Inc., *Tokyo*

Prentice-Hall of Southeast Asia Pte. Ltd., *Singapore*

Whitehall Books Limited, *Wellington, New Zealand*

Prentice-Hall, Inc., *Englewood Cliffs, New Jersey*

To Cathy–

*for always believing me when I
tell her this is the last one.*

CONTENTS

3

LOGIC-CIRCUIT DESIGN   55

4

FLIP-FLOPS   78

7

8

12

INTRODUCTION TO DIGITAL-COMPUTER OPERATION   **429**

13

INTRODUCTION TO THE MICROPROCESSOR
AND MICROCOMPUTER   **453**

PREFACE

This text presents a comprehensive up-to-date study of the principles and techniques used in modern digital systems and is intended for use in 2-year and 4-year technology programs.

Although a basic electronics background is helpful, a major portion of the material requires no prior electronics training. Hence, the text is well-suited for students of computer science or other nonelectronic technologies concerned with digital systems.

This second edition has undergone major revision in several areas to keep pace with the rapidly changing digital field. The new features added to this edition are:

1. Expanded material on IC logic families to include Schottky TTL, I²L and SOS devices.

2. Expanded coverage of semiconductor memories with more details on static and dynamic chips, and an introduction to charge-coupled and magnetic bubble devices.

3. Expanded introduction to digital computer operation (Chapter 12) to provide a better basis for the study of microprocessors.

4. A comprehensive new chapter on microprocessors and microcomputers discusses the structure and operation of a typical microcomputer and the internal structure of a microprocessor. This chapter should serve as an effective lead-in to a more detailed study in this area.

This edition, of course, retains all of the features that made the first edition so widely accepted. These include:

1. It is up-to-date in its coverage of digital IC families (TTL, MOS, CMOS, ECL, TRI-STATE LOGIC) and in its coverage of such topics as decoding, encoding, multiplexing, demultiplexing, strobing, data busing, and microprocessors as well as the standard topics.

2. It utilizes a block diagram approach to teach the basic digital and logic operations without confusing the reader with detailed circuit diagrams. In Chapter 8, the reader is introduced to the internal IC circuitry after he has a firm understanding of logic principles. At that point, he can interpret a logic block's input and output characteristics and "fit" it properly into a complete system.

3. Emphasis on the teaching of principles is emphasized, followed by *thoroughly explained* applications of the same principles so that they can be understood and applied to new applications. Among these applications are: the frequency counter, the digital clock, the keyboard entry encoder, the control sequencer, the digital voltmeter, the digital computer, and many others.

4. Extensive use of illustrations and practical illustrative examples is made to provide a vehicle to help the reader to understand the concepts involved.

5. Each chapter is followed by numerous questions and problems of various levels of complexity that give students the opportunity to test and examine his understanding of the material. Many of the problems also present modifications to, extension of, or new applications for the material covered in the chapter, thereby offering the student a chance to expand his knowledge of the material.

6. Chapter 9 explains and shows such applications of digital operations as decoding, encoding, multiplexing, data busing, and so on. These are areas that receive superficial treatment in many competing texts.

7. Chapters 2 and 3 show step-by-step procedures for both analysis and design of combinatorial logic circuits.

8. Chapter 7 is a comprehensive study of all types of binary counters and counter applications. Included are serial, parallel, variable-MOD, UP, DOWN, UP/DOWN, presettable, and shift counters.

9. The topics are presented in a clear, straightforward manner with no unnecessary side issues. A good instructor can use this as an effective base on which to add interesting side-lines to his course.

The author is indebted to Professor Les Laskowski, who provided invaluable assistance with the new material on microprocessors.

RONALD J. TOCCI

DIGITAL SYSTEMS
Principles and Applications

1

INTRODUCTORY CONCEPTS

In science, technology, business, and, in fact, in most other fields of endeavor, we are constantly dealing with *quantities*. These quantities are measured, monitored, recorded, manipulated arithmetically, observed, or in some other way utilized in most physical systems. It is important when dealing with various quantities that we be able to represent their values efficiently and accurately.

1.1 NUMERICAL REPRESENTATIONS

There are basically two ways of representing the numerical value of quantities: analog and digital.

Analog Representations

In *analog representation* one quantity is represented by another quantity which is directly proportional to the first quantity. An example of analog representation is an automobile speedometer, in which the deflection of the needle is proportional to the speed of the auto. The angular position of the needle represents the value of the auto's speed, and the needle follows any changes that occur as the auto speeds up or slows down.

Another example is the common room thermostat, in which the bending of the bimetallic strip is proportional to the room temperature. As the temperature changes gradually, the curvature of the strip changes proportionally.

Still another example of an analog quantity is found in the familiar audio microphone. In this device an output voltage is generated in proportion to the amplitude of the sound waves that impinge on the microphone. The variations in the output voltage follow the same variations as the input sound.

Analog quantities such as those cited above have an important characteristic: *they can gradually vary over a continuous range of values*. The automobile speed can have *any* value between zero and, say, 100 mph. Similarly, the microphone output might be anywhere within a range of zero to 10 mV (i.e., 1 mV, 2.3724 mV, 9.9999 mV).

Digital Representations

In *digital representation* the quantities are not represented by proportional quantities but by symbols called *digits*. As an example, consider the digital clock, which provides the time of day in the form of decimal digits which represent hours and minutes (and sometimes seconds). As we know, the time of day continuously changes, but the digital clock reading does not change continuously; rather, it changes in steps of one per minute (or per second). In other words, this digital representation of the time of day changes in *discrete* steps, as compared to the analog representation of time provided by a wristwatch, where the dial reading changes continuously.

The major difference between analog and digital quantities, then, can be simply stated as follows:

$$analog \equiv continuous$$

$$digital \equiv discrete \ (step \ by \ step)$$

Because of the discrete nature of digital representations, there is no ambiguity when reading the value of a digital quantity, whereas the value of an analog quantity is often open to interpretation.

EXAMPLE 1.1   Which of the following involve analog quantities and which involve digital quantities?
(a)   Resistor substitution box.
(b)   Current meter.
(c)   Temperature.
(d)   Sand grains on a beach.
(e)   Radio volume control.

Solution:
(a)   Digital.
(b)   Analog.
(c)   Analog.
(d)   Digital, since the number of grains can be only certain discrete (integer) values and not any value over a continuous range.
(e)   Analog.

1.2 DIGITAL AND ANALOG SYSTEMS

A *digital system* is a combination of devices (electrical, mechanical, photoelectric, etc.) arranged to perform certain functions in which quantities are represented digitally. In an analog system the physical quantities are principally analog in nature. Many practical systems are *hybrid*, in that both digital and analog quantities are present and continual conversion between the two types of representation takes place.

Some of the more common digital systems are digital computers and calculators, digital voltmeters, and numerically controlled machinery. In these systems the electrical and mechanical quantities change only in discrete steps. Examples of *analog systems* are analog computers, radio broadcast systems, and audio tape recording. In these systems the quantities change gradually over a continuous range. For example, in AM radio broadcasting you can tune your radio to any frequency over a continuous band from 535 to 1605 kHz.

Generally speaking, digital systems offer the advantages of greater speed and accuracy and the capability of memory. In addition, digital systems are less susceptible than analog systems to fluctuations in the characteristics of the system components and are generally more versatile in a wider range of applications.

In the real world most quantities are analog in nature, and it is these quantities which are often being measured or monitored or controlled. Thus, if the advantages of digital techniques are to be utilized, it is obvious that many hybrid systems must exist. Chief among these are industrial process control systems in which analog quantities such as temperature, pressure, liquid level, and flow rate are measured and controlled. Figure 1.1 shows the symbolic block diagram of such a system. As the diagram shows, the analog quantity is measured and the measured value is then converted to a digital quantity by an *analog-to-digital (A/D) converter*. The digital quantity is then operated on by the central processor, which is completely digital and whose output is then reconverted to an analog quantity in a *digital-to-analog (D/A) converter*. This analog output is fed to a controller, which

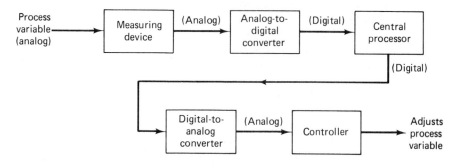

Figure 1.1 Block diagram of typical hybrid process control.

then exerts some type of effect on the process to adjust the value of the original measured analog quantity.

1.3 DIGITAL NUMBER SYSTEMS

Many number systems are in use in digital technology. The most common are the decimal, binary, octal, and hexadecimal systems. The decimal system is clearly the most familiar to us because it is a tool that we use every day. Examining some of its characteristics will help us to better understand the other systems.

Decimal System

The *decimal system* is composed of *10* numerals or symbols. These 10 symbols are 0, 1, 2, 3, 4, 5, 6, 7, 8, 9; using these symbols as *digits* of a number, we can express any quantity. The decimal system, also called the *base-10* system, because it has 10 digits, has evolved naturally as a result of the fact that man has 10 fingers. In fact, the word "digit" is the Latin word for "finger."

The decimal system is a *positional-value system* in which the value of a digit depends on its position. For example, consider the decimal number 453. We know that the digit 4 actually represents 4 *hundreds*, the 5 represents 5 *tens*, and the 3 represents 3 *units*. In essence, the 4 carries the most weight of the three digits; it is referred to as the *most significant digit (MSD)*. The 3 carries the least weight and is called the *least significant digit (LSD)*.

Consider another example, 27.35. This number is actually equal to two tens plus seven units plus three tenths plus five hundredths, or $2 \times 10 + 7 \times 1 + 3 \times 0.1 + 5 \times 0.01$. The decimal point is used to separate the integer and fractional parts of the number.

More rigorously, the various positions relative to the decimal point carry weights that can be expressed as powers of 10. This is illustrated in Figure 1.2,

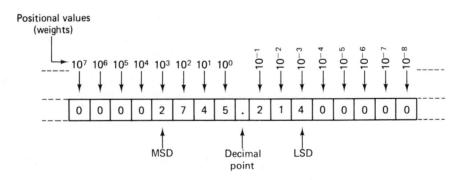

Figure 1.2   Decimal position values as powers of 10.

where the number 2745.214 is represented. The decimal point separates the positive powers of 10 from the negative powers. The number 2745.214 is thus equal to

$$(2 \times 10^{+3}) + (7 \times 10^{+2}) + (4 \times 10^1) + (5 \times 10^0)$$
$$+ (2 \times 10^{-1}) + (1 \times 10^{-2}) + (4 \times 10^{-3})$$

In general, any number is simply the sum of the products of each digit value times its positional value.

Decimal Counting

When counting in the decimal system, we start with 0 in the units position and take each symbol (digit) in progression until we reach 9. Then we add a 1 to the next higher position and start over with zero in the first position (see Figure 1.3). This process continues until the count of 99 is reached. Then we add a 1 to the third position and start over with zeros in the first two positions. The same pattern is followed continuously as high as we wish to count.

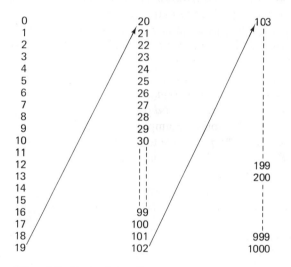

Figure 1.3   Decimal counting.

It is important to note that in decimal counting the units position (LSD) changes upward with each step in the count, the tens position changes upward every 10 steps in the count, the hundreds position changes upward every 100 steps in the count, and so on.

Another characteristic of the decimal system is that using only two decimal places we can count through $10^2 = 100$ different numbers (0 to 99).* With three

*Zero is counted as a number.

places we can count through 1000 numbers (0 to 999); and so on. In general, with N places or digits we can count through 10^N different numbers, starting with and including zero. The largest number will always be $10^N - 1$.

Binary System

Unfortunately, the decimal number system does not lend itself to convenient implementation in digital systems. For example, it is very difficult to design electronic equipment so that it can work with 10 different voltage levels (each one representing one decimal character, 0–9). On the other hand, it is very easy to design simple, accurate electronic circuits that operate with only two voltage levels. For this reason, almost all digital systems use the binary number system (base 2) as the basic number system of its operations, although other systems are often used in conjunction with binary.

In the *binary system* there are only two symbols or possible digit values, 0 and 1. Even so, this base-2 system can be used to represent any quantity that can be represented in decimal or other number systems. In general though, it will take a greater number of binary digits to express a given quantity.

All the statements made earlier concerning the decimal system are equally applicable to the binary system. The binary system is also a positional-value system, wherein each binary digit has its own value or weight expressed as powers of 2. This is illustrated in Figure 1.4. Here, places to the left of the *binary point* (counterpart of the decimal point) are positive powers of 2 and places to the right are negative powers of 2. The number 1011.101 is shown represented in the figure. To find its equivalent in the decimal system we simply take the sum of the products of each digit value (0 or 1) times its positional value.

$$1011.101_2 = (1 \times 2^3) + (0 \times 2^2) + (1 \times 2^1) + (1 \times 2^0)$$
$$+ (1 \times 2^{-1}) + (0 \times 2^{-2}) + (1 \times 2^{-3})$$
$$= 8 + 0 + 2 + 1 + 0.5 + 0 + 0.125$$
$$= 11.625_{10}$$

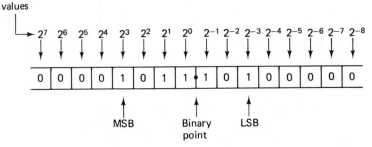

Figure 1.4 Binary position values as powers of 2.

Notice in the preceding operation that subscripts (2 and 10) were used to indicate the base in which the particular number is expressed. This convention is used to avoid confusion whenever more than one number system is being employed.

In the binary system, the term *binary digit* is often abbreviated to the term *bit* which we will use henceforth. Thus, in the number expressed in Figure 1.4 there are 4 bits to the left of the binary point, representing the integer part of the number, and 3 bits to the right of the binary point, representing the fractional part. The most significant bit (MSB) is the leftmost bit (largest weight). The least significant bit (LSB) is the rightmost bit (smallest weight). These are indicated in Figure 1.4.

Appendix I contains a table of values of powers of 2, both positive and negative, which can be utilized when converting between binary and decimal. More will be said later about such conversions.

Binary Counting

When counting in the binary system, we start with 0 in the first position (units position) and then proceed to 1 in the first position. Then we add a 1 to the next higher position (2^1, or two's position) and start over with zero in the first position. This is illustrated in Figure 1.5. When the count of 11_2 is reached, we add a 1 to the third position and start over with zeros in the first two positions. The process is followed continuously.

The binary counting sequence has an important characteristic, as shown in Figure 1.5. The units bit (LSB) changes either f om 0 to 1 or 1 to 0 with *each* count. The second bit (two's position) stays at 0 for two counts, then 1 for two

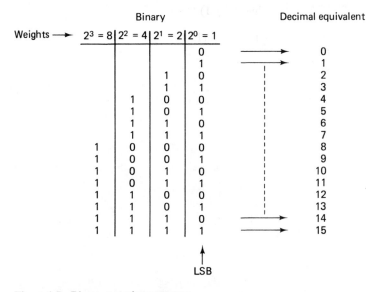

Figure 1.5 Binary counting sequence.

counts, then 0 for two counts, and so on, The third bit (four's position) stays 0 for four counts, then 1 for four counts; and so on. The fourth bit (eight's position) stays at 0 for eight counts, then at 1 for eight counts. If we wanted to count further we would add more places, and this pattern would continue with 0s and 1s alternating in groups of 2^{N-1}. For example, using a fifth binary place, the fifth bit would alternate sixteen 0s, then sixteen 1s, and so on.

As we saw for the decimal system, it is also true of the binary system that using N bits or places we can go through 2^N counts. For example, with 2 bits we can go through $2^2 = 4$ counts (00_2 through 11_2); with 4 bits we can go through $2^4 = 16$ counts (0000_2 through 1111_2); and so on. The last count will always be all 1s and is equal to $2^N - 1$ in the decimal system. For example, using 4 bits, the last count is $1111_2 = 2^4 - 1 = 15_{10}$.

EXAMPLE 1.2   What is the largest number that can be represented using 8 bits?

Solution: $2^N - 1 = 2^8 - 1 = 255_{10} = 11111111_2$.

1.4 REPRESENTING BINARY QUANTITIES

In digital systems the information that is being processed is usually present in binary form. Binary quantities can be represented by any device that has only two operating states or possible conditions. For example, a switch has only two states: open or closed. We can arbitrarily let an open switch represent binary 0 and a closed switch represent binary 1. With this assignment we can now represent any binary number as illustrated in Figure 1.6(a), where the states of the various switches represent 10010_2.

Another example is shown in Figure 1.6(b), where holes punched in paper are used to represent binary numbers. A punched hole is a binary 1 and absence of a hole is a binary 0.

There are numerous other devices which have only two operating states or can be operated in two extreme conditions. Among these are: light bulb (bright or dark), diode (conducting or nonconducting), relay (energized or deenergized),

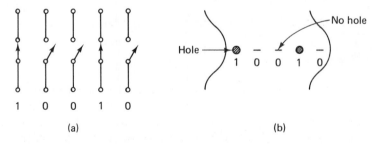

Figure 1.6   (a) Using switches and (b) punched paper to represent binary numbers.

transistor (cutoff or saturated), photocell (illuminated or dark), thermostat (open or closed), mechanical clutch (engaged or disengaged), and magnetic tape (magnetized or demagnetized).

In electronic digital systems, binary information is represented by voltages (or currents) that are present at the inputs and outputs of the various circuits. Typically, the binary 0 and 1 are represented by two nominal voltage levels. For example, zero volts (0 V) might represent binary 0, and +5 V might represent binary 1. In actuality, because of circuit variations, the 0 and 1 would be represented by voltage ranges. This is illustrated in Figure 1.7(a), where any voltage between 0 and 0.8 V represents a 0 and any voltage between 2 and 5 V represents a 1. All input and output signals will fall in either of these ranges. Figure 1.7(b) shows a typical digital waveform as it sequences through the binary value 01010.

We can now see another significant difference between digital and analog systems. In digital systems, the exact value of a voltage *is not* important; for example, a voltage of 3.6 V means the same as a voltage of 4.3 V. In analog systems, the exact value of a voltage *is* important. For instance, if the analog voltage is proportional to the temperature measured by a transducer, 3.6 V would represent a different temperature than would 4.3 V. In other words, the voltage value carries significant information. This characteristic means that the design of accurate analog circuitry is generally more difficult than the design of digital circuits.

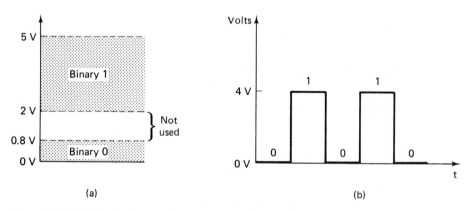

Figure 1.7   (a) Typical voltage assignments in digital system; (b) typical digital signal.

Parallel and Serial Transmission

It is generally necessary to transmit signals representing complete binary numbers consisting of several bits from one part of a system to another. There are basically two ways of doing this: parallel transmission and serial transmission.

In *parallel representation* or *transmission* of binary numbers, each bit is derived from a separate circuit output and transmitted over a separate line. Figure 1.8(a) illustrates the arrangement for a 5-bit number. Each circuit output represents one

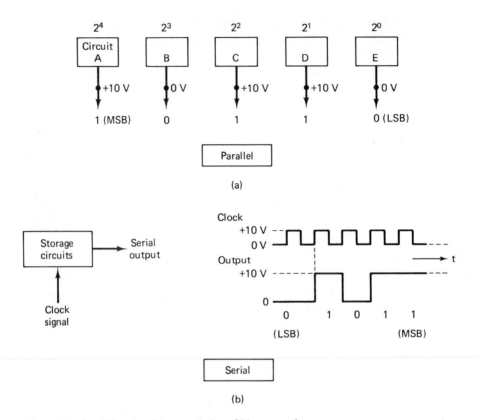

Figure 1.8 Parallel and serial transmission of binary numbers.

binary digit (bit) and can be either 0 V (binary 0) or 10 V (binary 1). The five circuit outputs are present simultaneously, so at any time the complete binary number is available at the outputs. In Figure 1.8(a) the binary number 10110 is represented.

In *serial transmission,* which is illustrated in Figure 1.8(b), only *one* signal output line is used to transmit the binary number. This output line transmits the various bits one at a time in sequence, generally starting with the least significant bit. It should be apparent that some sort of timing is needed to be able to distinguish among the various bits. In the figure a *clock signal* is used to provide the sequencing. A clock signal is an important part of most digital systems and is used to provide accurate timing of operations. In serial transmission, each time a clock pulse occurs, the output signal changes to the next bit of the binary number. In Figure 1.8(b) the output sequences through the bits 01011 (note that time increases from left to right). The actual binary number being transmitted is 11010.

The principal trade-off between serial and parallel transmission is one of speed versus simplicity. In general, when binary numbers are represented in parallel in a digital system the operation is much faster, because all the bits are present simultaneously—as opposed to serial representation, wherein the bits are present one at a time. This is particularly apparent for larger binary numbers, such as 36 bits, which are used in some computers. On the other hand, parallel representation generally requires more complex circuitry. Thus, as we shall see later, parallel digital systems are faster and more complex than serial systems, which are slower but simpler. In many digital systems, both types of representation are used in different parts of the system. In such cases there must be a method of converting between the parallel and serial representations. This conversion process will be discussed later.

1.5 PULSE CHARACTERISTICS

The binary information in a digital system is continually changing, often at a very fast rate. Thus, it becomes important to become familiar with certain characteristics of fast-changing digital signals. Some of these characteristics can be observed by looking at a single digital pulse, such as the one in Figure 1.9. The single pulse represents a digital signal that is initially at 0 V (binary 0) until a time when it makes a rapid jump to $+5$ V (binary 1), where it stays until a later time when it rapidly returns to 0 V.

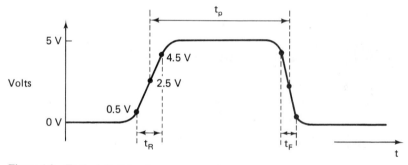

Figure 1.9 Typical digital pulse.

Rise Time t_R and Fall Time t_F

The pulse's *rise time* t_R is defined in Figure 1.9 as the time it takes for the signal to rise from its 10 per cent to its 90 per cent points. *Fall time*, t_F, is similarly defined for the negative-going portion of the waveform. In digital systems t_R and t_F are typically in the microsecond or nanosecond range. There are many digital circuits that will operate reliably only if the input waveforms have rise and fall times below a certain *maximum* value. As we shall see later, there are several methods for speeding up a pulse waveform (shortening its t_R and t_F).

Pulse Duration

The *pulse duration* t_p is the time between the 50 per cent points on the rising and falling edges. Values of t_p can occur over a very wide range, from nanoseconds up to seconds.

Pulse-Repetition Frequency

In many applications a repetitive train of digital pulses is used, such as the clock signal shown in Figure 1.8(b). The number of pulses that occur per second is called the *pulse-repetition frequency* (PRF) and is expressed in pulses per second (pps). This is the same as frequency (hertz), and both terms are used in pulse work. Figure 1.10 shows a periodic pulse train. The length of time between the rising (or falling) edges of successive pulses is called the period (T) and represents the length of one cycle of the waveform. The pulse-repetition frequency or rate can be calculated from

$$\text{PRF} = \frac{1}{T} \quad \text{or} \quad f = \frac{1}{T} \tag{1.1}$$

where PRF is in pps and f is in Hz when T is expressed in seconds.

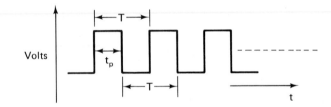

Figure 1.10   Repetitive pulse waveform.

EXAMPLE 1.3   Calculate the pulse-repetition frequency if T is 2 μs.

Solution:

$$\text{PRF} = \frac{1}{2\ \mu s} = \frac{1}{2 \times 10^{-6}} = \frac{10^6}{2} = 500,000 \text{ pps}$$

Duty Cycle

The *duty cycle* of a repetitive pulse train is defined as the ratio of pulse width t_p to period T in per cent. Stated mathematically,

$$\text{duty cycle} = \frac{t_p}{T} \times 100\% \tag{1.2}$$

For example, if a pulse waveform has a period of 4 μs and the width of one pulse is 1 μs, then the duty cycle is 25 per cent.

The special case of a 50 per cent duty cycle occurs quite often in pulse work. Here the pulse train is entirely symmetrical since it is low for half a period and high for the other half. This type of pulse waveform is often referred to as a *square-wave*.

Propagation Delay

A very important consideration in any digital system is the amount of delay that a circuit exhibits between its input and output. In other words, when the input to a circuit changes, there is always some delay before the circuit output begins to change in response to the input. This delay is referred to as *propagation delay*, t_{pd}, and is illustrated in Figure 1.11.

Propagation delays can become very significant in high-speed digital systems, in which signals are supposed to be present at certain points at certain times. Accumulated delays can often produce serious malfunctions. For this reason, systems designers must always be aware of the t_{pd} values for the individual circuits.

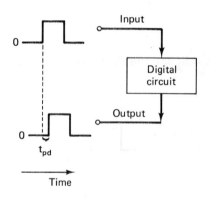

Figure 1.11   Propagation delay.

1.6   MEMORY

When an input signal is applied to most devices or circuits, the output somehow changes in response to the input, and when the input signal is removed, the output returns to its original state. These circuits do not exhibit the property of *memory*, since their outputs revert back to normal. In digital circuitry there are certain types of devices and circuits that do have memory. When an input is applied to such a circuit, the output will change its state, but it will remain in the new state even after the input is removed. This property of retaining its response to a momentary input is called memory. Figure 1.12 illustrates nonmemory and memory operation.

Memory devices and circuits play an important role in digital systems because they provide means for storing binary numbers either temporarily or permanently,

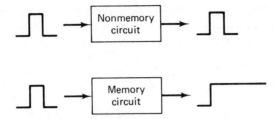

Figure 1.12   Comparison of nonmemory and memory operation.

with the ability to change the stored information at any time. As we shall see, the various memory elements include magnetic types and those which utilize electronic latching circuits (called *flip-flops*).

1.7   DIGITAL COMPUTERS

Digital techniques have found their way into innumerable areas of technology, but the area of automatic *digital computers* is by far the most notable and most extensive. Although digital computers affect some part of all of our lives, it is doubtful that many of us know exactly what a computer does. In simplest terms, *a computer is a system of hardware that performs arithmetic operations, manipulates data (usually in binary form), and makes decisions.*

For the most part, human beings can do whatever computers can do, but computers can do it with much greater speed and accuracy. This is in spite of the fact that computers perform all their calculations and operations one step at a time. For example, a human being can take a list of 10 numbers and find their sum all in one operation by listing the numbers one over the other and adding them column by column. A computer, on the other hand, can add numbers only two at a time so that adding this same list of numbers will take nine actual addition steps. Of course, the fact that the computer requires only a microsecond or less per step makes up for this apparent inefficiency.

A computer is faster and more accurate than people are, but unlike most people it has to be given a complete set of instructions that tell it *exactly* what to do at each step of its operation. This set of instructions is called a *program*, and is prepared by one or more persons for each job the computer is to do. These programs are placed in the computer's memory unit in binary-coded form, with each instruction having a unique code. The computer takes these instruction codes from memory *one at a time* and performs the operation called for by the code. Much more will be said about this later.

It would be impossible to list all of the applications that use computers today and will use computers in the future. Instead, we will look at three examples from the areas of business, industrial control, and scientific-engineering problem solving. These examples will not give a complete picture of how the computer performs in a given application, nor will they encompass all the areas in which computers are used today. Rather, they are intended to show why computers are so extremely useful.

Business Applications

A typical business task is the processing of paychecks for employees who work at an hourly rate. In a noncomputerized payroll system servicing a large company (i.e., 1000 employees) a substantial amount of paperwork and energy is expended in meeting the pay schedule. Calculations must be made concerning such items as regular and overtime hours worked, wage rate, gross earnings, federal and state taxes, social security, union dues, insurance deductions, and net earnings. Summaries must also be made on items such as earnings to date, taxes to date, social security accumulation, vacation, and sick leave. Each of the calculations would have to be performed manually (calculator) for each employee—consuming a large amount of time and energy.

By contrast, a computer payroll system requires much less time and manpower, is less prone to error, and provides a better record-keeping system. In such a system a master file is generated for each employee, containing all the necessary data (e.g., wage rate, tax exemptions, etc.). The master files of all employees, representing a massive amount of data, is stored on magnetic tape and is updated each payroll cycle. The computer is programmed to read the data from the master file tape, combine it with data for the pay period in question, and make all the necessary calculations. Some of the advantages of the computerized system are:

1. Payroll calculation time is shortened.
2. Accumulated earnings can be generated automatically.
3. Accurate payroll documents are prepared automatically.
4. Clerical effort is greatly reduced.
5. One master employee file is used to prepare all reports.
6. Data are readily available for budgeting control and cost analysis.

Science and Engineering Problems

Scientists and engineers use mathematics as a language that defines the operation of physical systems. In many cases, the mathematical relationships are extremely complex and must be evaluated for many different values of the system variables. A computer can evaluate these complex mathematical expressions at high speeds. In addition, it can perform repeated calculations using different sets of data, tabulate the results, and determine which sets of values produced the best results. In many cases, the computer can save an engineer hours, and even days, of tedious calculations, thereby providing more time for creative work.

Process Control Applications

Time is not a critical factor when a computer is used to process business data or do engineering calculations in the sense that the results are not required immediately (i.e., within a few milliseconds or seconds). Computers are often used in applications where the results of their calculations are required immediately to be used in controlling a process. These are called *real-time* applications.

An example of a real-time application can be found in industrial process control, which is used in industries such as paper mills, oil refineries, chemical plants, and many others. A non-computer-controlled system might involve one or more technicians located in a central control room. Instruments are used to monitor the various process parameters and send signals back to the control console, which displays pertinent readings on analog or digital meters. The technician records these readings, interprets them, and decides on what corrective actions, if any, are necessary. Corrections are made using various controls on the control console.

If the same system were under computer control, the measuring instruments would send their signals to the computer, which processes them and responds with appropriate control signals to be sent back to the process. Some advantages of the computer-controlled system are:

1. It reacts much faster and with greater precision to variations in the measured process parameters. Thus, serious conditions in the process can be detected and corrected much more quickly than can be done with human intervention.

2. Certain economic savings can be realized because the computer can be programmed to keep the process operating at maximum efficiency in the use of materials and energy.

3. Continuous records can be kept of the process parameter variations, thereby enabling certain trends to be detected prior to the occurrence of an actual problem. These records can be kept in much less space, since magnetic tape would replace a bulky paper-filing system.

These examples of what computers can do are really not indicative of the recent trends in computer usage that have been brought about by microprocessors and microcomputers. These developments have brought computers into many commercial products, such as automobiles, cash registers, traffic controllers, and copy machines, as well as into the home.

Major Parts of a Computer

There are several types of computer systems, but each can be broken down into the same functional units. Each unit performs specific functions, and all units function together to carry out the instructions given in the program. Figure 1.13

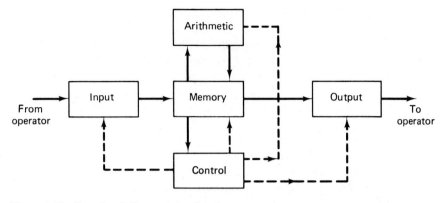

Figure 1.13   Functional diagram of a digital computer.

shows the five major functional parts of a digital computer and their interaction. The solid lines with arrows represent the flow of information. The dashed lines with arrows represent the flow of timing and control signals.

The major functions of each unit are:

1. *Input unit:* A complete set of instructions and data are fed into the computer system through this unit and into the memory unit, to be stored until needed. The information can come into the input unit by means of punched cards, magnetic tape, magnetic disc, and switches on the operator's console.

2. *Memory unit:* The memory stores the instructions and data received from the input unit. It stores the results of arithmetic operations received from the arithmetic unit. It also supplies information to the output unit.

3. *Control unit:* This unit takes instructions from the memory unit one at a time and interprets them. It then sends appropriate signals to all the other units to cause the specific instruction to be executed.

4. *Arithmetic unit:* All arithmetic calculations and decisions are performed in this unit, which then sends all results to the memory unit to be stored.

5. *Output unit:* This unit takes data from the memory unit and prints out, displays, or otherwise presents the information to the operator (or process, in the case of a process control computer).

How Many Types of Computers Are There?

The answer depends on the criteria used to classify them. Computers are often classified according to physical size, which usually, but not always, is also indicative of their relative capabilities. The *microcomputer* is the smallest and newest member of the computer family. It generally consists of several IC chips, including a *microprocessor* chip, *memory* chips, and *input/output interface* chips.* These chips are a result of tremendous advances in large-scale integration (LSI), which have also served to bring prices down so that it is possible to buy a fully assembled microcomputer now for less than $200.

Minicomputers are larger than microcomputers and have prices that go into the tens of thousands of dollars (including input/output peripheral equipment). Minis are widely used in industrial control systems, scientific applications for schools and research laboratories, and in business applications for small businesses. Although more expensive than microcomputers, minicomputers continue to be widely used because they are generally faster and possess more capabilities. These differences in speed and capability, however, are rapidly shrinking.

The largest computers are those found in large corporations, banks, universities, and scientific laboratories. These "maxicomputers" can cost as much as several million dollars and include complete systems of peripheral equipment such as magnetic tape units, magnetic disc units, punched card punchers and readers, keyboards, printers, and many more. Applications of maxicomputers range from computationally oriented science and engineering problem solving to the data-oriented business applications, where emphasis is on maintaining and updating large quantities of data and information.

1.8 FINAL COMMENTS

In this book we shall be dealing with devices and circuits that are used in the processing of information. With the advent of integrated-circuit technology, it has become obvious that the processing of information by digital techniques is generally more efficient and reliable than are analog methods. Digital methods offer the following features, which are advantageous in the processing of measurements or other data:

1. High speed of operation; for example, a simple addition operation can normally be performed in much less than 1 μs.

2. Large quantities of information can be stored in digital memory systems for indefinitely long periods of time.

3. The accuracy of the quantity representation is not generally affected by the stability of basic circuits or by normal amounts of noise. This is because of the nature of the binary system, which requires the circuits to operate in either of two states, where each state is defined over a fairly broad range. To illustrate, refer to Figure 1.14, where the quantity 4.25 is represented both by an analog circuit output and a digital (binary) output.

---

*Complete microcomputers are available on a single LSI chip with limited memory and input/output capacity.

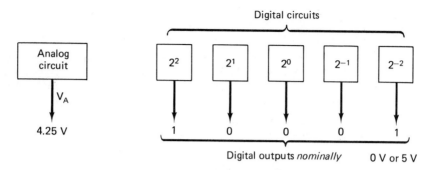

Figure 1.14 Analog and digital representation of 4.25.

The analog output represents the quantity by a voltage that is *ideally* equal to 4.25 V. However, the accuracy of this voltage is affected by stability and accuracy of the circuit elements and also by noise. The digital representation is simply the binary equivalent of 4.25, which is 100.01_2 and is present at the outputs of the various digital circuits. Each high digital output need only be within a specified range, such as 2.0 V to 5 V, to be considered a binary 1. Similarly, each low digital output need only by within a specified range, such as 0 V to 0.8 V, to be considered a binary 0. As such, the digital outputs, and therefore the whole quantity, are not as dependent on the accuracy of circuit elements or on the presence of small amounts of noise.

Furthermore, the number of significant places in the quantity being represented can easily be increased in the digital method simply by adding more circuits to represent the additional bits. On the other hand, analog methods are usually limited to about three or four significant places.

In this book we shall encounter many digital circuits which will be used to implement arithmetic operations such as those performed in computers and calculators. These circuits will often be referred to as *logic circuits*, because they will also be used to perform various decisions based on certain logic. Such logic circuits are also used in computers as well as in many other control system and data processing applications. We shall examine many of these applications in the course of this book.

QUESTIONS AND PROBLEMS

1.1 Which of the following involve analog quantities and which are digital?
  (a) Pressure gauge.
  (b) Atoms in a sample of material.
  (c) Radio tuning dial.
  (d) Aircraft altitude.
  (e) TV VHF channel selector.

1.2 Convert the following binary numbers to their equivalent decimal values:
(a) 11001
(b) $1001.1001_2 = $ _____ $_{10}$
(c) $10011011001.10110 = $ _____ $_{10}$

1.3 List the binary numbers in sequence from 0 to 128.

1.4 How high can we count in binary using 10 places?

1.5 How many bits are needed to count in binary up to 290?

1.6 Explain the major differences between serial and parallel transmission of binary data.

1.7 Consider the waveform in Figure 1.15. From the waveform diagram, determine:
(a) t_R (b) t_F (c) t_p

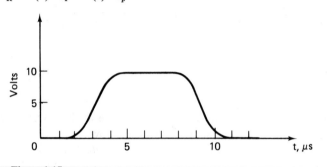

Figure 1.15

1.8 Sketch a binary waveform with the following characteristics: PRF = 1 kHz; 25 per cent duty cycle; −1 V binary 0 level; +6 V binary 1 level.

1.9 Explain the difference between memory and nonmemory circuits.

1.10 Define a digital computer in terms of what it does.

1.11 What are the five major parts of a computer and their functions?

1.12 Name the three types of computers and briefly describe the differences among them.

1.13 Summarize the major advantages of digital systems over analog systems.

LOGIC-CIRCUIT ANALYSIS
USING BOOLEAN ALGEBRA

Digital devices and digital circuits operate in the binary number system; that is, all the circuit variables are either 0 or 1 (low or high). This characteristic of digital devices makes it possible to use *Boolean algebra* as a mathematical tool for the analysis and design of digital circuits and systems. As we shall see, Boolean algebra differs significantly from the conventional algebra that we learned in high school and is generally easier to use and understand. The usefulness of this mathematical tool is universal in the digital field since it permits the design of a digital circuit or instrument to be performed in a logical manner. In addition, Boolean algebra allows an engineer or technician to easily follow the operation of someone else's design.

2.1 BOOLEAN CONSTANTS AND VARIABLES

A major difference between Boolean algebra and ordinary algebra is that in Boolean algebra the constants and variables are allowed to have only two possible values, 0 or 1. A Boolean variable is a quantity that may, at different times, be equal to either 0 or 1. The Boolean variables are often used to represent the voltage level present on a wire or at the input/output terminals of a circuit. For example, in a certain digital system the Boolean value of 0 might be assigned to any voltage in the range from 0 to 0.8 V while the Boolean value of 1 might be assigned to any voltage in the range 2 to 5 V.*

Thus, Boolean 0 and 1 do not represent actual numbers but instead represent

*Voltages between 0.8 and 2 V are undefined (neither 0 or 1) and under normal circumstances should not occur.

the state of a voltage variable or what is called its *logic level*. A voltage in a digital circuit is said to be at the logic level 0 or the logic level 1, depending on its actual numerical value. In the digital logic field several other terms are used synonomously with 0 and 1. Some of the more common ones are shown in Table 2.1.

Table 2.1

Logic 0	Logic 1
False	True
Off	On
Low	High
No	Yes
Open switch	Closed switch

Boolean algebra is used to express the effects that various digital circuits have on logic inputs and to manipulate logic variables for the purpose of determining the best method for performing a given circuit function. In all our work to follow we shall use letter symbols to represent logic variables. For example, A might represent a certain digital circuit output, and at any time we must have either $A = 0$ or $A = 1$: if not one, then the other.

The fact that only two values are possible makes Boolean algebra relatively easy to work with as compared to ordinary algebra. In Boolean algebra there are no fractions, decimals, negative numbers, square roots, cube roots, logarithms, imaginary numbers, and so on. In fact, in Boolean algebra there are only three basic operations.

2.2 BOOLEAN LOGIC OPERATIONS

The basic logic operations are:

1. *Logical addition*, which is also called *OR addition* or simply the *OR operation*. The common symbol for this operation is the plus sign ($+$).

2. *Logical multiplication*, which is also called *AND multiplication* or simply the *AND operation*. The common symbol for this operation is the multiplication sign ($\cdot$).

3. *Logical complementation or inversion*, which is also called the *NOT operation*. The common symbol for this operation is the overbar ($\overline{}$).

2.3 THE OR OPERATION

Let A and B represent two independent logic variables. When A and B are combined using OR addition, the result, x, can be expressed as

$$x = A + B$$

In this expression the $+$ sign does not stand for ordinary addition; it stands for OR addition, whose rules are given in the table shown in Figure 2.1(a).

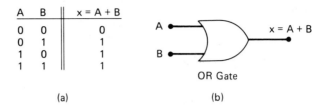

A	B	x = A + B
0	0	0
0	1	1
1	0	1
1	1	1

(a)

OR Gate

(b)

Figure 2.1   (a) Truth table defining the OR operation; (b) circuit symbol for a two-input OR gate.

It should be apparent from the table that except for the case where $A = B = 1$, the OR operation is the same as ordinary addition. However, for $A = B = 1$ the OR sum is 1 (not 2 as in ordinary addition). This is easy to remember if we recall that only 0 and 1 are possible values in Boolean algebra, so that the largest value we can get is 1. This same result is true if we have $x = A + B + C$, for the case where $A = B = C = 1$. That is,

$$x = 1 + 1 + 1 = 1$$

We can therefore say that in OR addition the result will be 1 if any one *or* more variables is a 1. This is also visible in the table in Figure 2.1(a). Incidentally, this table is called a *truth table*, which is a term borrowed from mathematical logic, where *true* and *false* are used instead of 1 and 0.

The expression $x = A + B$ can either be read as "x equals *A plus B*" or as "x equals *A OR B*." Both expressions are in common use. The key thing to remember is that the $+$ sign stands for OR addition, as defined by the truth table in Figure 2.1(a), and not for ordinary addition.

The OR Gate

In digital circuitry an *OR gate* is a circuit that has two or more inputs and whose output is equal to the OR sum of the inputs. Figure 2.1(b) shows the symbol for a two-input OR gate. The inputs A and B are logic voltage levels and the output x is a logic voltage level whose value is the result of the OR addition of A and B; that is, $x = A + B$. In other words, the OR gate operates such that its output is high (logic 1) if either input A or B or both are at a logic 1 level. The OR gate output will be low (logic 0) only if all its inputs are at logic 0.

This same idea can be extended to OR gates with more than two inputs. For example, Figure 2.2 shows a three-input OR gate and its accompanying truth table.* Note that there are eight cases shown in the table, whereas there were four

*Problem 2.1 at the end of the chapter shows how to construct a truth table so that all possible input cases are accounted for.

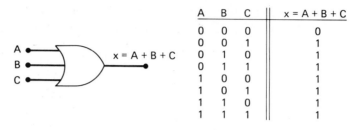

A	B	C	$x = A + B + C$
0	0	0	0
0	0	1	1
0	1	0	1
0	1	1	1
1	0	0	1
1	0	1	1
1	1	0	1
1	1	1	1

Figure 2.2   Symbol and truth table for a three-input OR gate.

cases for the two-input gate. In general, if N is the number of input variables, there will be 2^N possible cases, since each variable can take on either of two values. So for four inputs there will be $2^4 = 16$ possible cases to consider, and so on.

Examination of the truth table of Figure 2.2 shows again that the output is 1 for every case where one or more inputs is 1. This general principle of OR-gate operation is the same for any number of inputs.

Using the language of Boolean algebra, the output x can be expressed as $x = A + B + C$, where again it must be emphasized that the $+$ means OR addition. The output of any OR gate, then, can be expressed as the OR sum of its various inputs. We will put this to use when we subsequently analyze logic circuits.

Summary of the OR Operation

The important points to remember concerning the OR operation and OR gates are:

1. The OR operation produces a result of 1 when *any* of the input variables is 1.

2. The OR operation produces a result of 0 *only* when all the input variables are 0.

3. In OR addition, $1 + 1 = 1, 1 + 1 + 1 = 1$, and so on.

EXAMPLE 2.1   In many industrial control systems it is required to activate an output function whenever any one of several inputs is activated. For example, in a chemical process it may be desired that an alarm be activated whenever the process temperature exceeds a maximum value *or* whenever the pressure goes above a certain limit. Figure 2.3 is a block diagram of this situation. The temperature transducer circuit produces an output voltage proportional to the process temperature. This voltage, V_T, is compared to a temperature reference voltage, V_{TR}, in a voltage comparator circuit. The comparator output is normally a low voltage (logical 0), but it switches to a high voltage (logical 1) whenever V_T exceeds V_{TR}, indicating that process temperature is excessive. A similar arrangement is used for the pressure measurement, so its associated comparator output goes from low to high when the pressure is excessive.

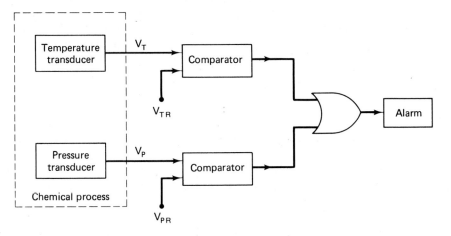

Figure 2.3 Example of the use of OR gate in alarm system.

Since we want the alarm to be activated when either temperature *or* pressure is too high, it should be apparent that the two comparator outputs can be fed to a two-input OR gate. The OR gate output thus goes high (1) for either alarm condition and will activate the alarm. This same idea can be obviously extended to situations with more than two process variables.

EXAMPLE 2.2 For the situation depicted in Figure 2.4, determine the waveform at the OR-gate output.

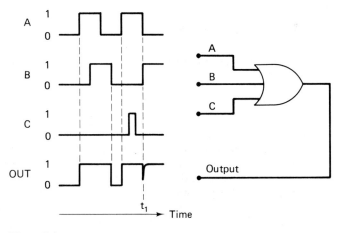

Figure 2.4

Solution: The three OR-gate inputs *A*, *B*, and *C* are varying, as shown by their waveform diagrams. The OR-gate output is determined by realizing that it will be high whenever *any* of the three inputs is at a high level. Using this reasoning, the OR-output waveform is as shown in the figure. Particular attention should be paid to what occurs at time t_1. The diagram shows that at that instant of time, input A is going from high to low while input B is going from low to high. Since these inputs are making their transitions at approximately the same time and since these transitions take a certain amount of time, there is a short interval of time when these OR-gate inputs are both in the undefined range between 0 and 1. When this occurs, the OR-gate output also becomes a value in this range, as evidenced by the glitch or spike on the output waveform at t_1. The occurrence of this glitch and its size (amplitude and width) depends on the speed with which the input transitions occur. It should be noted that if the C input had been sitting high while A and B were changing, the glitch would not occur because the high at C would keep the OR output high.

2.4 THE AND OPERATION

If two logic variables A and B are combined using AND multiplication, the result, x, can be expressed as

$$x = A \cdot B$$

In this expression the $\cdot$ sign stands for the Boolean operation of AND multiplication, whose rules are given in the truth table shown in Figure 2.5(a).

It should be apparent from the table that AND multiplication is *exactly* the same as ordinary multiplication. Whenever A or B are 0, their product is zero; when both A *and* B are 1, their product is 1. We can therefore say that in the AND operation the result will be 1 *only if* all the inputs are 1; for all other cases the result is 0.

The expression $x = A \cdot B$ is read "x equals A *and* B." The multiplication sign is sometimes omitted as in ordinary algebra, so the expression becomes $x = AB$. The key thing to remember is that the AND operation is the same as ordinary multiplication, where the variables can be either 0 or 1.

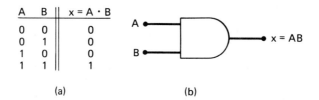

(a)   (b)

Figure 2.5   (a) Truth table for the AND operation; (b) AND gate symbol.

The AND Gate

A two-input AND gate is shown symbolically in Figure 2.5(b). The AND-gate output is equal to the AND product of the logic inputs; that is, $x = AB$. In other words, the AND gate is a circuit that operates such that its output is high only when all its inputs are high. For all other cases the AND gate output is low.

This same operation is characteristic of AND gates with more than two inputs. For example, a three-input AND gate and its accompanying truth table are shown in Figure 2.6. Once again, note that the gate output is 1 only for the case where $A = B = C = 1$. The expression for the output is $x = ABC$. For a four-input AND gate, the output is $x = ABCD$, and so on.

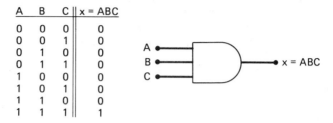

A	B	C	x = ABC
0	0	0	0
0	0	1	0
0	1	0	0
0	1	1	0
1	0	0	0
1	0	1	0
1	1	0	0
1	1	1	1

Figure 2.6   Truth table and symbol for a three-input AND gate.

Summary of the AND Operation

1. The AND operation is performed exactly like ordinary multiplication of 1s and 0s.

2. An output equal to 1 occurs only for the single case where all inputs are 1.

3. The output is 0 for any case where one or more inputs are 0.

EXAMPLE 2.3   Determine the output waveform for the AND gate shown in Figure 2.7.

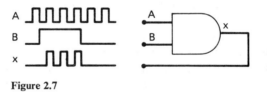

Figure 2.7

Solution: The output x will be at 1 only when A and B are both high at the same time. Using this fact, the x waveform can be determined as shown in the figure.

Notice that the x waveform is 0 whenever B is 0, regardless of the signal at A. Also notice that the x waveform is the same as A whenever B is 1. Thus, we can think of the B input as a *control* input whose logic level determines whether the A waveform gets through to the x output or not. In this situation, the AND gate is used as an *inhibit circuit*. We can say that $B = 0$ is the inhibit condition producing a 0 output. Conversely, $B = 1$ is the *enable* condition, which enables A to reach the output. This inhibit operation is an important application of AND gates.

2.5 THE NOT OPERATION

The NOT operation is unlike the OR or AND operations in that it can be performed on a single variable. For example, if the variable A is subjected to the NOT operation, the result x can be expressed as

$$x = \bar{A}$$

where the overbar represents the NOT operation. This expression is read "x equals NOT A" or "x equals the *inverse* of A" or "x equals the *complement* of A." Each of these are in common usage and they all indicate that the logic value of $x = \bar{A}$ is *opposite to* the logic value of A. The truth table in Figure 2.8(a) clarifies this for the two cases $A = 0$ and $A = 1$. That is,

$$\bar{1} = 0 \quad\quad \text{because NOT 1 is 0}$$

and

$$\bar{0} = 1 \quad\quad \text{because NOT 0 is 1}$$

The NOT operation is also referred to as *inversion* or *complementation,* and these terms will be used interchangeably throughout the text. Although we will use the overbar symbol exclusively to represent inversion, it is important to mention that another common symbol for inversion is the prime symbol ('). That is,

$$A' = \bar{A}$$

Both symbols should be recognized as NOT symbols.

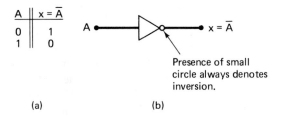

(a)   (b)

Figure 2.8   (a) Truth table and (b) symbol for the NOT circuit (inverter).

The NOT Circuit

Figure 2.8(b) shows the symbol for a NOT circuit, which is also called an *inverter.* This circuit always has only a single input, and its output logic level is always opposite to the logic level of this input. There is no such thing as an inverter with more than one input.

Summary of Boolean Operations

The rules for the OR, AND, and NOT operations may be summarized as follows:

$$\begin{array}{ccc} OR & AND & NOT \\ 0 + 0 = 0 & 0 \cdot 0 = 0 & \bar{0} = 1 \\ 0 + 1 = 1 & 0 \cdot 1 = 0 & \bar{1} = 0 \\ 1 + 0 = 1 & 1 \cdot 0 = 0 & \\ 1 + 1 = 1 & 1 \cdot 1 = 1 & \end{array}$$

2.6 DESCRIBING LOGIC CIRCUITS ALGEBRAICALLY

Any logic circuit, no matter how complex, may be completely described using the Boolean operations previously defined, because the OR gate, AND gate, and NOT circuit are the basic building blocks of digital systems. For example, consider the circuit in Figure 2.9. This circuit has three inputs, A, B, and C, and a single output, x. Utilizing the Boolean expression for each gate, we can easily determine the expression for the output.

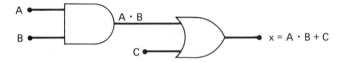

Figure 2.9 Logic circuit with its Boolean expression.

The expression for the AND-gate output is written $A \cdot B$. This AND output is connected as an input to the OR gate along with C, another input. The OR gate operates on its inputs, so its output is the OR sum of the inputs. Thus, we can express the OR output as $x = A \cdot B + C$. (This final expression could also be written as $x = C + A \cdot B$, since it does not matter which term of the OR sum is written first.)

Occasionally, there may be confusion as to which operation in an expression is performed first. The expression $A \cdot B + C$ can be interpreted in two different ways: (1) $A \cdot B$ is ORed with C, or (2) A is ANDed with the term $B + C$. To avoid this confusion, it will be understood that if an expression contains both AND and OR operations, the AND operations are performed first, unless there are *parentheses* in the expression, in which case the operation inside the parentheses is to be performed first. This is the same rule that is used in ordinary algebra to determine the hierarchy of operations.

To illustrate further, consider the circuit in Figure 2.10. The expression for the OR-gate output is simply $A + B$. This output serves as an input to the AND gate along with another input, C. Thus, we express the output of the AND gate as $x = (A + B) \cdot C$. Note the use of parentheses here to indicate that A and B are ORed *first*, before their OR sum is ANDed with C. Without the parentheses it

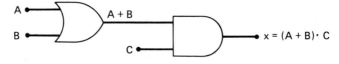

Figure 2.10   Logic circuit whose output expression requires parentheses.

would be interpreted *incorrectly*, since $A + B \cdot C$ means A is ORed with the
product $B \cdot C$.

Circuits Containing Inverters

Whenever an *inverter* (a NOT circuit) is present in a logic-circuit diagram, its
output expression is simply equal to the input expression with a bar over it. Figure
2.11 shows two examples using inverters. In Figure 2.11(a) the input A is fed
through an inverter, whose output is therefore $\bar{A}$. The inverter output is fed to an
OR gate together with B, so the OR output is equal to $\bar{A} + B$. Note that the bar
is only over the A, indicating that A is first inverted and then ORed with B.

In Figure 2.11(b) the output of the OR gate is equal to $A + B$ and is fed
through an inverter. The inverter output is therefore equal to $(\overline{A + B})$, since it
inverts the *complete* input expression. Note that the bar covers the entire expression $(A + B)$. This is important because, as will be shown later, the expressions
$(\overline{A + B})$ and $(\bar{A} + \bar{B})$ are *not* equivalent. The expression $(\overline{A + B})$ means that A
is ORed with B and then their OR sum is inverted, whereas the expression $(\bar{A} + \bar{B})$
indicates that A is inverted and B is inverted and the results are then ORed together.

Figure 2.12 shows two more examples, which should be studied carefully.
Note especially the use of *two* separate sets of parentheses in Figure 2.12(b). Also
notice in Figure 2.12(a) that the input variable C is connected as input to two
different gates.

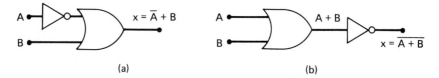

(a)   (b)

Figure 2.11   Circuits using inverters.

2.7   EVALUATING LOGIC-CIRCUIT OUTPUTS

Once the Boolean expression for a circuit output is obtained, the logic level of
the output can be determined for any values of circuit inputs. For example, suppose that we wanted to know the logic level of the output x for the circuit in Figure
2.12(a) for the case where $A = 0$, $B = 1$, $C = 1$, and $D = 0$. As in ordinary alge-

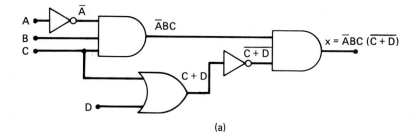

(a)

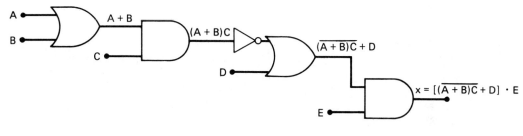

(b)

Figure 2.12 More examples.

bra, the value of x can be found by plugging the values of the variables into the expression and performing the indicated operations as follows:

$$x = \bar{A}BC(\overline{C + D})$$
$$= \bar{0} \cdot 1 \cdot 1 \cdot (\overline{1 + 0})$$
$$= 1 \cdot 1 \cdot 1 \cdot (\overline{1 + 0})$$
$$= 1 \cdot 1 \cdot 1 \cdot (\bar{1})$$
$$= 1 \cdot 1 \cdot 1 \cdot 0$$
$$= 0$$

As another illustration, let us evaluate the output of the circuit in Figure 2.12(b) for $A = 0$, $B = 0$, $C = 1$, $D = 1$, and $E = 1$.

$$x = [D + (\overline{A + B})\overline{C}] \cdot E$$
$$= [1 + (\overline{0 + 0}) \cdot 1] \cdot 1$$
$$= [1 + \overline{0 \cdot 1}] \cdot 1$$
$$= [1 + \bar{0}] \cdot 1$$
$$= [1 + 1] \cdot 1$$
$$= 1 \cdot 1$$
$$= 1$$

As a general rule, the following rules must always be followed when evaluating expressions such as the foregoing ones:

1. First, perform all inversions of single terms; that is, $\bar{0} = 1$ or $\bar{1} = 0$.
2. Then perform all operations within parentheses.
3. Perform an AND operation before an OR operation unless parentheses indicate otherwise.
4. If an expression has a bar over it, perform the operations of the expression first and then invert the result.

Determining Output Level from a Diagram

The output logic level for given input levels can also be determined directly from the circuit diagram without using the Boolean expression. This technique is often used by technicians during the troubleshooting or testing of a logic system since it also tells them what each gate output is supposed to be as well as the final output. To illustrate, the circuit of Figure 2.12(a) is redrawn in Figure 2.13 with the input levels $A = 0$, $B = 1$, $C = 1$, and $D = 0$. The procedure is to start from the inputs and to proceed through each inverter and gate, writing down each of their outputs in the process until the final output is reached.

In Figure 2.13 the first AND gate has *all* three inputs at the 1 level because the inverter changes the $A = 0$ to a 1. This condition produces a 1 at the AND output since $1 \cdot 1 \cdot 1 = 1$. The OR gate has inputs of 1 and 0, which produces a 1 output since $1 + 0 = 1$. This 1 is inverted to 0 and applied to the final AND gate along with the 1 from the first AND output. The 0 and 1 inputs to the final AND gate produce an output of 0, because $0 \cdot 1 = 0$.

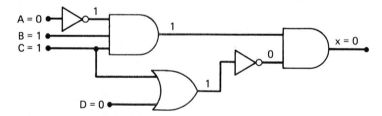

Figure 2.13 Determining output level from circuit diagram.

2.8 IMPLEMENTING CIRCUITS FROM BOOLEAN EXPRESSIONS

If the operation of a circuit is defined by a Boolean expression, a logic-circuit diagram can be implemented directly from that expression. For example, if we needed a circuit that was defined by $x = A \cdot B \cdot C$, we would immediately know that all that was needed was a three-input AND gate. If we needed a circuit that was defined by $x = A + \bar{B}$, we would use a two-input OR gate with an inverter on one of the inputs. The same reasoning used for these simple cases can be extended to more complex circuits.

Suppose that we wanted to construct a circuit whose output is $y = AC + B\bar{C} + \bar{A}BC$. This Boolean expression contains three terms $(AC, B\bar{C}, \bar{A}BC)$, which are ORed together. This tells us that a three-input OR gate is required with inputs that are equal to AC, $B\bar{C}$, and $\bar{A}BC$, respectively. This is illustrated in Figure 2.14(a), where a three-input OR gate is drawn with inputs labeled as AC, $B\bar{C}$, and $\bar{A}BC$.

Each OR-gate input is an AND product term, which means that an AND gate with appropriate inputs can be used to generate each of these terms. This is shown in Figure 2.14(b), which is the final circuit diagram. Note the use of inverters to produce the $\bar{A}$ and $\bar{C}$ terms required in the expression.

This same general approach can always be followed, although we shall find that there are some clever, more efficient techniques that can be employed. For now, however, this straightforward method will be used to minimize the number of new things that are to be learned.

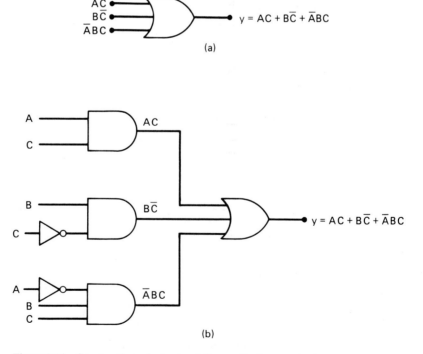

Figure 2.14 Constructing a logic circuit from a Boolean expression.

2.9 NOR GATES AND NAND GATES

Two other types of logic gates, NOR gates and NAND gates, are used extensively in digital circuitry. These gates actually combine the basic operations AND, OR, and NOT, which makes it relatively easy to describe them using the Boolean algebra operations learned previously.

The NOR Gate

Figure 2.15(a) shows the common symbol for a two-input *NOR gate*. The operation of the NOR gate is equivalent to the OR gate followed by an inverter shown in Figure 2.15(b), so the output expression for each is $x = \overline{A + B}$. In other words, the NOR gate *first* performs the OR operation on the inputs and then performs the NOT operation on the OR sum.

This is easy to remember because the NOR-gate symbol is just the OR symbol with a small circle on the output. This *small circle represents the inversion operation.*

The truth table in Figure 2.15(c) shows that the NOR output in each case is the inverse of the OR output. Whereas the OR output is *high* when *any* input is *high*, the NOR-gate output is *low* when any input is *high*. The same operation can be extended to NOR gates with more than two inputs.

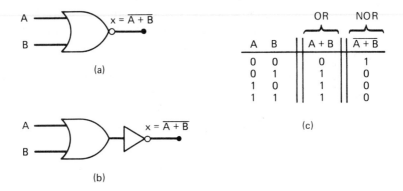

Figure 2.15   (a) NOR symbol; (b) equivalent circuit; (c) truth table.

EXAMPLE 2.4   Determine the Boolean expression for a three-input NOR gate followed by an inverter.

Solution: Refer to Figure 2.16, where the circuit diagram is shown. The expression at the NOR output is $(\overline{A + B + C})$, which is then fed through an inverter to produce

$$x = \overline{(\overline{A + B + C})}$$

The presence of the double inversion signs indicates that the quantity $(A + B + C)$ has been inverted and then inverted again. It should be clear that this simply results in the expression $(A + B + C)$ being unchanged. That is,

$$x = \overline{(\overline{A + B + C})} = A + B + C$$

Whenever two inversion bars are over the same variable or quantity, they cancel each other out, as in the example above. However, in cases such as $\overline{A} + \overline{B}$ the inverter bars do not cancel. This is because the smaller inverter bars invert the single variables A and B, respectively, while the wider bar inverts the quantity $(\overline{A} + \overline{B})$. Thus, $\overline{\overline{A} + \overline{B}} \neq A + B$. Similarly, $\overline{\overline{A}\overline{B}} \neq AB$.

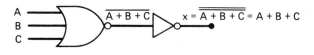

Figure 2.16

The NAND Gate

Figure 2.17(a) shows the common symbol for a two-input *NAND gate*. The operation of the NAND gate is equivalent to the AND gate followed by an inverter shown in Figure 2.17(b), so the output expression for each is $x = \overline{A \cdot B}$. Thus, the NAND gate *first* performs the AND operation on the inputs and then performs the NOT operation on the AND product. Once again, the order of these operations should be clear f om the NAND symbol.

The truth table in Figure 2.17(c) shows that the NAND-gate output in each case is the inverse of the AND output. Whereas the AND output is *high* only when *all* inputs are *high*, the NAND output is *low* only when *all* inputs are *high*. This same type of operation can be extended to NAND gates with more than two inputs.

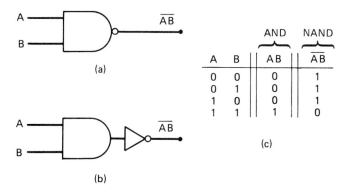

A	B	AND AB	NAND $\overline{AB}$
0	0	0	1
0	1	0	1
1	0	0	1
1	1	1	0

(c)

Figure 2.17　(a) NAND symbol; (b) equivalent circuit; (c) truth table.

The NAND gate and NOR gate are much more versatile than the AND and OR gates, since each can be used to produce any desired Boolean expression. We will be able to investigate this property later after we have studied some Boolean rules and theorems.

2.10 THE POWER OF BOOLEAN ALGEBRA

We have seen that Boolean algebra can be used to analyze a logic circuit and express its operation mathematically. This brings us to the most important application of Boolean algebra, the *simplification of logic circuits*. Once the Boolean expression for a logic circuit has been obtained, it may often be reduced to a simpler form by the use of certain Boolean theorems.* The simpler Boolean expression, since it is equivalent to the original, can be used in place of the original expression. This can often result in a considerable reduction in the number of gates and connections in the logic circuit.

For example, Figure 2.18(a) shows a logic circuit whose output expression $x = AB(\overline{A} + BC)$ can be obtained by the methods used earlier. By using Boolean theorems we will be able to simplify this expression to its equivalent, $x = AB\overline{C}$, which is implemented in Figure 2.18(b). It should be apparent that if both circuits are equivalent in the logic they perform, then the simpler circuit is more desirable from the standpoints of cost and size. The following sections will introduce a set of Boolean theorems and show how they can be applied to realize simplifications such as those in this example.

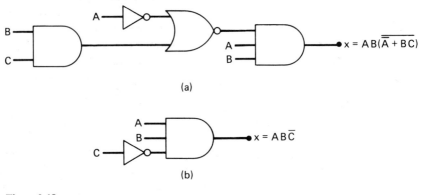

(a)

(b)

Figure 2.18

2.11 BOOLEAN THEOREMS

The first group of theorems is given in Figure 2.19. In each theorem x represents a logic variable that can be either 0 or 1. Each theorem is accompanied by its equivalent logic diagram, which will help to verify its validity. Any one of these

*"Theorem" is just another name for rule or law.

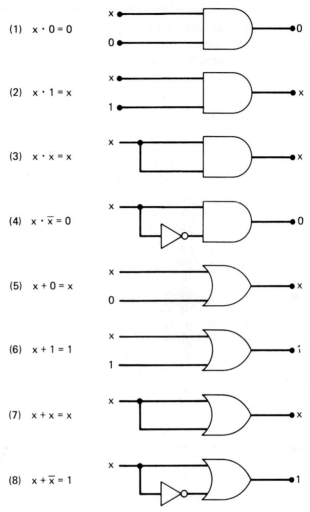

(1) $x \cdot 0 = 0$

(2) $x \cdot 1 = x$

(3) $x \cdot x = x$

(4) $x \cdot \overline{x} = 0$

(5) $x + 0 = x$

(6) $x + 1 = 1$

(7) $x + x = x$

(8) $x + \overline{x} = 1$

Figure 2.19   Single-variable theorems.

theorems, and in fact any Boolean theorem, may be proved simply by checking it for all possible values of the variables. The theorems in Figure 2.19 are single-variable theorems with x the only variable, so we have to check for the cases $x = 0$ and $x = 1$.

Theorem (1) states that if any variable is ANDed with 0, the result has to be 0. This is easy to remember because the AND operation is just like ordinary multiplication, where we know that anything multiplied by 0 is 0. We also know that the output of an AND gate will be 0 whenever any input is 0, regardless of the level on the other input.

Theorem (2) is also obvious by comparison with ordinary multiplication.

Theorem (3) can be proved by trying each case. If $x = 0$, then $0 \cdot 0 = 0$; if $x = 1$, then $1 \cdot 1 = 1$. Thus, $x \cdot x = x$.

Theorem (4) can be proved in the same manner. However, it can also be reasoned that at any time either x or its inverse $\bar{x}$ has to be at the 0 level, so their AND product always has to be 0.

Theorem (5) is straightforward since 0 *added* to anything does not affect its value, either in regular addition or in OR addition.

Theorem (6) states that if any variable is ORed with 1, the result will always be 1. Checking this for both values of x: $0 + 1 = 1$ and $1 + 1 = 1$. Equivalently, we can remember that an OR-gate output will be 1 when *any* input is 1, regardless of the value of the other input.

Theorem (7) can be proved by checking for both values of x: $0 + 0 = 0$ and $1 + 1 = 1$.

Theorem (8) can be proved similarly, or we can just reason that at any time either x or $\bar{x}$ has to be at the 1 level so that we are always ORing a 0 and a 1, which always results in 1.

Before introducing any more theorems, it should be pointed out that in applying theorems (1)–(8) the variable x may actually represent an expression containing more than one variable. For example, if we have $A\bar{B}(\overline{A\bar{B}})$, we can invoke theorem (4) by letting $x = A\bar{B}$. Thus, we can say that $A\bar{B}(\overline{A\bar{B}}) = 0$. The same idea can be applied to the use of any of these theorems.

Multivariable Theorems

The theorems presented next involve more than one variable:

(9) $x + y = y + x$

(10) $x \cdot y = y \cdot x$

(11) $x + (y + z) = (x + y) + z = x + y + z$

(12) $x(yz) = (xy)z = xyz$

(13) $x(y + z) = xy + xz$

(14) $x + xy = x$

(15) $x + \bar{x}y = x + y$

Theorems (9) and (10) are called the *commutative laws*. These laws indicate that the order in which we add or multiply two variables is unimportant; the result is the same.

Theorems (11) and (12) are the *associative laws*, which state that we can group the terms of a sum or a product any way we want. In other words, given the product $x \cdot y \cdot z$, we can first multiply x and y and then multiply the result by z; or we can first multiply y and z and then multiply the result by x. A similar idea applies to addition (the OR operation).

Theorem (13) is the *distributive law*, which states that an expression can be expanded by multiplying term by term just the same as in ordinary algebra. This theorem also indicates that we can factor an expression. That is, if we have a sum of two (or more) terms, each of which contains a common variable, the common variable can be factored out just like in ordinary algebra. For example, if we have the expression $A\bar{B}C + \bar{A}\bar{B}\bar{C}$, we can factor out the $\bar{B}$ variable:

$$A\bar{B}C + \bar{A}\bar{B}\bar{C} = \bar{B}(AC + \bar{A}\bar{C})$$

As another example, consider the expression $ABC + ABD$. Here the two terms have the variables A and B in common, so $A \cdot B$ can be factored out of both terms. That is,

$$ABC + ABD = AB(C + D)$$

This distributive law is very useful in many simplification applications, as the following examples illustrate.

EXAMPLE 2.5   Simplify the expression $y = A\bar{B}D + A\bar{B}\bar{D}$.

Solution: Factor out the common variables $A\bar{B}$ using theorem (13):

$$y = A\bar{B}(D + \bar{D})$$

Using theorem (8), the term in parentheses is equivalent to 1. Thus,

$$y = A\bar{B} \cdot 1$$
$$= A\bar{B} \quad \text{[using theorem (2)]}$$

EXAMPLE 2.6   Simplify $z = (\bar{A} + B)(A + B)$.

Solution: The expression can be expanded by multiplying out the terms [theorem (13)].

$$z = \bar{A} \cdot A + \bar{A} \cdot B + B \cdot A + B \cdot B$$

Invoking theorem (4), the term $\bar{A} \cdot A = 0$. Also, $B \cdot B = B$ [theorem (3)].

$$z = 0 + \bar{A} \cdot B + B \cdot A + B = \bar{A}B + AB + B$$

Factoring out the variable B [theorem (13)], we have

$$z = B(\bar{A} + A + 1)$$

Finally, using theorem (6),

$$z = B$$

Theorems (9)–(13) are easy to remember and use since they are exactly identical to those of ordinary algebra. Theorems (14) and (15), on the other hand, do not have any counterparts in ordinary algebra. Both of these are also useful in simplification techniques. Each can be proved by trying all possible cases for x and y. This is illustrated for theorem (14) as follows:

Case 1:   For $x = 0, y = 0,$

$$x + xy = x$$
$$0 + 0 \cdot 0 = 0$$
$$0 = 0$$

Case 2:   For $x = 0, y = 1,$

$$x + xy = x$$
$$0 + 0 \cdot 1 = 0$$
$$0 + 0 = 0$$
$$0 = 0$$

Case 3:   For $x = 1, y = 0,$

$$x + xy = x$$
$$1 + 1 \cdot 0 = 1$$
$$1 + 0 = 1$$
$$1 = 1$$

Case 4:   For $x = 1, y = 1,$

$$x + xy = x$$
$$1 + 1 \cdot 1 = 1$$
$$1 + 1 = 1$$
$$1 = 1$$

Theorem (14) can also be proved by factoring and using theorem (6) as follows:

$$x + xy = x(1 + y)$$
$$= x \cdot 1 \quad \text{[using theorem (6)]}$$
$$= x \quad \text{[using theorem (2)]}$$

EXAMPLE 2.7 Simplify $x = ACD + \bar{A}BCD.$

Solution: Factoring out the common variables CD, we have

$$x = CD(A + \bar{A}B)$$

Utilizing theorem (15), we can replace $A + \bar{A}B$ by $A + B$, so

$$x = CD(A + B)$$
$$= ACD + BCD$$

DeMorgan's Theorems

Two of the most important theorems of Boolean algebra were contributed by a great mathematician named DeMorgan. *DeMorgan's theorems* are extremely useful in simplifying expressions in which a product or sum of variables is complemented (inverted). The two theorems are:

(16) $\overline{(x + y)} = \bar{x} \cdot \bar{y}$

(17) $\overline{(x \cdot y)} = \bar{x} + \bar{y}$

Theorem (16) shows that when the OR sum of two variables $(x + y)$ is complemented, this is the same as if the two variables were individually complemented and then ANDed together. Stated differently, *the complement of an OR sum equals the AND product of the complements.*

Theorem (17) shows that when the product of two variables $(x \cdot y)$ is complemented, the result is equivalent to complementing the individual variables and then ORing the results. In other words, *the complement of an AND product is equal to the OR sum of the complements.* These two theorems can be easily proved by checking each one for all values of x and y and is left as an exercise for the reader.

DeMorgan's theorems are used whenever it is desired to modify an expression containing large inverter signs (inverters over more than one variable). When applying these theorems, remember that x and y can represent single variables or expressions. For example, let us apply them to $\overline{(A\bar{B} + C)}$. First note that we have the OR sum of $A\bar{B}$ and C, which is inverted. Invoking theorem (16), we can write this as

$$\overline{(A\bar{B} + C)} = \overline{(A\bar{B})} \cdot \bar{C}$$

Note that here we treated $A\bar{B}$ as x and C as y. The result can be further simplified since we have a product $A\bar{B}$ that is complemented. Using theorem (17), the expression becomes

$$\overline{A\bar{B}} \cdot \bar{C} = (\bar{A} + \bar{\bar{B}}) \cdot \bar{C}$$

Notice that we can replace $\bar{\bar{B}}$ by B, so we finally have

$$(\bar{A} + B) \cdot \bar{C} = \bar{A}\bar{C} + B\bar{C}$$

This final result contains only inverter signs that invert a single variable.

DeMorgan's theorems are easily extended to more than two variables. For example, it can be proved that

$$\overline{x + y + z} = \bar{x}\cdot\bar{y}\cdot\bar{z}$$

$$\overline{x\cdot y\cdot z} = \bar{x} + \bar{y} + \bar{z}$$

and so on for more variables.

Implications of DeMorgan's Theorems

Let us examine these theorems (16) and (17) from the standpoint of logic circuits. First, consider theorem (16),

$$\overline{x + y} = \bar{x}\cdot\bar{y}$$

The left-hand side of the equation can be viewed as the output of a NOR gate whose inputs are x and y. The right-hand side of the equation, on the other hand, is the result of first inverting both x and y and then putting them through an AND gate. These two representations are equivalent and are illustrated in Figure 2.20(a). What this means is that an AND gate with inverters on each of its inputs is equivalent to a NOR gate. In fact, both representations are used to represent the NOR function. When the AND gate with inverted inputs is used to represent the NOR function, it is usually drawn as shown in Figure 2.20(b), where the small ircles on the inputs represent the inverters.

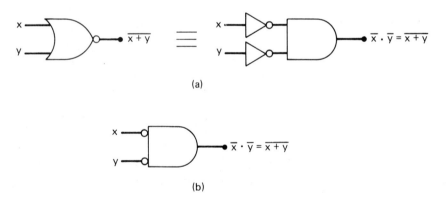

(a)

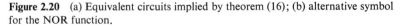

(b)

Figure 2.20   (a) Equivalent circuits implied by theorem (16); (b) alternative symbol for the NOR function.

Now consider theorem (17),

$$\overline{x\cdot y} = \bar{x} + \bar{y}$$

The left side of the equation can be implemented by a NAND gate with inputs x and y. The right side can be implemented by first inverting inputs x and y and then putting them through an OR gate. These two equivalent representations are shown in Figure 2.21(a). The OR gate with inverters on each of its inputs is equivalent to the NAND gate. In fact, both representations are used to represent the NAND function. When the OR gate with inverted inputs is used to represent the NAND function, it is usually drawn as shown in Figure 2.21(b), where the circles again represent inversion.

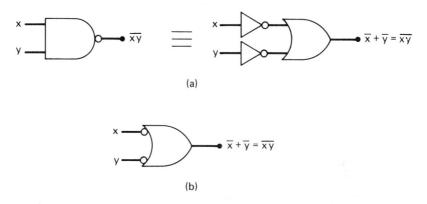

(a)

(b)

Figure 2.21 (a) Equivalent circuits implied by theorem (17); (b) alternative symbol for the NAND function.

EXAMPLE 2.8 Simplify the expression $z = \overline{(\bar{A} + C) \cdot (B + \bar{D})}$.

Solution: Using theorem (17), we can rewrite this as

$$z = \overline{(\bar{A} + C)} + \overline{(B + \bar{D})}$$

We can think of this as breaking the large inverter sign down the middle and changing the AND sign ($\cdot$) to an OR sign ($+$). Now the term $\overline{(\bar{A} + C)}$ can be simplified by applying theorem (16). Likewise, $\overline{(B + \bar{D})}$ can be simplified.

$$z = \overline{(\bar{A} + C)} + \overline{(B + \bar{D})}$$
$$= (\bar{\bar{A}} \cdot \bar{C}) + \bar{B} \cdot \bar{\bar{D}}$$

Here we have broken the larger inverter signs down the middle and replaced the ($+$) with a ($\cdot$). Canceling out the double inversions, we have finally

$$z = A\bar{C} + \bar{B}D$$

Example 2.8 points out that when using DeMorgan's theorems to reduce an expression, an inverter sign may be broken at any point in the expression and the operator at that point in the expression is changed to its opposite ($+$ is changed to $\cdot$, and vice versa). This procedure is continued until the expression is reduced to one in which only single variables are inverted. Two more examples are shown next. Study these carefully and note the procedure.

1. $z = \overline{A + \bar{B} \cdot C}$

   $= \bar{A} \cdot (\overline{\bar{B} \cdot C})$

   $= \bar{A} \cdot (\bar{\bar{B}} + \bar{C})$

   $= \bar{A} \cdot (B + \bar{C})$

2. $q = \overline{(A + BC) \cdot (D + EF)}$

   $= \overline{(A + BC)} + \overline{(D + EF)}$

   $= (\bar{A} \cdot \overline{BC}) + (\bar{D} \cdot \overline{EF})$

   $= [\bar{A} \cdot (\bar{B} + \bar{C})] + [\bar{D} \cdot (\bar{E} + \bar{F})]$

   $= \bar{A}\bar{B} + \bar{A}\bar{C} + \bar{D}\bar{E} + \bar{D}\bar{F}$

2.12 SIMPLIFYING LOGIC CIRCUITS

The theorems that have been presented can be used to simplify any Boolean expression. Unfortunately, it is not always obvious which theorems should be applied and in which order. Usually, if done correctly, any path will lead to the simplest result. Sometimes, two different paths may lead to different results that are equally simple. The following examples should be studied carefully to see some of the reasoning used in logic-circuit simplification. Even though there is no set procedure in applying these theorems, by studying these examples and the preceding ones and doing the problems at the end of the chapter you will eventually develop a facility for arriving at a reasonably good solution. For interested students, Karnaugh mapping, a graphical technique for simplifying logic circuits, is described and illustrated in Appendix II.

EXAMPLE 2.9 Simplify the logic circuit shown in Figure 2.22(a).

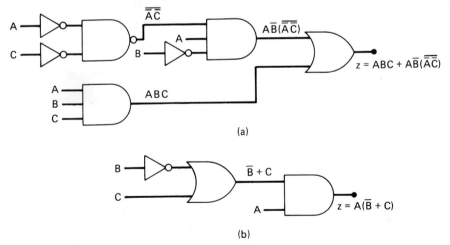

Figure 2.22

Solution: The first step is to determine the expression for the output. The result is

$$z = ABC + A\bar{B} \cdot (\overline{\overline{AC}})$$

Once the expression is determined, it is usually a good idea to break down all large inverter signs using DeMorgan's theorems and then multiply out all terms.

$$z = ABC + A\bar{B}(\overline{\bar{A}} + \bar{C}) \qquad \text{[theorem (17)]}$$
$$= ABC + A\bar{B}(A + C)$$
$$= ABC + A\bar{B}A + A\bar{B}C$$
$$= ABC + A\bar{B} + A\bar{B}C \qquad \text{[theorem (3)]}$$

When the expression is in this form, we should look for common variables among the various terms with the intention of factoring. The first and third terms above have AC in common, which can be factored out:

$$z = AC(B + \bar{B}) + A\bar{B}$$

Since $B + \bar{B} = 1$, then

$$z = AC(1) + A\bar{B}$$
$$= AC + A\bar{B}$$

We can now factor out A, which results in

$$z = A(C + \bar{B})$$

This result can be simplified no further. Its circuit implementation is shown in Figure 2.22(b). It is obvious that the circuit in (b) is a great deal simpler than the original circuit in (a).

EXAMPLE 2.10 Simplify the expression $z = ABC + AB\bar{C} + A\bar{B}C$.

Solution: We will look at two different ways to arrive at the same result.

Method 1: The first two terms in the expression have the variables AB in common. Thus,

$$z = AB(C + \bar{C}) + A\bar{B}C$$
$$= AB(1) + A\bar{B}C$$
$$= AB + A\bar{B}C$$

We can factor the variable A from both terms:

$$z = A(B + \bar{B}C)$$

Invoking theorem (15),

$$z = A(B + C)$$

Method 2: The original expression is $z = ABC + AB\bar{C} + A\bar{B}C$. The first two terms have the variables AB in common. The first and last terms have the variables

AC in common. How do we know whether to factor AB from the first two terms or AC from the two end terms? Actually, we can do both by using the ABC term *twice*. In other words, we can rewrite the expression as

$$z = ABC + AB\bar{C} + A\bar{B}C + ABC$$

where we have added an extra term ABC. This is valid and will not change the value of the expression since $ABC + ABC = ABC$ [theorem (7)]. Now we can factor AB from the first two terms and AC from the last two terms:

$$z = AB(C + \bar{C}) + AC(\bar{B} + B)$$
$$= AB \cdot 1 + AC \cdot 1$$
$$= AB + AC = A(B + C)$$

This is, of course, the same result as method 1. This trick of using the same term twice can always be used. In fact, the same term can be used more than twice if necessary.

EXAMPLE 2.11   Simplify $z = \bar{A}C(\overline{\bar{A}BD}) + \bar{A}B\bar{C}\bar{D} + A\bar{B}C$.

Solution:

$$z = \bar{A}C(A + \bar{B} + \bar{D}) + \bar{A}B\bar{C}\bar{D} + A\bar{B}C$$

Multiplying out,

$$z = \bar{A}CA + \bar{A}C\bar{B} + \bar{A}C\bar{D} + \bar{A}B\bar{C}\bar{D} + A\bar{B}C$$

Since $\bar{A} \cdot A = 0$, the first term is eliminated.

$$z = \bar{A}\bar{B}C + \bar{A}C\bar{D} + \bar{A}B\bar{C}\bar{D} + A\bar{B}C$$

Factor $\bar{B}C$ from the first and last terms. Also factor $\bar{A}\bar{D}$ from the second and third terms.

$$z = \bar{B}C(\bar{A} + A) + \bar{A}\bar{D}(C + B\bar{C})$$
$$= \bar{B}C + \bar{A}\bar{D}(C + B)   \text{[theorems (8) and (15)]}$$

This expression cannot be simplified further.

2.13 UNIVERSALITY OF NAND GATES AND NOR GATES

All Boolean expressions consist of various combinations of the basic operations of OR, AND, and INVERT. Therefore, any expression can be implemented using OR gates, AND gates, and inverters. It is possible, however, to implement any logic expression using only NAND gates and no other type of gate. This is because

NAND gates, in the proper combination, can be used to perform each of the Boolean operations OR, AND, and INVERT. This is demonstrated in Figure 2.23.

First, in Figure 2.23(a) we have a two-input NAND gate whose inputs are purposely connected together so that the variable A is applied to both. In this configuration, the NAND simply acts as an inverter, since its output is $x = \overline{A \cdot A} = \overline{A}$.

In Figure 2.23(b) we have two NAND gates connected so that the AND operation is performed. NAND gate 2 is used as an inverter to change $\overline{AB}$ to $\overline{\overline{AB}} = AB$, which is the desired AND function.

The OR operation can be implemented using NAND gates connected as shown in Figure 2.23(c). Here NAND gates 1 and 2 are used as inverters to invert the inputs, so the final output is $x = \overline{\overline{A} \cdot \overline{B}}$, which can be simplified to $x = A + B$ using DeMorgan's theorem.

In a similar manner, it can be shown that NOR gates can be arranged to implement any of the Boolean operations. This is illustrated in Figure 2.24. Part (a) shows that a NOR gate with its inputs connected together behaves as an inverter, since the output is $x = \overline{A + A} = \overline{A}$.

In Figure 2.24(b) two NOR gates are arranged so that the OR operation is performed. NOR gate 2 is used as an inverter to change $\overline{A + B}$ to $\overline{\overline{A + B}} = A + B$, which is the desired OR function.

The AND operation can be implemented with NOR gates as shown in Figure 2.24(c). Here NOR gates 1 and 2 are used as inverters to invert the inputs, so the final output is $x = \overline{\overline{A} + \overline{B}}$, which can be simplified to $x = A \cdot B$ by use of DeMorgan's theorem.

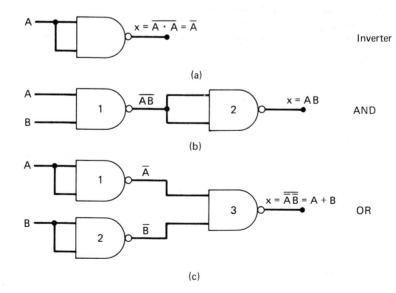

Figure 2.23   NAND gates can be used to implement any Boolean operation.

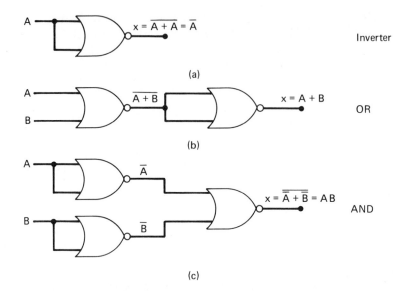

Figure 2.24 NOR gates used to implement Boolean operations.

The fact that NAND gates can be used to implement all the Boolean operations means that any logic circuit, no matter how complex, can be implemented using only NAND gates. The same is true for NOR gates. This characteristic of NAND and NOR gates is important. For example, it means that a company designing and producing a piece of digital equipment need stock only one type of gate. In Chapter 3 we will see how the design of any logic circuit can be implemented using only NAND or NOR gates.

2.14 EQUIVALENT LOGIC REPRESENTATIONS

We have seen how the same logic operation can be implemented in more than one way. For example, Figure 2.25(a) shows a NAND gate and 2.25b shows an OR gate with inverted inputs. Both circuits are equivalent and provide the same logic operation. In fact, the OR with inverted inputs is sometimes used to represent a NAND gate on logic schematics. The reason for the use of both representations is that the two can be interpreted differently, although equivalently.

The NAND-gate operation can be interpreted as: "the output goes LOW only if *all* inputs are HIGH." The operation of the circuit of Figure 2.25(b) can be interpreted as: "the output goes HIGH if *any* input is LOW." Obviously, these are two different ways of saying the same thing. The decision as to which representation should be used in a logic-circuit schematic depends on which of the output conditions is the *normal* condition and which is the *activated* condition. For example, if the output is normally HIGH during the circuit operation and goes LOW (is activated) only at certain special times, the representation of Figure 2.25(a) is

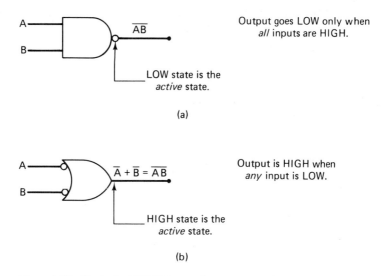

(a)

A$\overline{B}$

Output goes LOW only when *all* inputs are HIGH.

LOW state is the *active* state.

$\overline{A} + \overline{B} = \overline{AB}$

Output is HIGH when *any* input is LOW.

HIGH state is the *active* state.

(b)

Figure 2.25   Equivalent NAND-operation representations.

used. On the other hand, if the output is normally LOW and goes HIGH (is activated) only at certain times, the representation of Figure 2.25(b) is used.

In other words, when the LOW output state is the active state which causes other things in the circuit to happen, the representation in (a) is used. The small circle on the output indicates that the LOW state is the active output state. When the HIGH output state is the active state which causes other things to happen, (b) is used. The absence of a circle on the output indicates that the HIGH output state is the active state.

The same idea is used for other logic operations, such as the OR operation shown in Figure 2.26. Again, both representations are equivalent to the OR

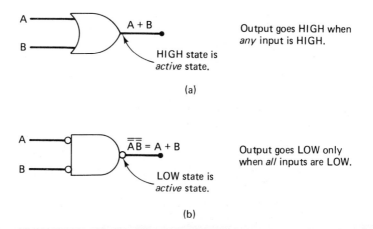

A + B

Output goes HIGH when *any* input is HIGH.

HIGH state is *active* state.

(a)

$\overline{\overline{A}\,\overline{B}} = A + B$

Output goes LOW only when *all* inputs are LOW.

LOW state is *active* state.

(b)

Figure 2.26   Equivalent OR-operation representations.

operation. The representation in (a) is used when the HIGH state is the active output state, and (b) is used when the LOW state is the active output state.

Figure 2.27 gives all the equivalent logic representations for the basic logic gates. Gates 1, 3, 6, and 8 (no circle on output) are used when the HIGH state is the active output state. Gates 2, 4, 5, and 7 are used when the LOW state is the active output state.

There is another important characteristic of the equivalent logic representations shown in Figure 2.27. Gates 1, 4, 5, and 8 all use AND logic, which means that *both* inputs have to be at their active levels to produce the output active level. For example, the operation of gate 8 can be interpreted as follows: the output goes HIGH only if *both* inputs are LOW. The output is active HIGH (no circle) and the inputs are active LOW (circles).

Similarly, gates 2, 3, 6, and 7 all use OR logic, which means that *either* input has to be at its active level to produce an active output level. For example, the operation of gate 2 can be interpreted as follows: output goes LOW when either input is LOW. The output and inputs are all active LOW.

Use of the convention described above can be very helpful to technicians who are troubleshooting or testing logic circuitry because they can tell which output level of a given circuit is significant, that is, which output level will cause something to happen in the circuits being driven. This information makes it easier to follow a complicated logic schematic. We will use this convention on logic schematics throughout the text whenever appropriate.

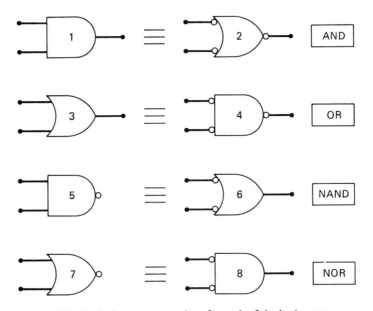

Figure 2.27 Equivalent representations for each of the basic gates.

QUESTIONS AND PROBLEMS

2.1 In Section 2.3 it was pointed out that if a digital circuit has N logic input variables, there will be 2^N different possible cases. If a truth table is to be constructed containing all the cases, there is a methodical way to list all the cases without missing any. Consider the truth table shown in Figure 2.28 for the four variables M, N, O, and P. Since $N = 4$, there are $2^4 = 16$ cases. All 16 cases can be listed if we momentarily consider the variables M, N, O, and P to represent a 4-bit binary number, with P being the LSB and M the MSB. If we now simply count in binary from 0000 to 1111 as described in Chapter 1, we will have all 16 cases. The same procedure is followed for any number of inputs. Add to the truth table the outputs for each case if it represents a four-input OR gate.

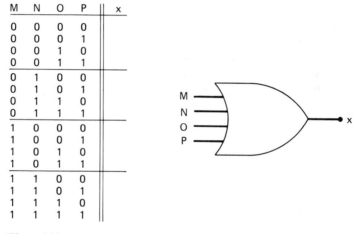

M	N	O	P	x
0	0	0	0	
0	0	0	1	
0	0	1	0	
0	0	1	1	
0	1	0	0	
0	1	0	1	
0	1	1	0	
0	1	1	1	
1	0	0	0	
1	0	0	1	
1	0	1	0	
1	0	1	1	
1	1	0	0	
1	1	0	1	
1	1	1	0	
1	1	1	1	

Figure 2.28

2.2 Explain how OR addition is different from ordinary addition.

2.3 What is the only condition under which an OR-gate output will be 0?

2.4 Determine the waveform at the output of the OR gate in Figure 2.29.

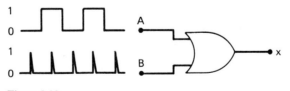

Figure 2.29

2.5 Write the Boolean expression for a six-input OR gate.

2.6 How is AND multiplication different from ordinary multiplication?

2.7 Under what conditions will the output of an AND gate be 0?

2.8   Write the expression for a four-input AND gate. Construct the complete truth table showing the output for all possible cases.

2.9   Change the OR gate in Figure 2.29 to an AND gate and determine the output waveform.

2.10 Is there such a thing as a three-input NOT gate?

2.11 Write the Boolean expression for the output x in Figure 2.30(a). Determine the value of x for all possible input conditions and list them in a truth table.

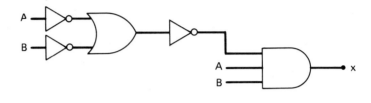

(a)

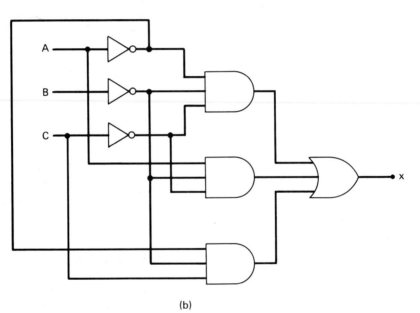

(b)

Figure 2.30

2.12 Repeat Problem 2.11 for the circuit in Figure 2.29(b).

2.13 Determine the output level of the circuit in Figure 2.12(b) for the following cases:
(a) $A = B = C = 1$, $D = 0$, $E = 1$
(b) $A = B = 0$, $C = 1$, $D = E = 0$.

2.14 For each of the following expressions, construct the corresponding logic circuit, using AND and OR gates and inverters:
(a) $x = \overline{AB(C + D)}$

(b) $z = \overline{(A + B + \overline{C}D\overline{E})} + \overline{B}C\overline{D}$

(c) $y = \overline{(M + N + \overline{P}Q)}$

2.15   How does a NOR gate differ from an OR gate?

2.16   Construct a truth table for a three-input NOR gate.

2.17   Under what conditions is the output of a NOR gate high?

2.18   Construct a truth table for a three-input NAND gate.

2.19   For what input conditions will a NAND output be low?

2.20   Modify the circuits that were constructed for Problem 2.14 so that NAND and NOR gates are used where appropriate.

2.21   Consider each of the following statements and for each indicate for which logic gate or gates (AND, OR, NAND, NOR) the statement is true:
(a)   Output is high *only* if all inputs are low.
(b)   Output will be low if the inputs are at different levels.
(c)   Output is high when both inputs are high.
(d)   Output is low *only* if all inputs are high.
(e)   All low inputs produce a high output.

2.22   Prove theorem (15) by trying all possible cases.

2.23   Prove both of DeMorgan's theorems by trying all cases.

2.24   Simplify the following terms using DeMorgan's theorems:
(a) $\overline{\overline{A}B\overline{C}}$ (b) $\overline{A + B\overline{C}}$ (c) $\overline{\overline{ABCD}}$

2.25   Simplify the circuits in Figure 2.30.

2.26   Simplify each of the following expressions:
(a) $x = RST + RS(\overline{T} + V)$
(b) $y = \overline{A}\overline{B}\overline{C} + AB\overline{C} + \overline{A}BC + A\overline{B}C$
(c) $z = (M + N)(\overline{M} + P)(\overline{N} + P)$
(d) $q = AB(\overline{B} + C) + C$

2.27   (a)   Using AND and OR gates, construct the logic circuit corresponding to the expression $x = AB + CD + EF$.
(b)   Replace each AND and OR gate in your circuit with its equivalent NAND-gate implementation (see Figure 2.23).
(c)   Write the expression for the revised circuit. Simplify it and compare it to the original expression.

2.28   (a)   Construct the logic circuit for $y = (A + B)(C + D)$.
(b)   Replace each AND and OR gate by its equivalent NOR-gate implementation (see Figure 2.24).
(c)   Write the expression for the revised circuit, simplify it, and compare it to the original.

2.29   Figure 2.31 is another example of an inhibit circuit. Input C is a 1-kHz squarewave. The inputs A and B control whether or not this squarewave passes through the AND gate to the output. Determine for which conditions on inputs A and B the output will be a squarewave.

2.30   Modify Figure 2.31 so that the signal at C is *inhibited* from reaching x only when A and B are simultaneously HIGH. (*Hint:* Use a separate gate to combine A and B.)

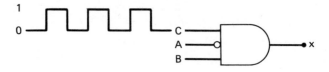

Figure 2.31

2.31 A NAND gate can also be used for inhibit operations, although the operation is not exactly the same as the AND gate inhibit operation. Change the gate in Figure 2.31 to a NAND gate and explain its operation for the various conditions on the *A* and *B* control inputs.

2.32 Figure 2.32 shows two different representations for the AND operation. Which of these should be used in a logic schematic if the output is driving a circuit that is activated by a LOW input?

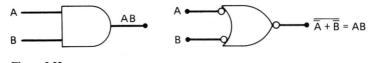

Figure 2.32

2.33 Determine which of the equivalent logic representations of Figure 2.27 should be used for each of the following applications:
(a) Provide a high voltage to turn on an indicator lamp whenever either of two logic inputs goes low.
(b) Produce a low voltage to turn off an indicator lamp when both input signals are simultaneously high.

3

LOGIC-CIRCUIT DESIGN

In Chapter 2 we saw how Boolean algebra is used to describe logic circuits and how it can be applied to the simplification of these circuits. In this chapter we will study the basic procedures used in *designing* logic circuits when the desired circuit requirements are given. The circuit requirements are normally given either in the form of a truth table showing the desired output level for all possible input combinations, or as a word statement describing the circuit operation. In this chapter we shall be concerned only with *combinatorial* logic networks, that is, logic networks that do not contain any feedback or memory. In combinatorial networks, the output (or outputs) at any instant of time depends only on the levels present at the various inputs at that instant of time.

3.1 SUM-OF-PRODUCTS AND PRODUCT-OF-SUMS EXPRESSIONS

Before beginning our study of logic design methods, we should first describe the two general forms of logic expressions that are used in these methods. The first general form for a logic expression is called the *sum-of-products* form. Some examples of the sum-of-products form (hereafter abbreviated S-of-P) are as follows:

1. $ABC + \bar{A}B\bar{C}$

2. $AB + \bar{A}B\bar{C} + \bar{C}\bar{D} + D$

3. $AB + CD + EF + GK + HL$

These examples essentially show that S-of-P expressions consist of the *sum* of two or more terms, where each term is a *product* of one or more variables.

The second general form is the *product-of-sums* form (hereafter abbreviated P-of-S). Several examples of this form are as follows:

1. $(A + B + C) \cdot (\bar{A} + \bar{B} + C)$
2. $(A + B) \cdot (\bar{A} + B + \bar{C}) \cdot (\bar{C} + \bar{D}) \cdot (D)$
3. $(A + B) \cdot (C + D) \cdot (E + F) \cdot (G + K) \cdot (H + L)$

From these examples it can be seen that the P-of-S expressions consist of the *product* of two or more terms in parentheses, where each of these terms is a *sum* of one or more variables.

3.2 DERIVING AN EXPRESSION FROM A TRUTH TABLE—SUM-OF-PRODUCTS SOLUTION

When the desired output level of a logic circuit is given for all possible input conditions, the results can be conveniently displayed in a truth table. The Boolean expression for the required circuit can then be derived from the truth table. For example, consider Figure 3.1(a), where a truth table is shown for a circuit that has two inputs, A and B, and output x. The table shows that output x is to be at the 1 level *only* for the case where $A = 0$ and $B = 1$. It now remains to determine what logic circuit will produce this desired operation. It should be apparent that one possible solution is that shown in Figure 3.1(b). Here an AND gate is used with inputs $\bar{A}$ and B, so $x = \bar{A} \cdot B$. Obviously x will be 1 *only if* both inputs to the AND gate are 1, namely $\bar{A} = 1$ (which means $A = 0$) and $B = 1$. For all other values of A and B, the output x has to be 0.

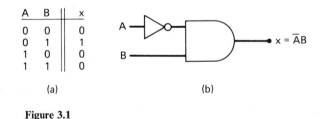

A	B	x
0	0	0
0	1	1
1	0	0
1	1	0

(a)   (b)

Figure 3.1

A similar approach can be used for other input conditions. For instance, if x were to be high only for the $A = 1$, $B = 0$ condition, the resulting circuit would be an AND gate with inputs A and $\bar{B}$. In other words, for any of the four possible input conditions we can generate a high x output by using an AND gate with appropriate inputs. The four different cases are shown in Figure 3.2. Each of the

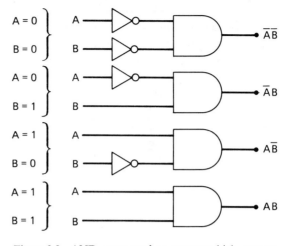

Figure 3.2 AND gates used to generate high outputs for each of the possible input conditions.

AND gates shown generates an output that is 1 *only* for the one given input condition and is 0 for all other conditions. It should be noted that the AND inputs are inverted or not inverted depending on the values that the variables have for the given condition. If the variable is 0 for the given condition, it is inverted before entering the AND gate.

Let us now consider the case shown in Figure 3.3(a), where we have a truth table which indicates that the output x is to be 1 for two different cases: $A = 0$, $B = 1$ and $A = 1$, $B = 0$. How can this be implemented? We know that the AND term $\bar{A} \cdot B$ will generate a 1 for the $A = 0$, $B = 1$ condition, and the AND term $A \cdot \bar{B}$ will generate a 1 for the $A = 1$, $B = 0$ condition. Since x has to be high for *either* condition, it should be clear that these terms should be ORed together to produce the desired output, x. This implementation is shown in Figure 3.3(b), where the resulting expression for the output is $x = \bar{A}B + A\bar{B}$.

In this example, an AND term is generated for each case in the table where the output x is to be a 1. The AND-gate outputs are then ORed together to produce the total output x, which will be high when either AND term is high. This

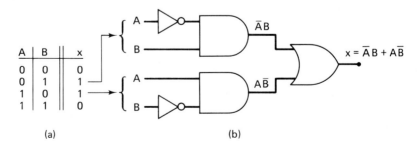

A	B	x
0	0	0
0	1	1
1	0	1
1	1	0

(a)

$x = \bar{A}B + A\bar{B}$

(b)

Figure 3.3

same procedure can be extended to examples with more than two inputs. Consider the following truth table for a three-input circuit:

A	B	C	x
0	0	0	0
0	0	1	0
0	1	0	1 $\longrightarrow \bar{A}B\bar{C}$
0	1	1	1 $\longrightarrow \bar{A}BC$
1	0	0	0
1	0	1	0
1	1	0	0
1	1	1	1 $\longrightarrow ABC$

Here there are three cases where the output x is to be 1. The required AND term for each of these cases is shown. Again, note that for each case where a variable is 0, it appears complemented in the AND term. The final expression for x is obtained by ORing the three AND terms. Thus,

$$x = \bar{A}B\bar{C} + \bar{A}BC + ABC$$

This expression can be implemented with three AND gates feeding an OR gate.

In all the preceding examples the expression for the output x was derived from the truth table in sum-of-products form. The general procedure for obtaining the output expression from a truth table in S-of-P form can be summarized as follows:

1. Write an AND term for each case in the table where the output is 1.
2. Each AND term contains each input variable in either inverted or non-inverted form. If the variable is 0 for that particular case in the table, it is inverted in the AND term.
3. All the AND terms are then ORed together to produce the final expression for the output.

Complete Design Problem

Once the output expression has been determined from the truth table in S-of-P form, it can easily be implemented using AND and OR gates. There will be one AND gate for each term in the expression and one OR gate, which is fed by the outputs of each AND gate. Usually, however, the S-of-P expression can be simplified using the techniques of Chapter 2, thereby resulting in a simpler circuit. The following example illustrates the complete design procedure.

EXAMPLE 3.1 Design a logic circuit that has three inputs, A, B, and C, and whose output will be high only when a majority of the inputs is high.

Solution: The first step is to set up the truth table based on the problem statement. The eight possible input cases are shown in Figure 3.4(a). Based on the problem statement, the output x should be a 1 whenever two or more inputs are 1. For all other cases the output should be 0. The next step is to write the AND terms for each case where $x = 1$, as shown in the figure. The S-of-P expression for x can then be written as

$$x = \bar{A}BC + A\bar{B}C + AB\bar{C} + ABC$$

This expression can be simplified in several ways. Perhaps the quickest way is to realize that the last term ABC has two variables in common with each of the other terms. Thus, we can use the ABC term to factor with each of the other terms. The expression is rewritten with the ABC term repeated three times (recall that this is legal in Boolean algebra).

$$x = \bar{A}BC + ABC + A\bar{B}C + ABC + AB\bar{C} + ABC$$

Factoring the appropriate pairs of terms, we have

$$x = BC(\bar{A} + A) + AC(\bar{B} + B) + AB(\bar{C} + C)$$

Since each term in parentheses is equal to 1, we have

$$x = BC + AC + AB$$

Further factoring of this expression will not produce any further simplification. The final expression is implemented in Figure 3.4(b). This final expression is also in S-of-P form, but it is much simpler than the original expression and requires a simpler circuit implementation, with fewer gates and fewer connections.

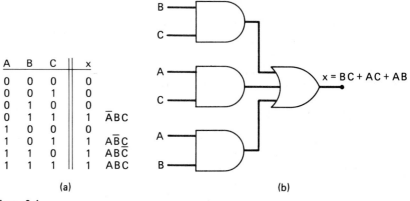

A	B	C	x	
0	0	0	0	
0	0	1	0	
0	1	0	0	
0	1	1	1	$\bar{A}BC$
1	0	0	0	
1	0	1	1	$A\bar{B}C$
1	1	0	1	$AB\bar{C}$
1	1	1	1	ABC

(a)

(b)

Figure 3.4

3.3 DERIVING AN EXPRESSION FROM A TRUTH TABLE—PRODUCT-OF-SUMS SOLUTION

The output expression for a truth table can also be derived in product-of-sums form. The procedure will be illustrated for the problem of Example 3.1. The truth table is redrawn in Figure 3.5(a), showing the desired values of output x. Also shown in the last column are the values of the inverted output, $\bar{x}$. We will first derive the S-of-P expression for $\bar{x}$ using the procedure of Section 3.2. Note that we write the AND terms that correspond to all cases where $\bar{x} = 1$. Thus, we have

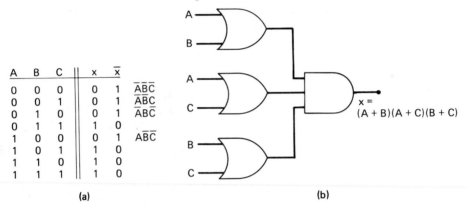

A	B	C	x	$\bar{x}$	
0	0	0	0	1	$\bar{A}\bar{B}\bar{C}$
0	0	1	0	1	$\bar{A}\bar{B}C$
0	1	0	0	1	$\bar{A}B\bar{C}$
0	1	1	1	0	
1	0	0	0	1	$A\bar{B}\bar{C}$
1	0	1	1	0	
1	1	0	1	0	
1	1	1	1	0	

(a)

(b) $x = (A + B)(A + C)(B + C)$

Figure 3.5

$$\bar{x} = \bar{A}\bar{B}\bar{C} + \bar{A}\bar{B}C + \bar{A}B\bar{C} + A\bar{B}\bar{C}$$

This expression can be simplified in a manner similar to that used in Example 3.1 and is left as an exercise for the reader. The result is

$$\bar{x} = \bar{B}\bar{A} + \bar{A}\bar{C} + \bar{B}\bar{C}$$

The next step is to invert both sides of the expression,

$$\bar{\bar{x}} = x = \overline{\bar{A}\bar{B} + \bar{A}\bar{C} + \bar{B}\bar{C}}$$

Using DeMorgan's theorems, the right side of the expression can be simplified as follows:

$$x = \overline{\bar{A}\bar{B} + \bar{A}\bar{C} + \bar{B}\bar{C}}$$
$$= \overline{\bar{A}\bar{B}} \cdot \overline{\bar{A}\bar{C}} \cdot \overline{\bar{B}\bar{C}}$$
$$= (\bar{\bar{A}} + \bar{\bar{B}}) \cdot (\bar{\bar{A}} + \bar{\bar{C}}) \cdot (\bar{\bar{B}} + \bar{\bar{C}})$$
$$= (A + B)(A + C)(B + C)$$

This expression is in the desired P-of-S form. Its implementation is shown in Figure 3.5(b). Each sum term is generated by an OR gate, and the outputs of each OR gate feed an AND gate. Comparing the P-of-S solution of Figure 3.5 with the S-of-P solution of Figure 3.4, we can see that both require the same number of gates and the same number of connections.

The procedure for obtaining the P-of-S solution can be summarized as follows:

1. Determine the S-of-P expression for the inverted output $\bar{x}$.
2. Simplify the expression for $\bar{x}$.
3. Invert the expression to solve for x and use DeMorgan's theorems to simplify it to product-of-sums form.

EXAMPLE 3.2   Three photocells are being illuminated by three different flashing lights. The lights are supposed to be flashing in sequence so that at no time should all three lights be on at the same time or off at the same time. Each photocell is used to monitor one of the lights, and each photocell is in a circuit that produces a low output voltage when the photocell is dark and a high output voltage when the photocell is illuminated. Design a logic circuit that has as its inputs the photocell-circuit outputs and which produces a high output whenever the three lights are *all* on or *all* off at the same time. Determine both a S-of-P solution and a P-of-S solution.

Solution: The first step is to set up the truth table. Let us call the three photocell-circuit outputs A, B, and C. The eight possible cases are listed in the table below. The desired output x and its inverse $\bar{x}$ are also shown. From the problem statement it should be clear that x has to be 1 only for the two cases $A = B = C = 0$ and $A = B = C = 1$ and 0 for all other cases. The values for $\bar{x}$, of course, are just the opposite.

A	B	C	x	$\bar{x}$
0	0	0	1	0
0	0	1	0	1
0	1	0	0	1
0	1	1	0	1
1	0	0	0	1
1	0	1	0	1
1	1	0	0	1
1	1	1	1	0

Sum-of-products solution:

$$x = \bar{A}\bar{B}\bar{C} + ABC$$

This expression for output x cannot be simplified further. Its implementation is shown in Figure 3.6(a).

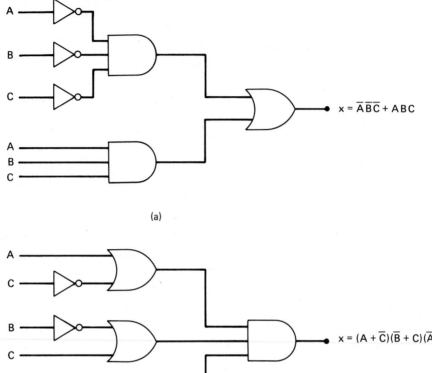

(a)

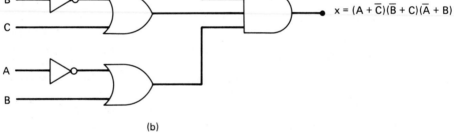

(b)

Figure 3.6   (a) S-of-P solution; (b) P-of-S solution for Example 3.2.

Product-of-sums solution:

$$\bar{x} = \underbrace{\bar{A}\bar{B}C}_{1} + \underbrace{\bar{A}B\bar{C}}_{2} + \underbrace{\bar{A}BC}_{3} + \underbrace{A\bar{B}\bar{C}}_{4} + \underbrace{A\bar{B}C}_{5} + \underbrace{AB\bar{C}}_{6}$$

There are several paths that can be followed in simplifying this expression. One possible path is as follows:

$$\bar{x} = \underbrace{\bar{A}C(\bar{B} + B)}_{1, 3} + \underbrace{B\bar{C}(\bar{A} + A)}_{2, 6} + \underbrace{A\bar{B}(\bar{C} + C)}_{4, 5}$$

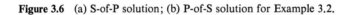

$$\bar{x} = \bar{A}C + B\bar{C} + A\bar{B}$$

This cannot be simplified further. Inverting both sides and applying DeMorgan's theorems,

$$x = \overline{\overline{A}C + B\overline{C} + A\overline{B}}$$
$$= \overline{\overline{A}C} \cdot \overline{B\overline{C}} \cdot \overline{A\overline{B}}$$
$$= (\overline{\overline{A}} + \overline{C}) \cdot (\overline{B} + \overline{\overline{C}}) \cdot (\overline{A} + \overline{\overline{B}})$$
$$= (A + \overline{C}) \cdot (\overline{B} + C) \cdot (\overline{A} + B)$$

This expression is the desired P-of-S solution. Its implementation is shown in Figure 3.6(b).

In Example 3.2 the P-of-S solution turned out to be somewhat more complex than the S-of-P solution. This can be attributed to the fact that there were only two cases where $x = 1$, whereas there were six cases where $\bar{x} = 1$. If it had been the other way around, with more cases where $x = 1$, the P-of-S solution would generally be simpler.

3.4 SUM-OF-PRODUCTS SOLUTION CONVERTED TO NAND GATES

It was shown in Chapter 2 that NAND gates can be used to implement any logic circuit. It turns out that a logic circuit which is in S-of-P form (ANDs followed by an OR) such as Figure 3.6(a) can easily be converted to all NAND gates without increasing the circuit complexity. To illustrate, let us convert the circuit of Figure 3.6(a) to NANDs. The original circuit is redrawn in Figure 3.7(a). To convert it to NANDs, each inverter, AND gate, and OR gate is replaced by its NAND-gate equivalent (see Figure 2.23). This results in the diagram of Figure 3.7(b). This diagram is a great deal more complex than the original circuit. However, it can be greatly simplified by eliminating the successive NAND inverters (NANDS 1 and 2 and NANDS 3 and 4), since two inversions in succession cancel each other out. This leaves us with the final diagram of Figure 3.7(c), which is the desired NAND implementation.

Comparing this final NAND circuit with the original circuit, it can be seen that they are identical in structure; that is, the NAND circuit could have been obtained simply by replacing each AND, OR, and inverter of the original circuit by a NAND gate. This characteristic is only true if the original circuit is in S-of-P form, that is, several AND gates feeding a single OR gate. The only exception to this is if the S-of-P circuit has a single-variable term, such as $x = A + BCD$. Here the A term will feed the OR gate directly. When converting to all NAND gates, a NAND inverter has to be placed after the A input. This is illustrated in

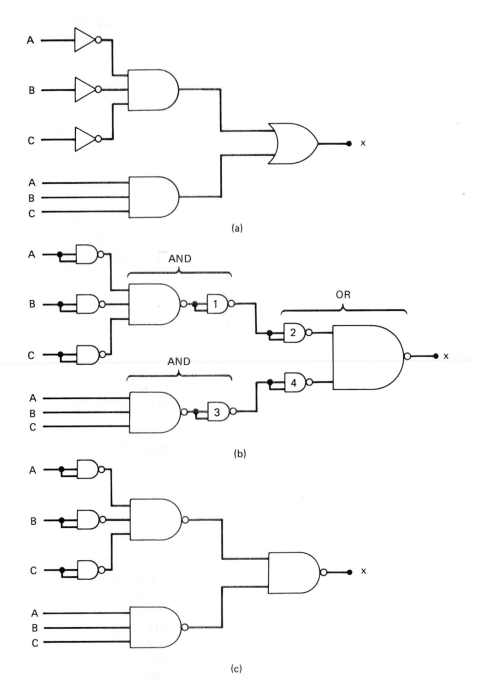

(a)

(b)

(c)

Figure 3.7 Converting a S-of-P circuit to all NAND gates.

Figure 3.8, which shows the S-of-P circuit and its equivalent NAND circuit for this case.

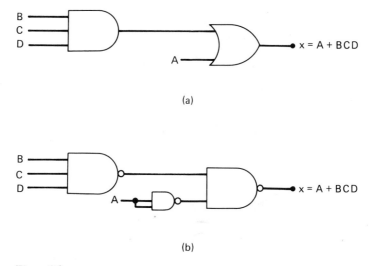

(a)

(b)

Figure 3.8

To summarize, when converting an S-of-P circuit to NANDs:

1. Replace all ANDs, ORs, and inverters with NAND gates.
2. Place a NAND inverter after any *single* variable that is feeding the final NAND gate.

The equivalencies of the circuits in Figures 3.7 and 3.8 can be easily proved by determining the output expression for each circuit and comparing them for equivalence.

3.5 PRODUCT-OF-SUMS SOLUTION CONVERTED TO NOR GATES

We have seen that a circuit implemented in S-of-P form can be easily converted to all NAND gates. In a similar manner, it can be shown that a P-of-S circuit can be easily converted to all NOR gates. To illustrate, refer to Figure 3.9(a), where the P-of-S circuit solution for Example 3.2 is redrawn. To convert it to NORs, each inverter, AND gate, and OR gate is replaced by its NOR-gate equivalent (see Figure 2.24). This results in the diagram of Figure 3.9(b). This diagram can be simplified by eliminating the successive NOR inverters, since two successive inversions cancel each other out. This leaves us with the final diagram of Figure 3.9(c), which is the desired NOR implementation.

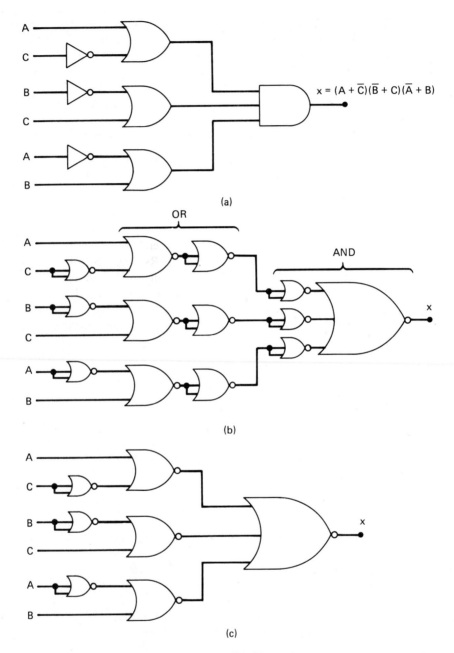

$$x = (A + \overline{C})(\overline{B} + C)(\overline{A} + B)$$

(a)

(b)

(c)

Figure 3.9 Converting a P-of-S circuit to all NOR gates.

Comparing this final NOR circuit with the original, it can be seen that each gate and inverter in the original circuit has essentially been replaced by a NOR gate. This characteristic is only true if the original circuit is in P-of-S form. The only exception is if the P-of-S circuit has a single variable feeding the AND gate, such as $x = A(B + C)(C + D)$. Here the A term will feed the AND gate directly. When converting to all NOR gates, a NOR inverter has to be placed after the A input.

To summarize, when converting a P-of-S circuit to all NORs:

1. Replace all ANDs, ORs, and inverters with NOR gates.

2. Place a NOR inverter after any *single* variable that is feeding the final NOR gate.

We have now seen how a S-of-P circuit can be easily converted to all NANDs and how a P-of-S circuit can be easily converted to all NORs. Actually, any circuit can be converted to all NANDs or all NORs simply by replacing each gate by its NAND or NOR equivalent. When designing a logic circuit that is to be implemented using NAND gates, the designer should perform the S-of-P solution (Section 3.2) and then convert it to NANDs. Likewise, if a NOR implementation is desired, the P-of-S solution should be performed (Section 3.3) and then converted to NORs.

3.6 EXCLUSIVE-OR AND EXCLUSIVE-NOR CIRCUITS

Two special logic circuits that occur quite often in digital systems are the *exclusive-OR* and *exclusive-NOR* circuits. Their description and operation will now be investigated.

Exclusive-OR

Consider the logic circuit of Figure 3.10(a). The output expression of this circuit is

$$x = \bar{A}B + A\bar{B}$$

The accompanying truth table shows that $x = 1$ for two cases: $A = 0$, $B = 1$ (the $\bar{A}B$ term) and $A = 1$, $B = 0$ (the $A\bar{B}$ term). In other words, *this circuit produces a high output whenever the two inputs are at opposite levels*. This is the exclusive-OR circuit, which will herafter be abbreviated EX-OR.

This particular combination of logic gates occurs quite often and is very useful in certain applications. In fact, the EX-OR circuit has been given a symbol of its own, shown in Figure 3.10(b). This symbol is assumed to contain all the logic

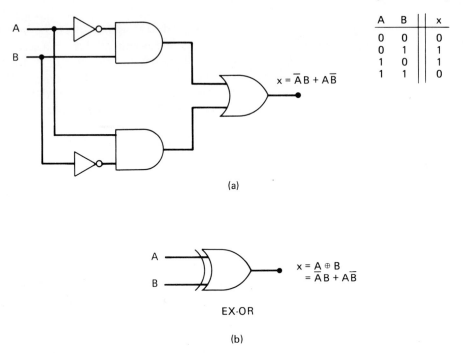

A	B	x
0	0	0
0	1	1
1	0	1
1	1	0

(a)

$x = \bar{A}B + A\bar{B}$

$x = A \oplus B$
$= \bar{A}B + A\bar{B}$

EX-OR

(b)

Figure 3.10   (a) Exlusive-OR circuit and truth table; (b) exclusive-OR gate symbol.

contained in the EX-OR circuit and therefore has the same logic expression and truth table. This EX-OR circuit is commonly referred to as an EX-OR *gate*, and we can consider it as another type of logic gate.

An EX-OR gate has only *two* inputs; there are no three-input or four-input EX-OR gates. The two inputs are combined such that $x = \bar{A}B + A\bar{B}$. A shorthand way that is sometimes used to indicate the EX-OR output expression is

$$x = A \oplus B$$

where the symbol $\oplus$ represents the EX-OR gate operation.

The characteristics of an EX-OR gate are summarized as follows:

1. It has only two inputs and its output is

$$x = \bar{A}B + A\bar{B} = A \oplus B$$

2. Its output is *high* only when the two inputs are at *opposite* levels.

Exclusive-NOR

The exclusive-NOR circuit (abbreviated EX-NOR) operates completely opposite to the EX-OR circuit. Figure 3.11(a) shows an EX-NOR circuit and its accompanying truth table. The output expression is

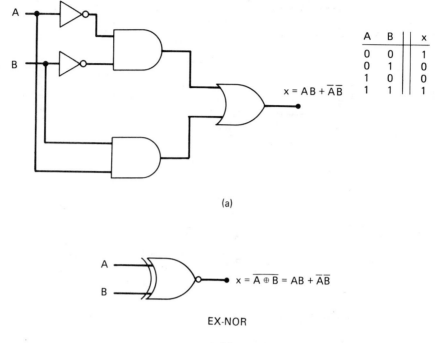

A	B	x
0	0	1
0	1	0
1	0	0
1	1	1

$x = AB + \bar{A}\bar{B}$

(a)

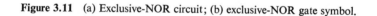

$x = \overline{A \oplus B} = AB + \bar{A}\bar{B}$

EX-NOR

(b)

Figure 3.11   (a) Exclusive-NOR circuit; (b) exclusive-NOR gate symbol.

$$x = AB + \bar{A}\bar{B}$$

which indicates along with the truth table that x will be 1 for two cases: $A = B = 1$ (the AB term) and $A = B = 0$ (the $\bar{A}\bar{B}$ term). In other words, *this circuit produces a high output whenever the two inputs are at the same level.**

It should be apparent that the output of the EX-NOR circuit is the exact inverse of the output of the EX-OR circuit. The symbol for an EX-NOR gate is obtained by simply adding a small circle at the output of the EX-OR symbol [Figure 3.11(b)].

The EX-NOR gate also has only *two* inputs, and it combines them such that its output is

$$x = AB + \bar{A}\bar{B}$$

A shorthand way to indicate the output expression of the EX-NOR is

$$x = \overline{A \oplus B}$$

which is simply the inverse of the EX-OR operation. The EX-NOR gate is summarized as follows:

*For this reason it is often called a "coincidence gate."

1. It has only two inputs and its output is

$$x = AB + \bar{A}\bar{B} = \overline{A \oplus B}$$

2. Its output is *high* only when the two inputs are at the same level.

It can be proved algebraically that the output of the EX-NOR is the exact inverse of an EX-OR:

$$\overline{AB + \bar{A}\bar{B}} = \overline{AB} \cdot \overline{\bar{A}\bar{B}} \qquad \text{(using DeMorgan's theorems)}$$
$$= (\bar{A} + \bar{B}) \cdot (A + B)$$
$$= \bar{A}A + \bar{A}B + A\bar{B} + B\bar{B} \qquad \text{(multiplying out)}$$

The first and last terms are 0 (remember that $x \cdot \bar{x} = 0$), so the end result is $\bar{A}B + A\bar{B}$, which is the EX-OR expression.

EXAMPLE 3.3 $x_1 x_0$ represents a 2-bit binary number that can have any value (00, 01, 10, or 11); for example, when $x_1 = 1$ and $x_0 = 0$, the binary number is 10, and so on. Similarly, $y_1 y_0$ represents another 2-bit binary number. Design a logic circuit, using x_1, x_0, y_1, and y_0 inputs, whose output will be *high* only when the two binary numbers $x_1 x_0$ and $y_1 y_0$ are *equal*.

Solution: The first step is to construct a truth table for the 16 (2^4) possible cases.

x_1	x_0	y_1	y_0	z (output)
0	0	0	0	1
0	0	0	1	0
0	0	1	0	0
0	0	1	1	0
0	1	0	0	0
0	1	0	1	1
0	1	1	0	0
0	1	1	1	0
1	0	0	0	0
1	0	0	1	0
1	0	1	0	1
1	0	1	1	0
1	1	0	0	0
1	1	0	1	0
1	1	1	0	0
1	1	1	1	1

The output z has to be high whenever the $x_1 x_0$ values match the $y_1 y_0$ values, that is, whenever $x_1 = y_1$ and $x_0 = y_0$. The table shows that there are four such cases.

We could now continue with the normal procedure, which would be to obtain a S-of-P expression for z, attempt to simplify it, and then implement the result. However, the nature of this problem makes it ideally suited for implementation using EX-NOR gates, and a little thought will produce a simple solution with minimum work. Refer to Figure 3.12; in this logic diagram x_1 and y_1 are fed to one EX-NOR gate and x_0 and y_0 are fed to another EX-NOR gate. The output of each EX-NOR will be high only when its inputs are equal. Thus, for $x_0 = y_0$ and $x_1 = y_1$ both EX-NOR outputs will be high. This is the condition we are looking for, because it means that the two 2-bit numbers are equal. The AND-gate output will be high only for this case, thereby producing the desired output.

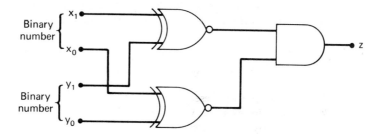

Figure 3.12 Circuit for detecting equality of two 2-bit binary numbers.

EXAMPLE 3.4 Simplify $x = \bar{A}BC + BC\bar{D} + A\bar{B}D + A\bar{C}D$.

Solution: Factoring,

$$x = BC(\bar{A} + \bar{D}) + AD(\bar{B} + \bar{C})$$

Using DeMorgan's theorem,

$$x = BC(\overline{AD}) + AD(\overline{BC})$$

This expression should be recognized as the EX-OR combination of the BC and AD terms. That is,

$$x = BC \oplus AD$$

This can be implemented using two two-input AND gates for the BC and AD terms and one EX-OR gate. The original expression for x requires four three-input AND gates, one four-input OR gate, and four inverters.

3.7 LOGIC CIRCUITS WITH MULTIPLE OUTPUTS

Many logic design problems involve more than one output. In other words, several outputs are generated from the same inputs. For these situations the outputs can be treated separately, with a complete design procedure followed for each output. Sometimes when this is done and the final output expressions are obtained, there may be some common terms among the expressions that may be shared. To illustrate, suppose that we have a two-output problem, where the output expressions are

$$x = AB + \bar{B}C$$
$$y = AB\bar{C} + \bar{A}B$$

Note that x contains an AB term. The y expression has an $AB\bar{C}$ term, which is the same as $(AB) \cdot \bar{C}$. Thus if AB is generated for the x output, it can be used to generate $AB\bar{C}$ for the y output.

This is illustrated in Figure 3.13. For this simple example, the savings realized by sharing common terms is not great (one gate-input connection is saved); however, in a more complex multiple-output problem, the savings can be considerable if the designer is clever in his use of "term sharing."

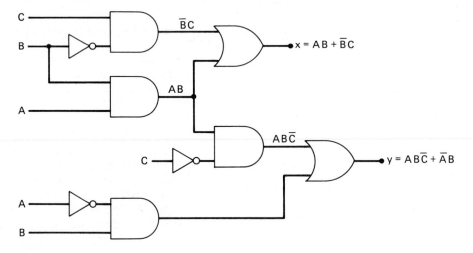

Figure 3.13 Example of term sharing.

3.8 DESIGNING WITHOUT A TRUTH TABLE

Occasionally a logic design problem is simple enough to perform without constructing a truth table, because the output expression can be written directly from the problem statement. The following examples illustrate.

EXAMPLE 3.5 Design a logic circuit with inputs A, B, C, and D such that the output is high *only* when both *C and D* are low *while* either *A or B* is high.

Solution: The output expression can be written as

$$x = (A + B) \cdot \overline{C + D}$$

Let us examine how this was obtained. First, note that A is ORed with B since either one has to be high for a high output. Next, note that the $\overline{C + D}$ term produces a high *only* if C and D are both low (this should be recognized as a NOR-gate term). The two terms $A + B$ and $\overline{C + D}$ are then ANDed since both conditions have to exist to produce a high output.

EXAMPLE 3.6 Design a circuit whose output goes high only when either A or B is low while C and D are not both high.

Solution:

$$x = \overline{AB} \cdot \overline{CD}$$

or

$$x = (\bar{A} + \bar{B}) \cdot (\bar{C} + \bar{D})$$

EXAMPLE 3.7 Design a logic circuit with two inputs A and B and two outputs X and Y which operates as follows:

1. $X = Y = 0$ while $A = 0$, regardless of the value of B.
2. If a positive pulse occurs at A, the pulse will also appear at output X if $B = 1$, or at output Y if $B = 0$.

Solution: Since both X and Y outputs must be 0 as long as $A = 0$, it should be clear that X and Y must be derived from the outputs of separate AND gates that have A as one input. The other inputs to the AND gates must be chosen to produce the operation in step 2. To produce a high pulse at output X when A pulses high requires that B also be 1. Thus, A and B should feed an AND gate to produce X. Similarly, A and $\bar{B}$ should feed an AND gate to produce Y, since Y should follow the pulse at A whenever $B = 0$ (or equivalently, $\bar{B} = 1$). Figure 3.14 shows the final circuit. Check it out to see that it satisfies the problem statement.

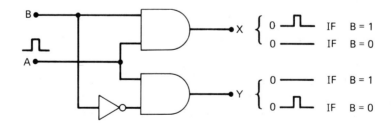

Figure 3.14

The preceding examples may not seem as apparent now as they will after you have gained more experience in logic work. It should be obvious, though, that considerable work can be saved in problems of this type if the designer can simply write down the desired circuit operation in Boolean form. Of course, many design problems are too complex for this and must be performed using the methods described previously.

QUESTIONS AND PROBLEMS

3.1   Construct a logic circuit whose output is high only when $A = 1$, $B = 0$, and $C = 1$.

3.2   Construct a logic circuit whose output is high only when $A = 0$, $B = 1$, and $C = 1$.

3.3   Construct a logic circuit whose output is high for *either* of the conditions of Problems 3.1 and 3.2.

3.4   Design the S-of-P logic circuit corresponding to the following truth table:

A	B	C	x
0	0	0	1
0	0	1	0
0	1	0	1
0	1	1	1
1	0	0	1
1	0	1	0
1	1	0	0
1	1	1	1

3.5   Design the S-of-P circuit for a logic circuit whose output is *high only* when a majority of inputs A, B, and C are *low*.

3.6   Develop the P-of-S solution for the truth table of Problem 3.4. Compare the circuit complexity with the S-of-P solution obtained previously.

3.7   Develop the P-of-S solution for the circuit of Problem 3.5.

3.8   Design a logic circuit whose output is *low* whenever the inputs A, B, and C are all at the same level. Use either S-of-P or P-of-S, whichever you think is best.

3.9   A 4-bit binary number is represented as $A_3A_2A_1A_0$, where A_3, A_2, A_1, and A_0 represent the individual bits with A_0 equal to the LSB. Design a logic circuit that will determine whenever the binary number is greater than 6.

3.10 A certain logic application requires a circuit with logic expression $x = A + BCD + \bar{B}\bar{C}D + \bar{A}\bar{B}C\bar{D}$.
   (a) Implement this expression using only NAND gates.
   (b) Implement this expression using only NOR gates.

3.11 (a) Implement the following expression using all NOR gates: $x = B(B + C + \bar{D})(\bar{A} + \bar{C} + D)$.
   (b) Implement this expression using all NAND gates.

3.12 Implement your design of Problem 3.5 using either all NAND or all NOR gates, whichever is easier.

3.13 Implement your design of Problem 3.6 using either all NAND or all NOR gates, whichever is easier.

3.14 Determine the output waveform for the circuit of Figure 3.15.

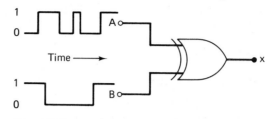

Figure 3.15

3.15 Apply the waveforms in Figure 3.15 to an EX-NOR circuit and draw the output waveform.

3.16 In Example 3.2 a logic circuit was designed to detect whenever three flashing lights got out of synchronization. Redo this design problem utilizing EX-ORs and one other gate.

3.17 Figure 3.16 represents a multiplier circuit that takes two 2-bit binary numbers x_1x_0 and y_1y_0 and produces an output binary number $Z_3Z_2Z_1Z_0$ that is equal to the arithmetic product of the two input numbers. Design the logic circuit for the multiplier. (*Hint:* The logic circuit will have four inputs and four outputs.)

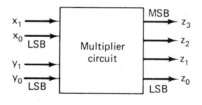

Figure 3.16

3.18 Figure 3.17 represents a *relative magnitude detector* that takes two 3-bit binary numbers $x_2x_1x_0$ and $y_2y_1y_0$ and determines whether they are equal and, if not, which one is larger. There are three outputs, defined as follows:

(a) $M = 1$ only if the two input numbers are equal.
(b) $N = 1$ only if $x_2x_1x_0$ is greater than $y_2y_1y_0$.
(c) $P = 1$ only if $y_2y_1y_0$ is greater than $x_2x_1x_0$.

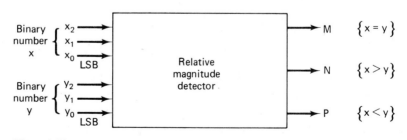

Figure 3.17

Design the logic circuitry for the comparator. The circuit has *six* inputs and *three* outputs and is therefore much too complex to handle using the truth-table approach. Refer to Example 3.3 as a hint to how you might start to solve this problem.

3.19   Design a logic circuit whose output is high whenever A and B are both high, while C and D are either both low or both high. Try to do this without using a truth table. Then check your result by constructing a truth table for your circuit to see if it agrees with the problem statement.

3.20   Four large tanks at a chemical plant contain different liquids being heated. Liquid-level sensors are being used to detect whenever the level in tanks A and B rises above a predetermined level. Temperature sensors in tanks C and D detect when the temperature in these tanks drops below a prescribed temperature limit. Assume that the liquid-level sensor outputs A and B are *low* when the level is satisfactory and *high* when the level is too high. Also, the temperature-sensor outputs C and D are *low* when the temperature is satisfactory and *high* when the temperature is too low. Design a logic circuit that will detect whenever the level in tank A or tank B is too high at the same time that the temperature in either tank C or tank D is too low.

3.21   Figure 3.18 shows the intersection of a main highway with a secondary access road. Vehicle-detection sensors are placed along lanes C and D (main road) and lanes A and B (access road). These sensor outputs are low (0) when no vehicle is present and high (1) when a vehicle is present. The intersection traffic light is to be controlled according to the following logic:
(a)   The E-W traffic light will be green whenever *both* lanes C and D are occupied.
(b)   The E-W light will be green whenever *either* C or D are occupied but lanes A and B are not *both* occupied.
(c)   The N-S light will be green whenever *both* lanes A and B are occupied but C and D are not *both* occupied.
(d)   The N-S light will also be green when *either* A or B is occupied while C and D are *both* vacant.
(e)   The E-W will be green when *no* vehicles are present. Using the sensor outputs A, B, C, and D as inputs, design a logic circuit to control the traffic light. There should be two outputs, N/S and E/W, which go high when the corresponding light is to be *green*. Simplify the circuit as much as possible and show *all* steps.

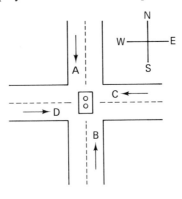

Figure 3.18

3.22 Design a logic circuit with two inputs, *A* and *B*, and two outputs, *X* and *Y*, so that is operates as follows:
(1)   *X* and *Y* are both HIGH as long as *A* is HIGH, regardless of the level of *B*.
(2)   If *A* pulses LOW, the LOW will appear at *X* if *B* = 0 or at *Y* if *B* = 1.

3.23 Design a logic circuit with inputs *A* and *B* and output *X* such that *X* will equal *A* whenever *B* = 0 and *X* will equal *Ā* whenever *B* = 1. This circuit is called a *controlled inverter* because *B* determines whether or not the input will be inverted.

3.24 Determine the output waveform of the circuit in Figure 3.19 for the two cases *B* = 0 and *B* = 1. This illustrates how an EX-OR can be used as a controlled inverter.

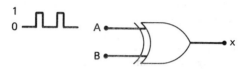

Figure 3.19

FLIP-FLOPS

The logic circuits considered thus far have been combinatorial circuits whose output levels at any instant of time are dependent upon the levels present at the inputs at that time. Any prior input-level conditions have no effect on the present outputs because combinatorial logic circuits have no memory. Most digital systems are made up of both combinatorial circuits and memory elements.

Figure 4.1 shows a block diagram of a general digital system that combines combinatorial logic gates with memory devices. The combinatorial portion accepts logic signals from external inputs and from the outputs of the memory elements. The combinatorial circuit operates on these inputs to produce various outputs, some of which are used to determine the binary values to be stored in the memory elements. The outputs of some of the memory elements, in turn, go to the inputs of logic gates in the combinatorial circuits. This process indicates that the external outputs of a digital system are a function of both its external inputs and the information stored in its memory elements.

The most widely used memory element is the flip-flop, which we shall study thoroughly in this chapter. The *flip-flop* (abbreviated FF) is a logic circuit with two outputs, which are the inverse of each other. Figure 4.2 indicates these outputs as Q and $\bar{Q}$ (actually any letter could be used, but Q is the most common). The Q output is called the *normal* FF output and $\bar{Q}$ is the *inverted* FF output. When a FF is said to be in the high (1) state or the low (0) state, this is the condition at the Q output. Of course, the $\bar{Q}$ output is always the inverse of Q.

There are two possible operating states for the FF: (1) $Q = 0, \bar{Q} = 1$; and (2) $Q = 1, \bar{Q} = 0$. The FF has one or more inputs, which are used to cause the FF to switch back and forth between these two states. As we shall see, once an input signal causes a FF to go to a given state, the FF will remain in that state even after that input signal is terminated. This is its *memory* characteristic.

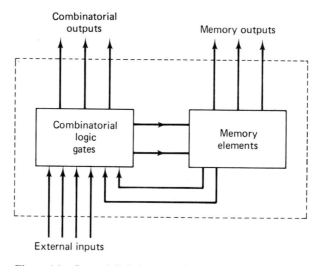

Figure 4.1 General digital system diagram.

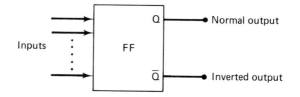

Figure 4.2 General flip-flop symbol.

The flip-flop, incidentally, is known by several other names, including *bistable multivibrator*, *latch*, and *binary*, but we will generally use flip-flop because it is the most common designation in the digital field. Other memory elements, which we will study in a later chapter, are also used in digital systems, but flip-flops are the most versatile because of their high speed of operation, the ease with which information can be stored into and read out of them, and the ease with which they can be interconnected with logic gates.

4.1 FLIP-FLOPS FROM LOGIC GATES

A basic FF circuit can be constructed from two NOR gates connected as shown in Figure 4.3. Notice that the output of NOR-1 serves as one of the inputs to NOR-2, and vice versa. The two outputs are Q and $\bar{Q}$, which are always the inverse of one another during normal operation. The two inputs are labeled SET and CLEAR, for reasons that will soon become apparent.

Let us begin our analysis of the NOR FF circuit by making both inputs low

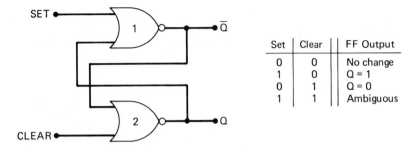

Set	Clear	FF Output
0	0	No change
1	0	Q = 1
0	1	Q = 0
1	1	Ambiguous

Figure 4.3   NOR gate flip-flop.

(SET = CLEAR = 0). In this situation we cannot determine what the Q and $\bar{Q}$ output values are since there are *two*, equally likely, possibilities: (1) $Q = 0$, $\bar{Q} = 1$, and (2) $Q = 1$, $\bar{Q} = 0$. To verify this, let us first assume that Q is 0. This 0 and the 0 from the SET input produce a 1 at the NOR-1 output, thereby making $\bar{Q} = 1$ (recall that a NOR output is 1 *only* when all its inputs are 0). This $\bar{Q} = 1$ is fed to the NOR-2 input, thereby producing a 0 at its output; so $Q = 0$, as was originally assumed.

Now, let us assume instead that $Q = 1$. This 1 applied to NOR-1 produces a 0 at its output, so $\bar{Q} = 0$. The $\bar{Q} = 0$ is fed to NOR-2 together with the 0 from the CLEAR input, thereby producing a 1 at its output; so $Q = 1$, as was originally assumed.

Thus, with SET and CLEAR both 0, the FF outputs can be in either state. Actually, the FF output state will depend on what has previously occurred at the inputs. The SET = CLEAR = 0 condition will *not* affect the FF outputs; they will simply remain in whatever state they happen to be in at the time. This is the first case shown in the table in Figure 4.3 and represents the "normal" state. In other words, the SET and CLEAR inputs are normally in the 0 state.

To make the FF go to a specific state, we must put a 1 on the appropriate input. To make $Q = 1$, we must apply a 1 to the SET while keeping CLEAR = 0. The 1 at the SET input causes NOR-1 to go to 0, so $\bar{Q} = 0$. This 0 fed to NOR-2 along with CLEAR = 0 causes NOR-2 to produce $Q = 1$. Thus, a 1 on the SET input (while CLEAR = 0) will always produce $Q = 1$ ($\bar{Q} = 0$). This 1 need only be present long enough to allow the gates to respond and pass the signal. When SET returns to 0, so that both inputs are 0, the FF will remain in the $Q = 1$ state. This is the second case listed in the table of Figure 4.3.

Clearly, since the circuit is completely symmetrical, it can be seen that a 1 applied to the CLEAR input while SET = 0 will produce a FF output state of $Q = 0$ ($\bar{Q} = 1$). When the CLEAR input returns to 0, the FF will remain in the $Q = 0$ state. This is the third case listed in the table of Figure 4.3.

The last case to consider is SET = CLEAR = 1. This condition will produce 0 at the output of both NOR gates, so $Q = 0$ and $\bar{Q} = 0$. This is obviously an illegal condition if it is desired that the FF outputs be the inverse of each other.

Furthermore, when the inputs are returned to 0, the FF output state will depend on which input reaches 0 first. This makes the SET = CLEAR = 1 condition ambiguous. For these reasons the latter case is never purposely used during the operation of this type of FF.

Summary of NOR FF

1. SET = 1, CLEAR = 0 always produces $Q = 1$, regardless of the prior state of FF output. This is called *setting* the FF to the 1 state or high state.

2. SET = 0, CLEAR = 1 always produces $Q = 0$, regardless of the prior FF output state. This is called *clearing* the FF to the 0 state or low state.

3. SET = 0, CLEAR = 0 does not affect the FF state. It remains in its prior state. This is the normal resting state of the FF inputs.

4. SET = 1, CLEAR = 1 is ambiguous and should not be used.

5. The 1s on the SET or CLEAR, which are used to cause the FF output to change states, can be either a constant level (dc) or a momentary pulse. When SET and CLEAR both return to 0, the Q output maintains its new state.

The NAND FF

Another basic FF circuit can be constructed using cross-coupled NAND gates as shown in Figure 4.4(a). The operation of this FF can be analyzed in the manner employed for the NOR FF. The operation can be summarized as follows:

1. SET = 1, CLEAR = 1 has no affect on the FF outputs. This is the normal resting state.

2. To set the FF to the $Q = 1$ state requires that SET = 0 while CLEAR = 1.

3. To clear the FF to the $Q = 0$ state requires that CLEAR = 0 while SET = 1.

4. The SET = 0, CLEAR = 0 condition is ambiguous and should not be used.

Comparing the NAND FF of Figure 4.4(a) with the NOR FF, we can see that they operate in basically the same manner except for the following difference: the NOR FF inputs are normally 0 and must be pulsed to the 1 state to change the state of the FF outputs; the NAND FF inputs are normally 1 and must be pulsed to 0 to change the FF output state.

Stated another way, the NOR FF has active HIGH inputs and the NAND FF has active LOW inputs. This difference is easier to see if we redraw the NAND FF using the equivalent representation for NAND gates (Section 2.14) as shown in

Figure 4.4(c). Here the circles on the SET and CLEAR inputs indicate normally HIGH inputs that are activated by a LOW logic level.

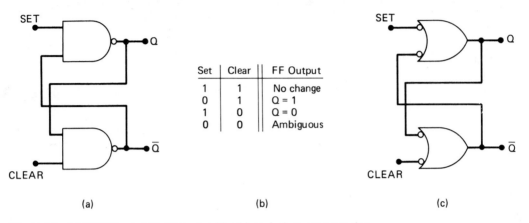

Set	Clear	FF Output
1	1	No change
0	1	Q = 1
1	0	Q = 0
0	0	Ambiguous

(a)   (b)   (c)

Figure 4.4   (a) NAND-gate FF; (b) truth table; (c) equivalent representation.

4.2 SET-CLEAR FLIP-FLOPS

The two FF circuits described above are examples of *SET-CLEAR (S-C) flip-flops.* The general logic symbols used to represent the S-C FF are shown in Figure 4.5 with their corresponding truth tables. Figure 4.5(a) represents the S-C FF that

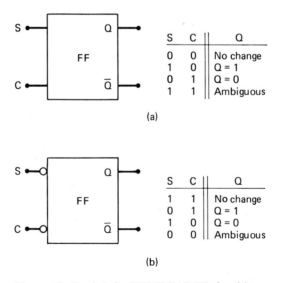

S	C	Q
0	0	No change
1	0	Q = 1
0	1	Q = 0
1	1	Ambiguous

(a)

S	C	Q
1	1	No change
0	1	Q = 1
1	0	Q = 0
0	0	Ambiguous

(b)

Figure 4.5   Symbols for SET-CLEAR FFs for: (a) type that responds to high inputs; (b) type that responds to low inputs.

responds to *high* levels on its S and C inputs, such as the NOR-gate FF of Figure
4.3. Figure 4.5(b) represents the S-C FF that responds to *low* levels on its S and C
inputs, such as the NAND-gate FF. Note the small circles shown on the S and C
inputs to indicate that this FF responds to 0s on these inputs. We shall generally
use these block symbols to represent S-C FFs instead of showing the complete
internal circuitry.

EXAMPLE 4.1   Each of the FFs in Figure 4.6 is initially in the low state ($Q = 0$).
Determine the FF output in response to the inputs shown.

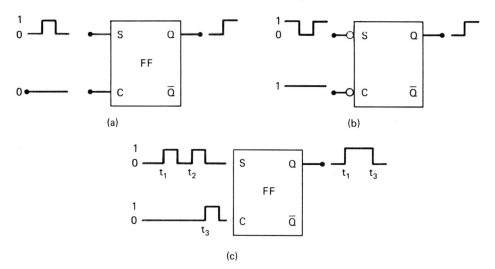

Figure 4.6

Solution:
(a)   This FF responds to *high* inputs. Therefore, Q will go to 1 when S jumps from
0 to 1. Q will remain high when the S input jumps back to 0.
(b)   This FF responds to *low* inputs. Therefore, nothing will happen as long as
$S = C = 1$. But when S goes low, this will set the FF to the $Q = 1$ state, where
it will remain even after S has returned to 1.
(c)   This FF responds to *high* inputs. The first pulse at S will set $Q = 1$. The second
pulse at S will have no effect, since $Q = 1$ already. The subsequent pulse at C will
clear the FF back to the $Q = 0$ state.

The S-C flip flop forms the basis for many other types of flip-flop circuits,
which will be described subsequently. By itself the S-C FF is useful as a memory
element to store information. To illustrate, Figure 4.7 shows a simple burglar-
alarm circuit using a S-C FF. The FF is initially in the $Q = 0$ state, so the alarm
is deactivated. With the photocell illuminated, the transistor is saturated, so
$V_x \approx 0$, applying a low to the S input. If the photocell goes dark, the transistor is
turned off and $V_x = 6$ V. This applies a high to the S input, causing the FF to go
to the $Q = 1$ state, thereby activating the alarm. The FF, once it has been set to

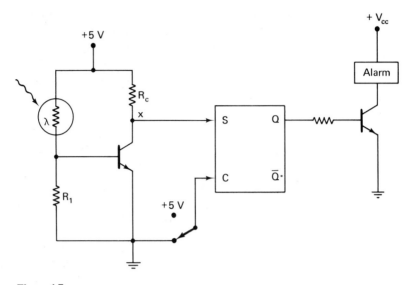

Figure 4.7

$Q = 1$, will stay there (acts as memory) even if the photocell immediately goes light again. Thus, the alarm will persist until the FF is cleared to the $Q = 0$ state by applying a momentary 1 to the C input.

It should be mentioned here that the term "RESET" is commonly used to mean the same thing as "CLEAR." In other words, the CLEAR input is also called the RESET input and the operation of clearing a FF is also called *resetting* a FF. These terms are both in common use and will be used interchangeably throughout the text.

4.3 CLOCK SIGNALS

Most digital systems operate as *synchronous sequential systems*. What this means is that the sequence of operations that takes place is synchronized by a *master clock signal*, which generates periodic pulses that are distributed to all parts of the system. This clock signal is usually one of the forms shown in Figure 4.8; very often it is a squarewave (50 per cent duty cycle), such as the one shown in Figure 4.8(b).

The clock signal is the signal that causes things to happen at regularly spaced intervals. In particular, operations in the system are made to take place at times when the clock signal is making a transition from 0 to 1 or from 1 to 0. These transition times are pointed out in Figure 4.8. The 0-to-1 transition is called the *rising edge* or *positive-going edge* of the clock signal; the 1-to-0 transition is called the *falling edge* or *negative-going edge* of the clock signal.

The synchronizing action of the clock signal is the result of using *clocked*

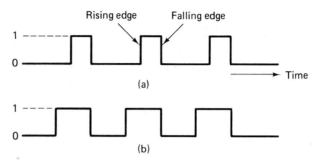

Figure 4.8 Clock signals.

flip-flops, which are designed to change states on either (but not both) the rising edge or the falling edge of the clock signal. In other words, the clocked FFs will change states at the appropriate clock transition and will rest between successive clock pulses. The frequency of the clock pulses is generally determined by how long it takes the FFs and gates in the circuit to respond to the level changes initiated by the clock pulse, that is, the propogation delays of the various logic circuits. In Section 4.4 we will begin the study of the various types of clocked flip-flops, which are being used extensively in most digital systems.

4.4 THE CLOCKED S-C FLIP-FLOP

Figure 4.9(a) shows the logic symbol for a *clocked S-C flip-flop* which is triggered by the positive-going edge of the clock signal. This means that the FF will change states *only* when a signal applied to its clock input (abbreviated *CLK*) makes a transition from 0 to 1. The S and C inputs control the state of the FF in the same manner as described earlier for the basic (unclocked) S-C flop-flop, but the FF does not respond to these inputs until the occurrence of the rising edge of the clock signal. This is illustrated by the waveforms in Figure 4.9(b), which can be analyzed as follows:

1. Initially all inputs are 0 and the Q output is 0.

2. When the rising edge of the first clock pulse occurs (point a), the S and C inputs are both 0, so the FF is not affected and remains in the $Q = 0$ state.

3. At the occurrence of the rising edge of the second clock pulse (point c), the S input is now high, with C still low. Thus, the FF sets to the 1 state at the rising edge of this clock pulse.

4. When the third clock pulse makes its positive transition (point e), it finds that $S = 0$ and $C = 1$, which causes the FF to clear to the 0 state.

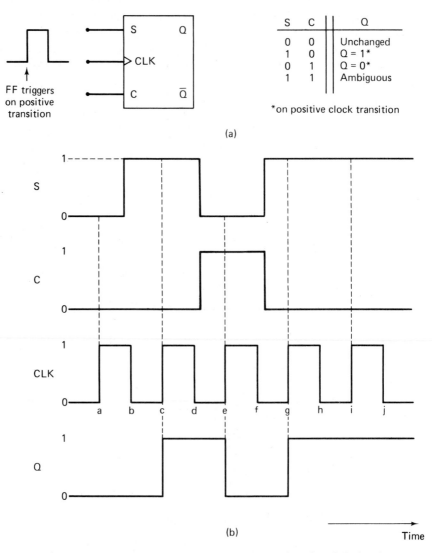

S	C	Q
0	0	Unchanged
1	0	Q = 1*
0	1	Q = 0*
1	1	Ambiguous

*on positive clock transition

(a)

(b)   Time

Figure 4.9 (a) Clocked S-C FF that responds to positive-going edge of clock pulse; (b) waveforms.

5. The fourth pulse sets the FF once again to the $Q = 1$ state (point *g*) because $S = 1$ and $C = 0$ when its positive edge occurs.

6. The fifth pulse also finds that $S = 1$ and $C = 0$ when it makes its positive-going transition. However, Q is already high, so it remains in that state.

7. The $S = C = 1$ condition should not be used, because it results in an ambiguous condition.

It should be noted from these waveforms that the FF is not affected by the negative-going edge of the clock pulses. Also, note that the S and C levels have no effect on the FF, except upon the occurrence of a positive-going transition of the clock signal. The S and C inputs are essentially *control* inputs, which control which state the FF will go to when the clock pulse occurs; the CLK input is the *trigger* input, which causes the FF to change states according to what the S and C inputs are when the clock edge occurs.

The symbol for the edge-triggered S-C FF shows a small triangle on the CLK input. This triangle is used to indicate that the CLK input responds only to a signal *transition* and is not affected by levels. The triangle symbol is not used by all logic circuit manufacturers, but it is gaining wider acceptance, so we will use it throughout the text.

Figure 4.10 shows the symbol for a clocked S-C flip-flop that triggers on the *negative*-going transition at its CLK input. The small circle and triangle on the CLK input indicates that this FF will trigger only when the CLK input goes from 1 to 0. This FF operates in the same manner as the positive-edge FF of Figure 4.9 except that the output can change states only on the falling edge of the clock pulses (points $b, d, f, h,$ and j). Both positive-edge and negative-edge triggering FFs are used in digital systems.

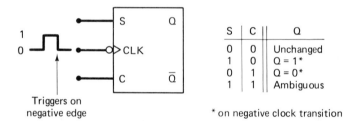

S	C	Q
0	0	Unchanged
1	0	Q = 1*
0	1	Q = 0*
1	1	Ambiguous

Triggers on
negative edge

* on negative clock transition

Figure 4.10   Clocked S-C FF that triggers on negative-going transactions.

4.5 THE CLOCKED J-K FLIP-FLOP

Figure 4.11(a) shows a *clocked J-K flip-flop* that is triggered by the positive-going edge of the clock signal. The J and K inputs control the state of the FF in the same way as the S and C inputs do for the clocked S-C FF except for one major difference: *the $J = K = 1$ condition does not result in an ambiguous output.* For this 11 condition, the FF will always go to the opposite state upon the positive transition of the clock signal. This is called the *toggle* mode of operation. In this mode, if both J and K are left high, the FF will change states (toggle) for each clock pulse. The operation of this FF is illustrated by the waveforms in Figure 4.11(b), which can be analyzed as follows:

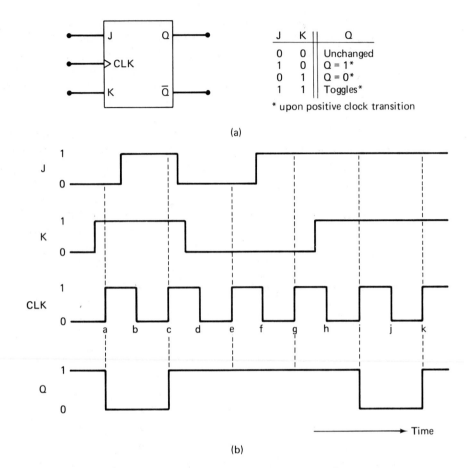

J	K		Q
0	0		Unchanged
1	0		Q = 1*
0	1		Q = 0*
1	1		Toggles*

* upon positive clock transition

(a)

(b)

Figure 4.11   (a) Clocked J-K FF that responds to positive edge of clock; (b) wave-forms.

1. Initially all inputs are 0 and the Q output is 1.

2. When the positive-going edge of the first clock pulse occurs (point *a*) the $J = 0$, $K = 1$ condition exists. Thus, the FF will be cleared to the $Q = 0$ state.

3. The second clock pulse finds $J = K = 1$ when it makes its positive transition (point *c*). This causes the FF to toggle to its opposite state, $Q = 1$.

4. At point *e* on the clock waveform, J and K are both 0, so the FF does not change states on this transition.

5. At point *g*, $J = 1$ and $K = 0$. This is the condition that sets Q to the 1 state. However, it is already 1, so it will remain there.

6. At point $i, J = K = 1$, so the FF toggles to its opposite state. The same thing occurs at point *k*.

It should also be noted from these waveforms that the FF is not affected by the negative-going edge of the clock pulses. Also, the *J* and *K* input levels have no effect except upon the occurrence of the positive-going edge of the clock signal. The *J* and *K* inputs by themselves cannot cause the FF to change states.

Figure 4.12 shows the symbol for a clocked J-K flip-flop that triggers on the negative-going clock-signal transitions. The small circle on the *CLK* input indicates that this FF will trigger when the *CLK* input goes from 1 to 0. This FF operates in the same manner as the positive-edge FF of Figure 4.11 except that the output can change states only on negative-going clock-signal transitions (points *b, d, f, h,* and *j*). Both polarities of edge-triggered J-K FFs are in common usage.

The J-K FF is much more versatile than the S-C FF because it has no ambiguous states. The $J = K = 1$ condition, which produces the toggling operation, finds extensive use in all types of binary counters. In essence, the J-K FF can do anything the S-C FF can do *plus* it has the toggle mode of operation. Because of this, the J-K FF presently enjoys widespread use in almost all modern digital systems.

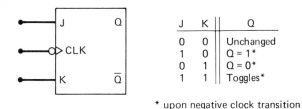

J	K		Q
0	0		Unchanged
1	0		Q = 1*
0	1		Q = 0*
1	1		Toggles*

* upon negative clock transition

Figure 4.12 J-K FF that triggers on negative-going transitions.

4.6 THE CLOCKED D FLIP-FLOP

Figure 4.13(a) shows the symbol for a *clocked D-type flip-flop* which triggers on positive transitions at the *CLK* input. The *D* input is a single control input which determines the state of the FF according to the accompanying truth table. Essentially, the FF output *Q* will go to the same state that is present on the *D* input whenever a positive transition occurs at the *CLK* input. This is illustrated by the waveforms in Figure 4.13(b).

Note that each time a positive transition occurs on the *CLK* input, the *Q* output takes on the same value as the level present at the *D* input. The negative transitions at the *CLK* input have no effect. The levels present at the *D* input have no effect until a positive clock transition occurs.

Negative-edge-triggered D FFs are also available and operate similarly except that they trigger on the negative-going clock transitions. The symbol for the negative-edge-triggered D FF has a small circle on the *CLK* input.

The D FF is used principally in the transfer of binary data, as illustrated in

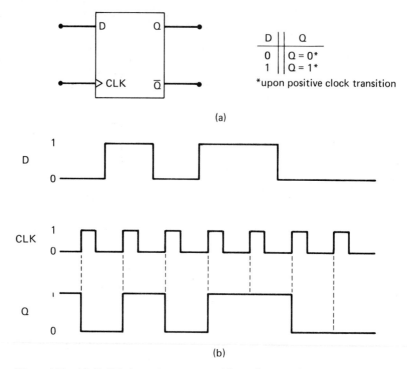

(a)

(b)

Figure 4.13 (a) D FF that triggers on positive-going transitions; (b) wave-forms.

Figure 4.14. Here the outputs *X*, *Y*, and *Z* of a combinatorial logic circuit are to be transferred to FFs *Q*1, *Q*2, and *Q*3 for storage. Using the D FFs, the levels present at *X*, *Y*, and *Z* will be transferred to Q_1, Q_2, and Q_3 respectively, upon application of a pulse to the common *CLK* inputs. The FFs can store these values for subsequent processing. This is an example of *parallel* transfer of binary data; the three bits *X*, *Y*, and *Z* are all transferred simultaneously.

S-C and J-K flip-flops can easily be modified to operate as D flip-flops. This is illustrated in Figure 4.15. Either of these arrangements may be used to perform the data-transfer operation when D FFs are not available. The reader should verify that these arrangements will act as D FFs.

D-Type Latch

There is another type of clocked flip-flop similar to the edge-triggered D flip-flop. It is called a *D latch* and has the structure shown in Figure 4.16(a). Basically, it consists of a NAND FF with a gating arrangement on its inputs. It operates as follows:

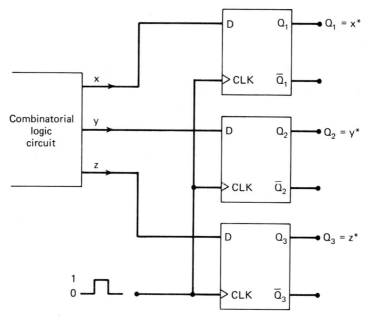

* on positive transition

Figure 4.14 Parallel transfer of binary data using D FFs.

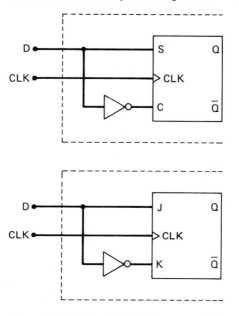

Figure 4.15 S-C and J-K FFs arranged to operate as D FFs.

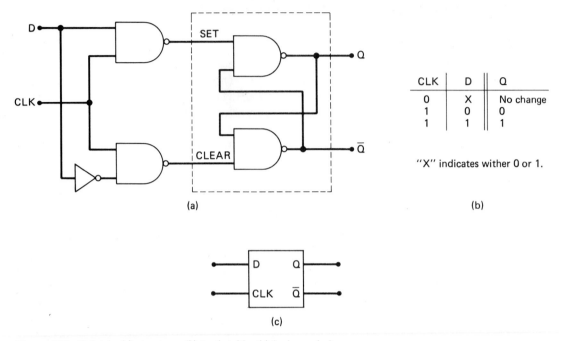

"X" indicates wither 0 or 1.

CLK	D	Q
0	X	No change
1	0	0
1	1	1

(a)   (b)

(c)

Figure 4.16   D latch: (a) structure; (b) truth table; (c) logic symbol.

1. When the *CLK* input is LOW, the *D* input has no effect, since the SET and CLEAR inputs of the NAND FF are kept HIGH.

2. When the *CLK* goes HIGH, the level present on the D input will produce a LOW at the appropriate NAND FF input needed to cause Q to have the same state as *D*. In other words, as long as *CLK* is HIGH, the Q output will take on the value of the *D* input. In fact, if *D* changes while *CLK* is HIGH, Q will follow the change exactly.

This last characteristic is what makes the D latch different from the edge-triggered D flip-flop. As we saw, the edge-triggered D FF can only change states when the correct *CLK* edge occurs. The *D* input has no effect at any other time. The latch, however, can change states when *CLK* is HIGH if *D* changes. This operation is summarized in the table of Figure 4.16(b).

The logic symbol for the D latch is given in Figure 4.16(c). It is the same as the symbol for the edge-triggered D FF except for the absence of the triangle on the *CLK* input, since the latch is not edge-triggered. Some logic circuit manufacturers do not use the triangle to indicate edge-triggering, so their logic symbols are the same for the latch and the edge-triggered FF.

4.7 SYNCHRONOUS AND ASYNCHRONOUS FF INPUTS

For the clocked flip-flops that we have been studying, the *S, C, J, K,* and *D* inputs have been referred to as *control* inputs. These inputs are also called *synchronous inputs*, because their effect on the FF output is synchronized with the *CLK* input. As we have seen, the synchronous control inputs must be used in conjunction with a clock signal to trigger the FF.

Most clocked FFs also have one or more *asynchronous inputs* which operate independently of the synchronous inputs and clock input. These asynchronous inputs can be used to set the FF to the 1 state or clear the FF to the 0 state at any time, regardless of the conditions at the other inputs. Stated in another way, the asynchronous inputs are *override* inputs, which can be used to override all the other inputs in order to place the FF in one state or the other.

Figure 4.17 shows a clocked J-K FF with DC SET and DC CLEAR inputs. These asynchronous inputs are activated by a 0 level, as indicated by the small circles on the FF symbol. The accompanying truth table indicates how these inputs operate. A low on the DC SET input *immediately* sets Q to the 1 state. A low on the DC CLEAR *immediately* clears Q to the 0 state. Simultaneous low levels on DC SET and DC CLEAR are forbidden since it leads to an ambiguous condition. When neither of these inputs are low, the FF is free to respond to the *J, K,* and *CLK* inputs, as previously described.

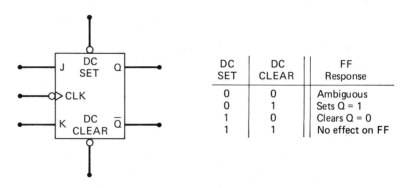

DC SET	DC CLEAR	FF Response
0	0	Ambiguous
0	1	Sets Q = 1
1	0	Clears Q = 0
1	1	No effect on FF

Figure 4.17 Clocked J-K FF with asynchronous inputs.

It is important to realize that these asynchronous inputs respond to dc levels. This means that if a constant 0 is held on the DC SET input, the FF will remain in the $Q = 1$ state regardless of what is occurring at the other inputs. Similarly, a constant low on the DC CLEAR input holds the FF in the $Q = 0$ state. Thus, the asynchronous inputs can be used to hold the FF in a particular state for any desired interval. Most often, however, the asynchronous inputs are used to set or clear the FF to the desired state by application of a momentary pulse.

Some FFs have asynchronous inputs that are activated by 1s rather than by 0s. For these FFs the small circle on the DC SET and DC CLEAR inputs is not shown. Other commonly used designations for asynchronous inputs are PRESET (same as DC SET) and RESET (same as DC CLEAR). In our use of FFs in the following material, the asynchronous inputs will not be shown on the FF symbol unless they are being used in the particular application.

The operation of the asynchronous inputs is illustrated in Figure 4.18. The synchronous inputs (J, K) are held HIGH as *CLK* pulses are applied. The FF will toggle on each negative-going transition as long as the PRESET and CLEAR inputs are both HIGH when the *CLK* edge occurs. This happens at points a, c, d, and f. The Q output can also change states in response to a LOW at PRESET or CLEAR. The LOW on PRESET causes Q to go to the 1 state *immediately*, regardless of the state of *CLK* (point b). Similarly, the LOW at CLEAR causes Q to go to 0 immediately (point e). Note that this LOW at CLEAR prevents the fourth *CLK* edge from toggling Q; this is the override feature.

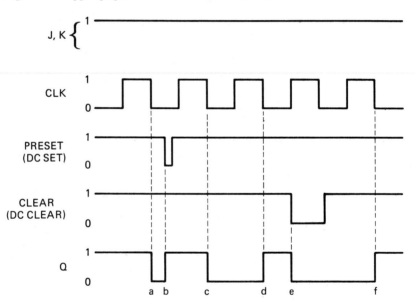

Figure 4.18 Asynchronous inputs can cause FF to change states immediately without waiting for *CLK* transition.

4.8 FF OPERATING CHARACTERISTICS

The e are several operating characteristics of flip-flops that are usually specified on manufacturers' data sheets. These characteristics or parameters are often needed by the logic circuit designer to determine whether a particular FF will be satisfactory for a given application.

Propagation Delays

Whenever a signal is to change the state of a FF's output, there is a delay from the time the signal is applied to the time when the output makes its change. Figure 4.19 illustrates the propagation delays that occur in response to a positive transition on the *CLK* input. Note that these delays are measured between the 50 per cent points on the input and output waveforms. The same types of delays occur in response to signals on a FF's asynchronous inputs (PRESET and CLEAR). The manufacturers' data sheets usually specify propagation delays in response to all triggering inputs.

Modern IC flip-flops have propagation delays that range from a few nanoseconds to around 1 μs: The values of t_{PLH} and t_{PHL} are generally not the same and they increase in direct proportion to the number of loads being driven by the Q output. FF propagation delays play an important part in certain situations which we will encounter later.

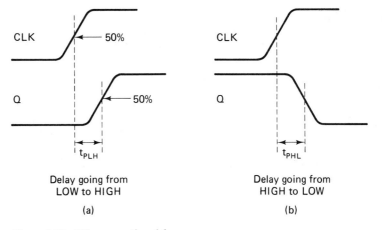

Delay going from   Delay going from
LOW to HIGH   HIGH to LOW

(a)   (b)

Figure 4.19   FF propagation delays.

Set-Up and Hold Times

The set-up time, t_S, is the amount of time that the control inputs (*S, C, J, K*, and *D*) must be held stable prior to the occurrence of the triggering edge of the *CLK* input in order for the FF to respond reliably. This is illustrated in Figure 4.20 for a D flip-flop. Here the *D* input has to have gone to 1 for a time, t_S, before the negative-going *CLK* transition in order for the FF to go to the 1 state.

The hold time, t_H, is the amount of time that the control inputs have to be held stable after the triggering edge of the *CLK* input. Figure 4.20 shows the *D* input being held high for a time, t_H, after the *CLK* goes from 1 to 0. If this requirement is not met, proper FF triggering cannot be assured.

In modern integrated-circuit FFs, the values of t_S and t_H are typically in the nanosecond region. Set-up times generally fall in the range 5 to 50 ns and hold

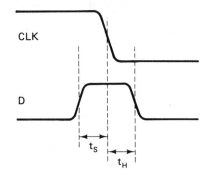

Figure 4.20 Set-up and hold time.

times are usually less than 10 ns. Many edge-triggered FFs in the TTL IC family
have $t_H = 0$, which means that the control inputs can change state at the same
time as the *CLK* input transition. The *master/slave* type of FF, discussed in Section
4.9, actually has *negative* t_H values, indicating that the control inputs can change
state *before* the clock transition occurs.

The t_S and t_H parameters are particularly important in applications where the
control inputs are changing at approximately the same time as the *CLK* input.
These applications include certain types of counters and shift registers, which will
be subsequently discussed. The t_S and t_H parameters are characteristics of all
clocked FFs, including the S-C, D and the J-K types.

Maximum Frequency

A flip-flop can have clock pulses applied to its *CLK* input at a rate that must not
exceed f_{max}, its maximum frequency limit. At clock frequencies above f_{max}, the FF
would be unable to respond quickly enough and its output would be distorted.
Modern IC FFs have f_{max} values ranging from 1 MHz up to hundreds of MHz.

4.9 MASTER/SLAVE FLIP-FLOPS

In digital systems the outputs of flip-flops are often connected, directly or through
logic gates, to the inputs of other flip-flops. Figure 4.21 is an illustration of this.
The output of FF Q_1 acts as the *J* input of FF Q_2, and both FFs are triggered by
the same clock pulse at their *CLK* inputs. Let us assume that initially $Q_1 = 1$ and
$Q_2 = 0$. Since Q_1 has both its *J* and *K* inputs high, it will toggle to the 0 state on
the negative transition of the clock pulse. Q_2 has a 1 on its *J* input via Q_1 *prior to*
the negative transition of the clock pulse. However, when the clock pulse goes
LOW, Q_1 will also go LOW, so the *J* input of Q_2 will be changing from 1 to 0 while
Q_2 is being clocked. This is called a *race condition* and can sometimes lead to
unpredictable triggering of Q_2.

Recall from our discussion of set-up and hold times (t_S and t_H) that edge-triggered FFs require that the control inputs (*S, C* or *J, K* or *D*) remain stable for a

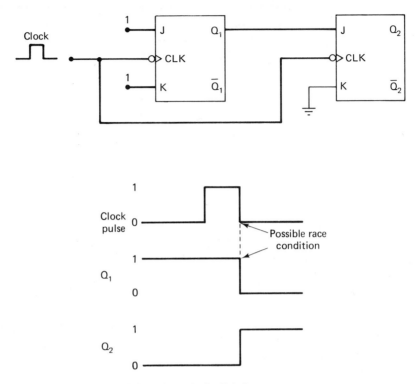

Figure 4.21   Race conditions that exist in digital systems.

time t_H after the clock makes its transition. If the t_H requirement for the FF is very small or zero, the race situation in Figure 4.21 might not cause a problem. This is because Q_1 will actually go low a short time after the clock goes low, owing to the inherent delay of FF Q_1 in responding to the clock pulse. Thus, Q_2 will respond properly by going to the 1 state on the falling edge of the clock.

What this means is that for the *edge-triggered* S-C, J-K, and D flip-flops, the race condition is not a problem as long as the control inputs do not change before the clock input makes its transition. In some applications, however, this cannot be reliably guaranteed, or the situation is too close to being marginal. For this reason, another group of clocked FFs have been developed, called master/slave FFs. The S-C, J-K, and D types of FFs are available in both edge-triggered and master/slave versions. The major difference between the two versions will be explained in the following paragraphs.

Set-Clear Master/Slave Flip-Flops

There are various arrangements for *master/slave* (hereafter abbreviated M/S) *flip-flops*. One common arrangement, shown in Figure 4.22, actually includes two S-C FFs (i.e., NAND FFs). One is called the *master*, the other is the *slave*, and both are triggered by low levels at their S and C inputs.

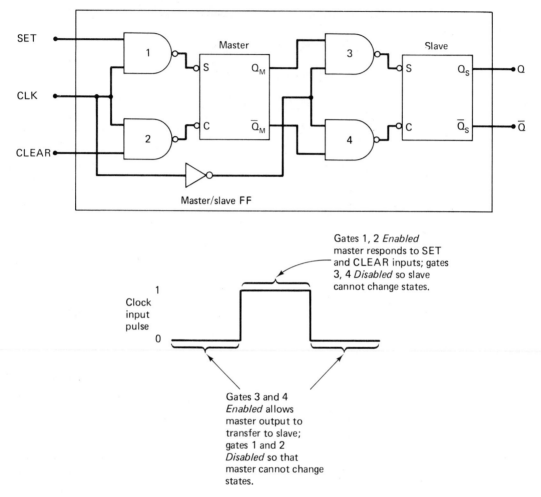

Figure 4.22 Master/slave S-C FF.

The master FF is triggered by the externally applied SET and CLEAR inputs. However, the master will respond to these inputs only when the *CLK* input is at its 1 level, because of NAND gates 1 and 2. When the *CLK* is low, these two gates are essentially *disabled*, meaning that their outputs will be forced high regardless of the other inputs. These 1s applied to the *S* and *C* of the master will have no effect on it.

The slave FF is triggered by the outputs of the master FF, Q_m and $\overline{Q_m}$. However, the slave will respond only when the *CLK* input is at its 0 level, because of NAND gates 3 and 4. Note that the *CLK* input is inverted before it is applied to these gates. Thus, when the *CLK* input is high, the *S* and *C* inputs of the slave are high and the slave FF cannot change states.

The operation of this master/slave arrangement can be explained as follows:

1.   While the *CLK* input is *high*, gates 3 and 4 are disabled, so the slave cannot change states. Since the slave outputs are the actual FF outputs Q and $\bar{Q}$, the FF cannot change states while the *CLK* is high.

2.   While the *CLK* is high, gates 1 and 2 are *enabled*, which means they will allow the SET and CLEAR inputs to activate the inputs of the master. Thus, the master can change states according to the levels of the SET and CLEAR inputs as long as *CLK* remains high.

3.   When the *CLK* goes low, gates 1 and 2 are disabled, thereby preventing the master from changing states regardless of what is happening at the SET and CLEAR inputs. At the same time, gates 3 and 4 are enabled, so Q_m and $\overline{Q_m}$ from the master are allowed to activate the S and C inputs of the slave. If $Q_m = 1$, the slave is set to the 1 state; if $Q_m = 0$, the slave is cleared to 0. Thus, the slave output (the overall FF output) takes on the present state of the master and remains there, since the master cannot change states while the *CLK* is low.

4.   When the *CLK* goes high again, the master is free to respond to the SET and CLEAR inputs; the slave remains in its previous state, since gates 3 and 4 are disabled.

The operation described above is actually somewhat more complicated. The various NAND gates are designed so that when the *CLK* input is going from 1 to 0, NAND gates 1 and 2 are disabled *before* gates 3 and 4 are enabled. This is done to ensure that the master and slave will not be affected by any changes in the SET and CLEAR inputs which occur at the same time as the *CLK* transition. This is important in avoiding the race problem.

It is important to understand that the slave FF changes states only when the *CLK* goes from 1 to 0, and it remains there for the duration of the time that the *CLK* stays at 0. Thus, the overall FF outputs Q and $\bar{Q}$ change state only on the 1 to 0 transition of the *CLK*. On the other hand, the master can change states at any time while the *CLK* is 1, depending on what occurs at the SET and CLEAR inputs. In fact, the master may change states more than once while the *CLK* is 1 if the SET and CLEAR inputs change. Once the *CLK* goes low, however, the current state of the master is then transferred to the slave output.

The preceding operation is illustrated by the waveforms in Figure 4.23. Initially, the master and slave outputs are 0. Examination of these waveforms shows that the master will respond to the SET and CLEAR inputs whenever the *CLK* input is high (note times $t_1, t_3, t_4,$ and t_6). Notice that during the second *CLK* pulse, the master changes states twice (at t_3 and t_4) since the SET and CLEAR inputs changed. The slave output can only change on the negative transitions of the *CLK* (times $t_2, t_5,$ and t_7), and it always goes to whatever the state of the master is at the occurrence of the negative transition of the *CLK*.

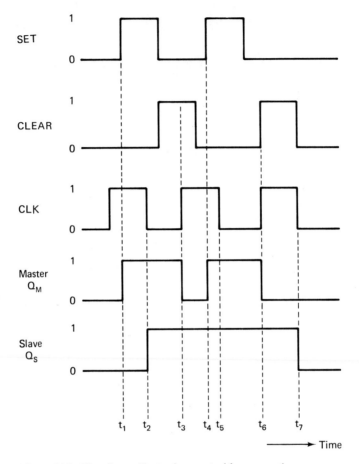

Figure 4.23 Waveforms illustrating master/slave operation.

Difference Between Edge-Triggered and Master/Slave Flip-Flops

The major difference between the edge-triggered FFs examined earlier and the master/slave FF can best be illustrated by considering the waveforms shown in Figure 4.24. There the outputs of an edge-triggered FF and a M/S FF are shown for the same S, C, and CLK inputs. The two FFs respond in the same manner at t_1 and t_2 when the CLK makes its negative-going transitions. At t_3 the SET input and CLK are both going from HIGH to LOW. The edge-triggered FF will respond unpredictably unless the fall of the SET input is delayed beyond the falling edge of the CLK by at least an amount t_H. The M/S FF, however, will be set to the 1 state at t_3 because the 1 at the SET input was stored in the master while the CLK was HIGH, and when the CLK went LOW, this 1 was transferred to the slave output.

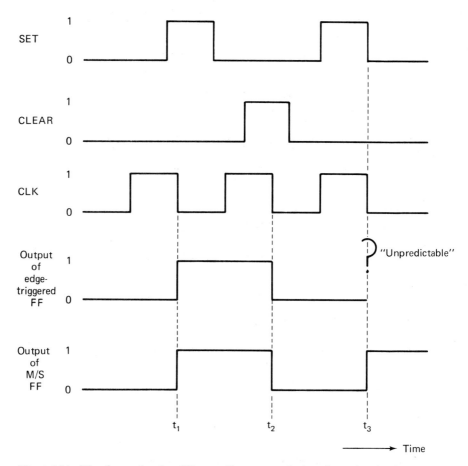

Figure 4.24 Waveforms showing difference in response of edge-triggered and M/S FFs.

In other words, the M/S FF responded to the SET = 1 condition, which was present *prior to* the *CLK* falling edge.

The situation at t_3 demonstrates the chief advantage of M/S FFs over the edge-triggered type. The M/S will provide reliable triggering even if the control inputs are changing at the same time that the *CLK* transition occurs. For this reason, M/S FFs must be used in applications where race situations such as the ones in Figures 4.21 and 4.24 can occur.

J-K and D-type FFs are also available in master/slave types as well as edge-triggered operation. The M/S operation of these FFs is exactly like the S-C M/S except for the difference in how the control inputs affect the output. We will distinguish between edge-triggered and M/S FFs by placing an M/S designation on the symbols for the master/slave type. Most M/S FFs are designed to trigger on the negative-going *CLK* transitions.

A few words should be added here concerning J-K *master/slave FFs* and how they respond in certain situations. They operate basically on the same master/slave principle described earlier, but they have a unique property which can cause problems if it is not understood. This property is stated as follows: *A J-K master/slave FF will trigger on the negative-going* CLK *transition and will go to a state determined by **any** HIGHs that occur on the J and K inputs while the* CLK *is HIGH.* What this means is that *any* HIGHs that occur on the *J* and/or *K* inputs while the *CLK* is HIGH will determine how the FF output responds when the *CLK* goes LOW. If *either* the *J* or *K*, but not both, go HIGH sometime during the *CLK* HIGH interval, the FF will SET or CLEAR accordingly on the negative-going transition. If *both J* and *K* go HIGH during the *CLK* HIGH interval, *but not necessarily at the same time*, the FF will toggle when the *CLK* goes LOW. In other words, the J-K M/S remembers *all* the HIGHs that occur on the *J* and *K* inputs during the *CLK* HIGH interval and responds accordingly when the CLK goes from HIGH to LOW. Thus, to avoid problems, the *J-K* inputs should not be allowed to change while *CLK* is HIGH.

4.10 FLIP-FLOP OPERATIONS: TRANSFER OPERATION

An operation that occurs very frequently in digital systems is the transfer of information from one FF or group of FFs to another FF or group of FFs. Figure 4.25 illustrates how this operation is performed using S-C, J-K, and D FFs. In each case, the logic value stored in FF *A* is transferred to FF *B* upon the occurrence of the negative transition of the TRANSFER pulse. That is, the *B* output after the pulse occurs will be the same as what the *A* output was prior to the pulse.

The transfer operations in Figure 4.25 are examples of *synchronous transfer* since the synchronous and *CLK* inputs are used to perform the transfer. A transfer operation can also be obtained using the asynchronous inputs of a FF. Figure 4.26 shows how an *asynchronous transfer* can be obtained using the DC SET and DC CLEAR inputs of any type of FF. Here, the asynchronous inputs respond to LOW levels (note the small circles on the symbol). When the TRANSFER ENABLE line is held LOW, the two NAND outputs are kept HIGH, with no effect on the FF inputs. When the TRANSFER ENABLE line is made HIGH, one of the NAND outputs will go LOW, depending on the state of the *A* and $\bar{A}$ outputs. This LOW will either set or clear FF *B* to the same state as FF *A*. This asynchronous transfer is done independently of the synchronous and *CLK* inputs of the FF. Asynchronous transfer is also called *jam transfer* because the data can be "jammed" into FF *B* even if the synchronous inputs are active.

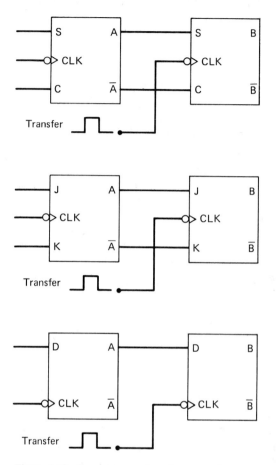

Figure 4.25 Synchronous transfer operation performed by various types of FFs.

Parallel Transfer of Information

Sets of FFs used to store a particular group of 0s and 1s are called *registers*. A common operation in digital systems involves transferring the information stored in one group of FFs to another group of FFs. Figure 4.27 illustrates this operation using D-type FFs. Register X consists of FFs X_1, X_2, and X_3; register Y consists of FFs Y_1, Y_2, and Y_3. Upon application of the TRANSFER pulse, the value stored in X_1 is transferred to Y_1, X_2 to Y_2, and X_3 to Y_3. The transfer of the contents of the X register into the Y register is a synchronous transfer. It is also referred to as a *parallel* transfer, since the contents of X_1, X_2, and X_3 are transferred *simultaneously* into Y_1, Y_2, and Y_3. If a *serial* transfer were performed, the contents of the X register would be transferred to the Y register one bit at a time. This will be examined in Section 4.11.

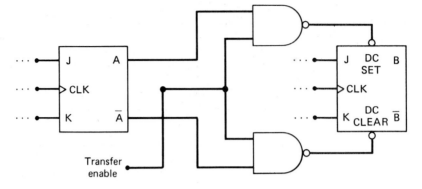

Figure 4.26   Asynchronous transfer operation.

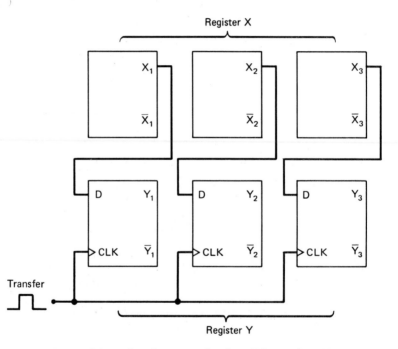

Figure 4.27   Parallel transfer of contents of register X into register Y.

4.11   FLIP-FLOP OPERATIONS: SHIFT REGISTERS

Shift registers are used to transfer the contents of one register into a second register one bit at a time. Before examining this operation of *serial* transfer, let us look at the operation of a basic shift register. Figure 4.28(a) shows four J-K FFs wired as a 4-bit shift register. Note that the FFs are connected so that the output of X_3

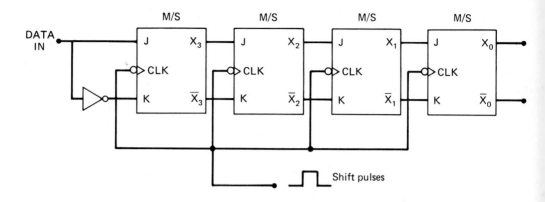

(a)

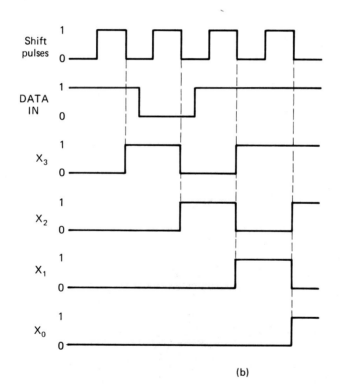

(b)

Figure 4.28 Four-bit shift register.

transfers into X_2, X_2 into X_1, and X_1 into X_0. What this means is that upon the occurrence of the shift pulse, each FF takes on the value stored previously in the FF on its left. FF X_3 takes on a value determined by the conditions present on its J and K inputs when the shift pulse occurs. For now, we will assume that X_3's J and K inputs are fed by the DATA IN waveform shown in Figure 4.28(b). We will also assume that all FFs are in the 0 state before shift pulses are applied.

The waveform diagrams show how the input information is essentially shifted from left to right from FF to FF. When the first shift pulse occurs, it finds that all FFs have $J = 0$, $K = 1$ except for X_3, which has $J = 1$, $K = 0$. Thus, on the falling edge of this pulse, only X_3 goes to 1. When the second shift pulse occurs, it finds that X_2 has $J = 1$, $K = 0$, and all other FFs have $J = 0$, $K = 1$, so X_2 goes HIGH and all other FFs go or stay LOW. Similar reasoning is used to determine the FF outputs for the succeeding shift pulses.

Possible Race Problem

The shift register exhibits the race condition discussed in Section 4.9 because there are times when the J, K inputs to a FF are changing at approximately the same time as its *CLK* input. For example, just prior to the falling edge of the second shift pulse, the J input of X_2 is 1, since $X_3 = 1$. On the falling edge X_3 will go LOW, so X_2's J input goes LOW at the same time as its *CLK* transition. However, since X_2 is a master/slave FF, it will trigger reliably according to what its J value was just prior to the *CLK* transition. Thus, the possible race problem is avoided. It is important to realize, then, that these simultaneous transitions on the J, K and *CLK* inputs can be neglected when analyzing this type of circuit.

Serial Transfer Between Registers

Figure 4.29(a) shows the two 3-bit registers once again. This time they are wired so that the contents of the X register is transferred serially (shifted) into register Y. Both registers are connected as shift registers. X_0, the last FF of register X, is connected to the inputs of Y_2, the first FF of register Y. Thus, as the shift pulses are applied, the information transfer takes place as follows: $X_2 \rightarrow X_1 \rightarrow X_0 \rightarrow Y_2 \rightarrow Y_1 \rightarrow Y_0$. The X_2 FF will go to states determined by its J, K inputs. For now, $J = 0$ and $K = 1$ will be used, so X_2 will go LOW on the first pulse and will remain there.

To illustrate, let us assume that before any shift pulses are applied, the contents of the X register is 1   0   1 (that is, $X_2 = 1$, $X_1 = 0$, $X_0 = 1$) and the Y register is at 0   0   0. Refer to the table in Figure 4.29(b), which shows how the states of each FF change as shift pulses are applied. The following points should be noted:

1. On the falling edge of each pulse, each FF takes on the value that was stored in the FF on its left prior to the occurrence of the pulse.

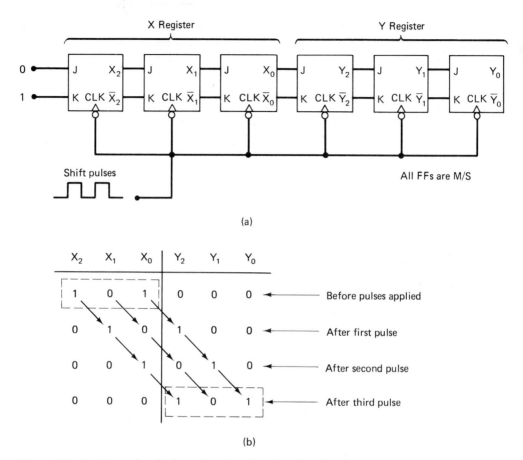

Figure 4.29   Serial transfer of information from X register into Y register.

2. After *three* pulses, the 1 that was initially in X_2 is in Y_2, the 0 initially in X_1 is in Y_1, and the 1 initially in X_0 is in Y_0. In other words, the 101 stored in the X register has now been shifted into the Y register. The X register is at 000.

3. The complete transfer of the three bits of data requires three shift pulses.

Parallel versus Serial Transfer

In parallel transfer, all the information is transferred simultaneously upon the occurrence of a single transfer command pulse (Figure 4.27), no matter how many bits are being transferred. In serial transfer, as exemplified by Figure 4.29, the complete transfer of N bits of information requires N clock pulses (3 bits requires three pulses, 4 bits requires four pulses, etc.). Parallel transfer, then, is obviously much faster than serial transfer using shift registers.

In parallel transfer, the output of each FF in register X is connected to a corresponding FF input in register Y. In serial transfer, only the last FF in register X is connected to register Y. In general, then, parallel transfer requires more interconnections between the sending register (X) and the receiving register (Y) than does serial transfer. This difference becomes more obvious when a greater number of bits of information are being transferred. This is an important consideration when the sending and receiving registers are remote from each other, since it determines how many lines (wires) are needed in the transmission of the information.

The choice of either parallel or serial transmission depends on the particular system application and specifications. Often, a combination of the two types is used to take advantage of the *speed* of parallel transfer and the *economy and simplicity* of serial transfer. More will be said later about information transfer.

4.12 FLIP-FLOP OPERATIONS: COUNTING

Refer to Figure 4.30(a). Each FF has its J and K inputs at the 1 level, so it will change states (toggle) whenever the signal on its *CLK* input goes from HIGH to LOW. The train of clock pulses are applied only to the *CLK* input of FF X_0. Output X_0 is connected to the *CLK* input of FF X_1 and output X_1 is connected to the *CLK* input of FF X_2. The waveforms in Figure 4.30(b) show how the FFs change states as the pulses are applied. The following important points should be noted:

1. FF X_0 toggles on the negative-going transition of each input clock pulse. Thus, the X_0 output waveform has a frequency that is exactly $\frac{1}{2}$ of the clock pulse frequency.

2. FF X_1 toggles each time the X_0 output goes from HIGH to LOW. The X_1 waveform has a frequency equal to exactly $\frac{1}{2}$ the frequency of the X_0 output and therefore $\frac{1}{4}$ of the clock frequency.

3. FF X_2 toggles each time the X_1 output goes from HIGH to LOW. Thus, the X_2 waveform has $\frac{1}{2}$ the frequency of X_1 and therefore $\frac{1}{8}$ of the clock frequency.

As described above, each FF divides the frequency of its input by *2*. Thus, if we were to add a fourth FF to the chain, it would have a frequency equal to $\frac{1}{16}$ of the clock frequency; and so on. Using the appropriate number of FFs, this circuit could divide a frequency by any power of 2. Specifically, using N FFs would produce an output frequency from the last FF which is equal to $\frac{1}{2^N}$ of the input frequency.

Counting Operation

In addition to functioning as a frequency divider, the circuit of Figure 4.30 also operates as a *binary counter*. This can be demonstrated by examining the sequence of states of the FFs after the occurrence of each clock pulse. Figure 4.31 presents the results in tabular form. Let the $X_2 X_1 X_0$ values represent a binary number where

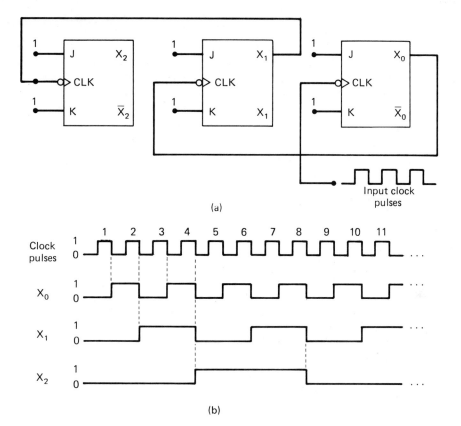

(a)

(b)

Figure 4.30 J-K FFs wired as a 3-bit binary counter (MOD-8).

X_2 is the 2^2 position, X_1 is the 2^1 position, and X_0 is the 2^0 position. The first eight $X_2 X_1 X_0$ states in the table should be recognized as the binary counting sequence from 000 to 111. After the first pulse the FFs are in the 001 state ($X_2 = 0$, $X_1 = 0$, $X_0 = 1$), which represents 001_2 (equivalent to decimal 1); after the second pulse the FFs represent 010_2, which is equivalent to 2_{10}; after three pulses, $011_2 = 3_{10}$; after four pulses, $100_2 = 4_{10}$; and so on until after seven pulses, $111_2 = 7_{10}$. On the eighth pulse the FFs return to the 000 state, and the binary sequence repeats itself for succeeding pulses.

Thus, for the first seven input pulses, the circuit functions as a binary counter in which the states of the FFs represent a binary number equivalent to the number of pulses that have occurred. This counter can count as high as $111_2 = 7_{10}$, and then it returns to 000.

MOD Number

The counter of Figure 4.30 has $2^3 = 8$ different states (000 through 111). It would be referred to as a *MOD-8 counter*, where the MOD number indicates the number of states in the counting sequence. If a fourth FF were added, the sequence of

2^2	2^1	2^0	
X_2	X_1	X_0	
0	0	0	Before applying clock pulses
0	0	1	After pulse #1
0	1	0	After pulse #2
0	1	1	After pulse #3
1	0	0	After pulse #4
1	0	1	After pulse #5
1	1	0	After pulse #6
1	1	1	After pulse #7
0	0	0	After pulse #8 recycles to 000
0	0	1	After pulse #9
0	1	0	After pulse #10
0	1	1	After pulse #11
.	.	.	. . .
.	.	.	. . .
.	.	.	. . .

Figure 4.31 Sequence of FF states shows binary counting sequence.

states would count in binary from 0000 to 1111, a total of 16 states. This would be called a *MOD-16 counter*. In general, if N FFs are connected in the arrangement of Figure 4.30, the counter will have 2^N different states, so it is a MOD-2^N counter. It would be capable of counting up to $2^N - 1$ before returning to its zero state.

The MOD number of a counter also indicates the frequency division obtained from the last FF. For instance, a 4-bit counter has 4 FFs, each representing one binary digit (bit), so it is a MOD-2^4 = MOD-16 counter. It can therefore count up to 15 ($= 2^4 - 1$). It can also be used to divide the input pulse frequency by a factor of 16 (the MOD number).

We have only looked at the basic FF binary counter. We will examine counters in much more detail in Chapter 7.

4.13 THE ONE-SHOT

A digital circuit that is somewhat related to the FF is the *one-shot* (abbreviated OS). Like the FF, the OS has two outputs, Q and $\bar{Q}$, which are inverses of each other. Unlike the FF, the OS has only one *stable* output state (normally $Q = 0$, $\bar{Q} = 1$), where it remains until it is triggered by an input signal. Once triggered, the OS outputs switch to the opposite state ($Q = 1$, $\bar{Q} = 0$). It remains in this *quasi-stable* state for a fixed period of time, t_p, which is usually determined by an RC time constant which is connected to the OS. After a time t_p, the OS outputs return to their stable resting state until triggered again.

Figure 4.32(a) shows the logic symbol for a OS. The value of t_p is usually

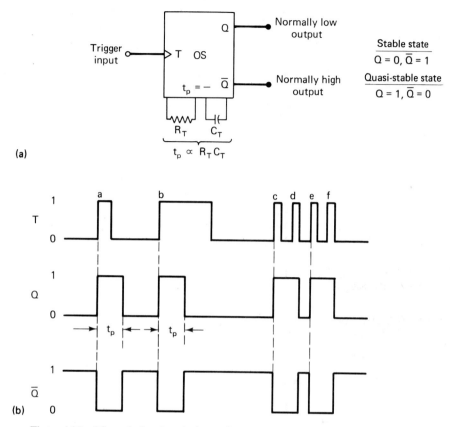

Figure 4.32   OS symbol and typical waveforms.

indicated somewhere on the OS symbol. In practice, t_p can vary from several nanoseconds up to several tens of seconds. The waveforms in the figure illustrate the OS operation. This particular OS is triggered by positive-going *transitions*, and the important points to note are:

1. The OS outputs go to the quasi-stable state for a time t_p whenever a positive-going trigger occurs (points *a, b, c, e*).

2. The input pulse duration has no effect on the OS operation. It responds only to the positive transition.

3. If a second trigger pulse occurs while the OS has already been triggered to its quasi-stable state (points *d* and *f*), it will have no effect on the OS outputs. The OS must return to its stable state before it can be retriggered.*

*A special OS called a "retriggerable" OS can be retriggered while it is in the quasi-stable state. Each new trigger pulse begins a new t_p interval.

4. The value of t_p is proportional to $R_T \times C_T$. This allows the OS user to choose values as desired.

The OS is also referred to as a *monostable multivibrator* or simply a *monostable*, because it has only one stable state. The major applications of the OS are as pulse generators, time-delay elements, or timing circuits. Some of these applications will be introduced in the problems at the end of the chapter.

4.14 ANALYSIS OF FLIP-FLOP CIRCUITS

In most applications FFs and combinatorial logic circuits are used together to perform many different functions. The analysis of such circuits is done on a step-by-step basis, but there is no general procedure that can be used in all cases. Through the experience gained from doing logic-circuit analysis and troubleshooting one will eventually develop good analysis techniques. The example that follows is designed to illustrate how knowledge of the operation of the individual logic elements is used to determine overall circuit operation.

EXAMPLE 4.2   Consider the circuit of Figure 4.33. Initially, all the FFs and the OS are in the 0 state before the clock pulses are applied. These pulses have a repetition rate of 1 kHz. Determine the waveforms at X, Y, Z, W, $\bar{Q}$, A, and B for 16 cycles of the clock input.

Solution: Initially the FFs and OS are in the 0 state, so $X = Y = Z = W = Q = 0$. The inputs to the NAND gate are $X = 0$, $\bar{Y} = 1$, and $\bar{Z} = 1$, so its output $A = 1$. The inputs to the OR gate are $W = 0$ and $\bar{Q} = 1$, so its output is $B = 1$.

As long as B remains HIGH, FF Z will have its J and K inputs both HIGH, so it will operate in the toggle mode. FFs X and Y are kept in the toggle mode since their J and K inputs are held at the 1 level permanently. It should be clear, then, that FFs X, Y, and Z will operate as a 3-bit counter as long as B stays HIGH. Thus, FF Z will toggle on the negative transition of each input clock pulse, FF Y will toggle on the negative transition of the Z output, and FF X will toggle on the negative transition of FF Y. This operation continues for clock pulses 1–4.

When clock pulse 4 makes its negative transition, Z goes LOW, Y goes LOW, and X goes HIGH. Therefore, the inputs to the NAND gate become $X = 1$, $\bar{Y} = 1$, and $\bar{Z} = 1$, causing the NAND output A to go LOW. This negative transition at A will trigger the OS to its quasi-stable state, so $\bar{Q}$ goes LOW for 3.5 ms. This negative transition of $\bar{Q}$ will not affect FF W since this D-type FF triggers on positive transitions at its CLK input. With $\bar{Q}$ and W both LOW, the OR gate output B goes LOW, so FF Z now has $J = 1$, $K = 0$. This means that Z will be *set* to the 1 state on the *next* clock pulse and will remain in this state for each successive clock pulse as long as $J = 1$, $K = 0$. Thus, Z goes HIGH on clock pulse 5 and stays HIGH for clock pulses 6 and 7. (*Note:* A goes back HIGH when Z goes HIGH on pulse 5 because $\bar{Z} = 0$.)

After 3.5 ms the OS returns to its stable state with $Q = 0$, $\bar{Q} = 1$. This positive

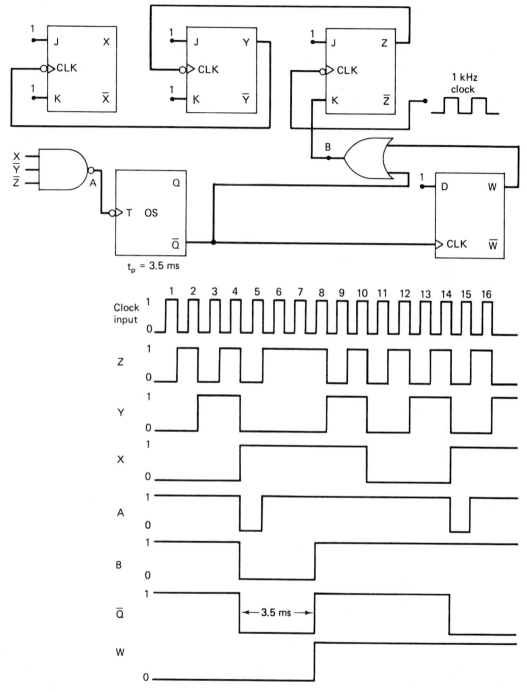

Figure 4.33

transition of $\bar{Q}$ causes FF W to go HIGH since $D = 1$. The 1 at W produces a 1 at B so that FF Z is back in the toggle mode with $J = K = 1$. The counter will now operate properly in response to all succeeding clock pulses.

On the negative transition of clock pulse 14, the NAND output A gain goes LOW, thereby triggering the OS. The OS, however, will not affect the K input of Z since W is keeping the OR output HIGH. Also, since its D input is HIGH, FF W will remain in the $W = 1$ state indefinitely allowing the counter to count normally for all succeeding clock pulses.

4.15   FLIP-FLOP SUMMARY

1. NOR-Gate FF

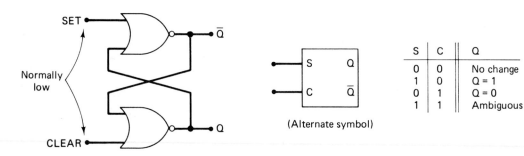

S	C	Q
0	0	No change
1	0	Q = 1
0	1	Q = 0
1	1	Ambiguous

(Alternate symbol)

Figure 4.34

2. NAND-Gate FF

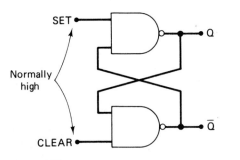

S	C	Q
0	0	Ambiguous
1	0	Q = 0
0	1	Q = 1
1	1	No change

(Alternate symbol)

Figure 4.35

3. Edge-Triggered Clocked S-C FF

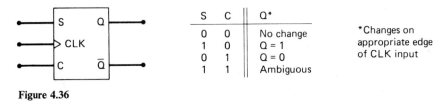

S	C	Q*
0	0	No change
1	0	Q = 1
0	1	Q = 0
1	1	Ambiguous

*Changes on appropriate edge of CLK input

Figure 4.36

4. Edge-Triggered J-K FF

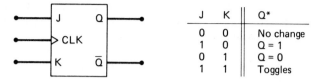

J	K	Q*
0	0	No change
1	0	Q = 1
0	1	Q = 0
1	1	Toggles

*Changes on appropriate edge of CLK

Figure 4.37

5. Edge-Triggered D FF

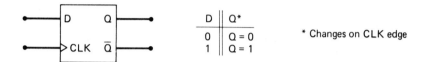

D	Q*
0	Q = 0
1	Q = 1

* Changes on CLK edge

Figure 4.38

6. D-Type Latch

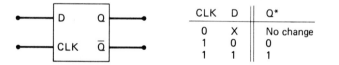

CLK	D	Q*
0	X	No change
1	0	0
1	1	1

*Q follows D input while CLK is HIGH

Figure 4.39

7. Asynchronous FF Inputs

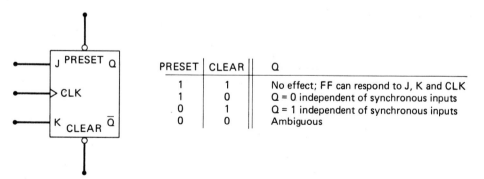

PRESET	CLEAR	Q
1	1	No effect; FF can respond to J, K and CLK
1	0	Q = 0 independent of synchronous inputs
0	1	Q = 1 independent of synchronous inputs
0	0	Ambiguous

Figure 4.40

8. Master/Slave FFs   Same as edge-triggered FFs except that control inputs do not have to remain stable as *CLK* triggering edge occurs.

QUESTIONS AND PROBLEMS

4.1 A NOR-gate FF has waveforms w and x of Figure 4.41(a) applied to its S and C inputs, respectively. Assume initially that $Q = 0$ and determine the Q and $\bar{Q}$ output waveforms in response to these inputs.

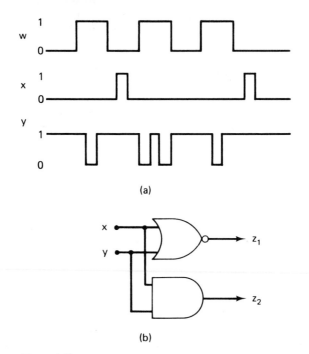

(a)

(b)

Figure 4.41

4.2 Waveforms w and y of Figure 4.41(a) are applied to the S and C inputs of the FF of Figure 4.4. Assume that $Q = 1$ initially and determine the Q and $\bar{Q}$ waveforms.

4.3 Why shouldn't the x and y waveforms in Figure 4.41 be applied to the respective S and C inputs of the FFs of Figure 4.5?

4.4 The x and y waveforms of Figure 4.41(a) are applied to the gates in Figure 4.41(b). The z_1 output signal is applied to the S input of the FF of Figure 4.5(a) and z_2 is applied to the C input. Determine the Q waveform. Assume that $Q = 0$ initially.

4.5 Change the NOR gate to an OR gate, the AND gate to a NAND gate, and repeat Problem 4.4 for the FF of Figure 4.5(b).

4.6 Apply the S, C, and CLK waveforms of Figure 4.9 to the FF of Figure 4.10 and determine the Q waveform.

4.7 A *toggle* FF is one that has a single input and operates such that the FF output changes state for each pulse applied to its input. The clocked S-C FF can be wired to operate in the toggle mode, as shown in Figure 4.42. The waveform applied to

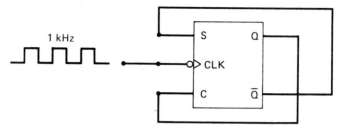

Figure 4.42

the *CLK* input is a 1-kHz squarewave. Verify that this arrangement operates in the manner described, and then determine the *Q* output waveform. Assume that $Q = 0$ initially.

4.8   A certain clocked S-C FF has set-up and hold times specified typically as $t_S = 30$ ns and $t_H = 5$ ns.
(a)   What is the minimum amount of time that the *S* or *C* inputs have to be HIGH before the *CLK* input transition occurs? What is the maximum?
(b)   How long must the *S* or *C* inputs remain HIGH once the *CLK* transition occurs?

4.9   Apply the *J, K,* and *CLK* waveforms of Figure 4.11 to the FF of Figure 4.12. Assume that $Q = 1$ initially and determine the *Q* waveform.

4.10   Explain the major difference between the S-C and J-K flip-flops.

4-11   Show how the J-K FF can be operated as a toggle FF. Apply a 10-kHz squarewave to its input and determine its output waveform.

4.12   A D-type FF is sometimes used to *delay* a binary waveform so that the binary information appears at the output a certain amount of time after it appears at the *D* input. Determine the *Q* waveform in Figure 4.43 and compare it to the input waveform. Note that it is delayed from the input by one clock period. How can a delay of two clock periods be obtained?

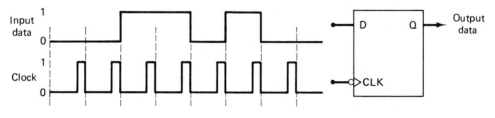

Figure 4.43

4.13   Explain the difference between synchronous and asynchronous FF inputs.

4.14   Determine the *Q* waveform for the FF in Figure 4.44. Assume that $Q = 0$ initially and remember that the asynchronous inputs over-ride all other inputs.

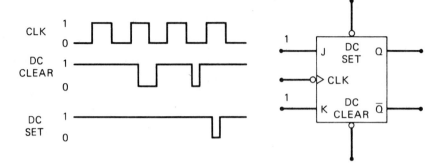

Figure 4.44

41.5 What is the major difference in the operation of edge-triggered FFs and master/slave FFs?

4.16 The waveforms shown in Figure 4.45 are to be applied to three different FFs: (a) positive-edge triggered J-K; (b) negative-edge triggered J-K; and (c) a M/S J-K. Draw the Q waveform response for each of these FFs, assuming that $Q = 0$ initially.

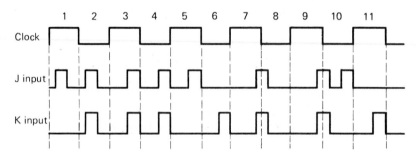

Figure 4.45

4.17 Answer true or false:

(a) The DC SET input will affect the FF output only while the *CLK* input is LOW.

(b) Edge-triggered FFs are sensitive to what the synchronous control inputs are when the *CLK* transition occurs.

(c) In a M/S S-C FF the Q output goes HIGH as soon as the S input goes HIGH.

(d) Asynchronous transfer of binary data uses the *CLK* input to perform the transfer.

(e) Parallel transfer from one FF register to another can only be done using D-type FFs.

(f) Parallel transfer is generally faster than serial transfer.

(g) A FF's synchronous control inputs cannot be used without also using the *CLK* input.

(h) The DC SET or DC CLEAR inputs are effective only while a FFs *CLK* input is LOW and its control inputs are LOW.

(i) A M/S J-K FF is not affected by any variations of the *J* or *K* inputs that occur while the *CLK* input is HIGH.

(j) Serial transfer from one register to another generally requires fewer register interconnections than parallel transfer.

4.18 Show the circuit arrangement required to perform parallel transfer from register *X* to register *Y* using three J-K FFs in each register.

4.19 A *recirculating* shift register is a shift register that keeps the binary information circulating through the register as clock pulses are applied. The shift register of Figure 4.28 can be made into a circulating register by connecting X_0 to the DATA IN line. No external inputs are used. Assume that this circulating register starts out with 1010 stored in it (that is, $X_3 = 1$, $X_2 = 0$, $X_1 = 1$, and $X_0 = 0$). List the sequence of states that the register FFs go through as eight shift pulses are applied.

4.20 Refer to Figure 4.29, where a 3-bit number stored in register *X* is serially shifted into register *Y*. How could the circuit be modified so that at the end of the transfer operation, the original number stored in *X* is present in both registers? (*Hint:* See Problem 19.)

4.21 Refer to the binary counter of Figure 4.30. Change it by connecting $\bar{X}_0$ to the *CLK* of FF X_1 and $\bar{X}_1$ to the *CLK* of FF X_2. Start out with all FFs in the 1 state and draw the various FF output waveforms (X_0, X_1, X_2) for 16 input pulses. Then list the sequence of FF states as was done in Figure 4.31. This counter is called a *down counter* because it counts down from 111 to 000.

4.22 How many FFs would be required to count from 0 to 127_{10}?

4.23 How many FFs are in a MOD-32 counter? If a 640-kHz clock signal is applied to this counter, what is the frequency at the output of the last FF (the one farthest from the clock input signal)?

4.24 Show how clocked S-C FFs can be used in a counter such as that in Figure 4.30. (*Hint:* See Problem 4.7.)

4.25 How many stable states does a one-shot have?

4.26 Determine the waveforms at Q_1, Q_2, and Q_3 in response to the single input pulse in Figure 4.46. Each OS triggers on negative-going transitions.

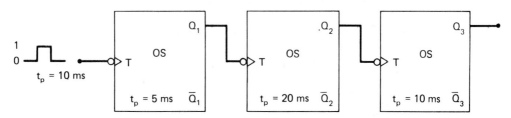

Figure 4.46

4.27 Show how one-shots may be used to take an input pulse having $t_p = 10\ \mu s$ and produce an output pulse with the same t_p but occurring 100 ms later? This is called a *pulse-delay circuit.*

4.28 Consider the circuit of Figure 4.47. Initially all FFs are in the 0 state. The circuit operation begins with a momentary start pulse applied to the DC SET inputs of FFs *X* and *Y*. Determine the waveforms at *A, B, C, X, Y, Z,* and *W* for 20 cycles of the clock pulses after the start pulse.

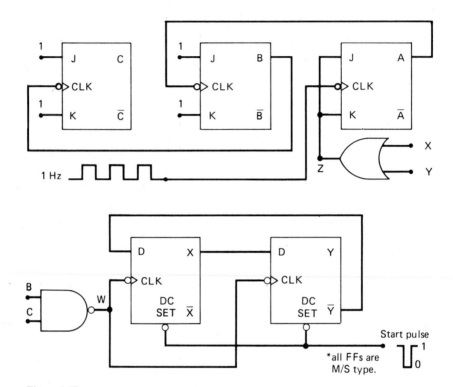

Figure 4.47

4.29 Figure 4.48(a) shows a situation that sometimes occurs in a digital system. A push-button switch is being used to provide a pulse at the output of the AND gate when $A = 1$. The gate output should be normally LOW (switch open) and goes HIGH for the period of time when the switch is depressed. However, owing to the phenomenon of "switch bounce," the gate output will jump back and forth between HIGH and LOW for several milliseconds, as shown in the figure. Switch bounce is the term used to describe the fact that the contracts of a mechanical switch usually vibrate open and closed for a few milliseconds (< 10 ms) beofre reaching a stable closed position. In many cases the waveform at *X* is undesirable, because the extra bounce pulses will cause erroneous operation of the circuit being driven. Several types of "de-bounce" circuits can be used to eliminate the effects of switch bounce. One such circuit is shown in Figure 4.48(b). Analyze and explain its operation; pay particular attention to the waveforms at *Q* and *Z*.

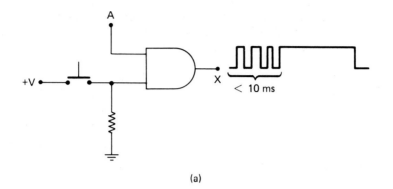

(a)

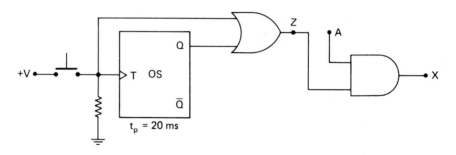

(b)

Figure 4.48

5

NUMBER SYSTEMS AND CODES

In digital systems numerical information is usually represented in the binary number system (or another related binary code). The binary system was introduced in Chapter 1, where its similarities to the decimal system were pointed out. Several other systems for representing numerical data are also important in digital systems, notably the *octal* (base-8), *hexadecimal* (base-16), *binary-coded-decimal* (BCD), and *excess-3* representations. The various number systems, their relationships, and arithmetic operations will be studied in this chapter.

5.1 BINARY-TO-DECIMAL CONVERSIONS

As explained earlier, the binary number system is a positional system where each binary digit (bit) carries a certain weight based on its position relative to the binary point. Any binary number can be converted to its decimal equivalent simply by summing together the weights of the various positions in the binary number which contain a 1. To illustrate:

$$1 \quad 1 \quad 0 \quad 1 \quad 1 \quad \text{(binary)}$$
$$2^4 + 2^3 + 0 + 2^1 + 2^0 = 16 + 8 + 2 + 1$$
$$= 27_{10} \quad \text{(decimal)}$$

The same method is used for binary numbers that contain a fractional part:

$$1\ 0\ 1.\ 1\ 0\ 1 = 2^2 + 2^0 + 2^{-1} + 2^{-3}$$
$$= 4 + 1 + 0.5 + 0.125$$
$$= 5.625_{10}$$

The following conversions should be performed and verified by the reader:

(1) $100110_2 = 38_{10}$

(2) $0.110001_2 = 0.765625_{10}$

(3) $11110011.0101_2 = 243.3125_{10}$

5.2 DECIMAL-TO-BINARY CONVERSIONS

There are several ways to convert a decimal number to its equivalent binary-system representation. A method that is convenient for small numbers is the reverse of the process described in Section 5.1. The decimal number is simply expressed as a sum of powers of 2 and then 1s and 0s are written in the appropriate bit positions. To illustrate:

$$13_{10} = 8 + 4 + 1 = 2^3 + 2^2 + 0 + 2^0$$
$$= 1 \quad 1 \quad 0 \quad 1_2$$

Another example:

$$25.375_{10} = 16 + 8 + 1 + 0.25 + 0.125$$
$$= 2^4 + 2^3 + 0 + 0 + 2^0 + 0 + 2^{-2} + 2^{-3}$$
$$= 1 \quad 1 \quad 0 \quad 0 \quad 1 \cdot 0 \quad 1 \quad 1_2$$

For larger decimal numbers, the above method is laborious. A more convenient method entails separate conversion of the integer and fractional parts. For example, take the decimal number 25.375, which was converted above. The first step is to convert the integer portion, 25. This conversion is accomplished by repeatedly *dividing* 25 by 2 and writing down the remainders after each division until a quotient of zero is obtained:

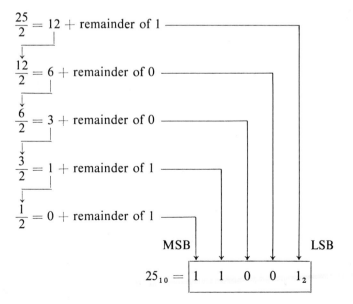

$$\frac{25}{2} = 12 + \text{remainder of } 1$$

$$\frac{12}{2} = 6 + \text{remainder of } 0$$

$$\frac{6}{2} = 3 + \text{remainder of } 0$$

$$\frac{3}{2} = 1 + \text{remainder of } 1$$

$$\frac{1}{2} = 0 + \text{remainder of } 1$$

$$\text{MSB} \qquad\qquad \text{LSB}$$

$$25_{10} = \boxed{1 \quad 1 \quad 0 \quad 0 \quad 1}_2$$

The desired binary conversion is obtained by writing down the remainders as shown above. Note that the *first* remainder is the LSB and the *last* remainder is the MSB.

The fractional part of the number (0.375) is converted to binary by repeatedly *multiplying* it by 2 and recording any carries in the integer position:

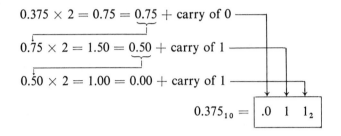

$$0.375 \times 2 = 0.75 = 0.75 + \text{carry of } 0$$

$$0.75 \times 2 = 1.50 = 0.50 + \text{carry of } 1$$

$$0.50 \times 2 = 1.00 = 0.00 + \text{carry of } 1$$

$$0.375_{10} = \boxed{.0 \quad 1 \quad 1_2}$$

Note that the repeated multiplications continue until a product of exactly 1.00 is reached,* since further multiplication results in all zeros. Notice here that the *first* carry is written in the first position to the right of the binary point.

Finally, the complete conversion for 25.375 can be written as the combination of the integer and fraction conversions:

$$25.375_{10} = 11001.011_2$$

The reader should apply this method to verify the following conversion:

$$632.85_{10} \approx 1001111000.11011_2$$

5.3   BINARY ADDITION

The addition of two binary numbers is performed in exactly the same manner as the addition of decimal numbers. In fact, binary addition is simpler since there are fewer cases to learn. Let us first review decimal addition:

$$
\begin{array}{rccc}
  & 3 & 7 & 6 \quad \text{LSD} \\
+ & 4 & 6 & 1 \\
\hline
  & 8 & 3 & 7
\end{array}
$$

The least-significant-digit (LSD) position is operated on first producing a sum of 7. The digits in the second position are then added to produce a sum of 13, which

---

*Most of the time 1.00 will not occur and the process is terminated after a suitable number of places in the binary fractional number is reached. In such cases the binary number is an approximation to the decimal number.

produces a *carry* of 1 into the third position. This produces a sum of 8 in the third position.

The same general steps are followed in binary addition. However, there are only four cases that can occur in adding the two binary digits (bits) in any position. They are:

$$0 + 0 = 0$$
$$1 + 0 = 1$$
$$1 + 1 = 0 + \text{carry of 1 into next position}$$
$$1 + 1 + 1 = 1 + \text{carry of 1 into next position}$$

The last case occurs when the two bits in a certain position are 1 and there is a carry from the previous position. Here are several examples of the addition of two binary numbers:

```
  011 (3)        1001 (9)       11.011 (3.375)
+ 110 (6)      + 1111 (15)    + 10.110 (2.750)
 1001 (9)       11000 (24)     110.001 (6.125)
```

It is not necessary to consider the addition of more than two binary numbers at a time because in all digital systems the circuitry that actually performs the addition can only handle two numbers at a time. When more than two numbers are to be added, the first two are added together and then their sum is added to the third number; and so on. This is not a serious drawback since modern digital machines can typically perform an addition operation in less than 1 μs.

Addition is the most important arithmetic operation in digital systems. As we shall see, the operations of subtraction, multiplication, and division as they are performed in most modern digital computers and calculators actually use only addition as their basic operation.

Binary Addition versus Logical OR Addition

It is important to understand the difference between the OR-addition operation and binary addition. OR addition is the Boolean *logic* operation performed by an OR gate, which produces a 1 output whenever any one or more inputs are 1. Binary addition is an *arithmetic* operation that produces an arithmetic sum of two binary numbers. Although the + sign is used for both OR addition and binary addition, the meaning of the + sign will usually be apparent from the context in which it is being used. The major differences between OR addition and binary addition can be summarized as follows:

OR Addition	*Binary Addition*
$1 + 1 = 1$	$1 + 1 = 0 + \text{carry of 1}$
$1 + 1 + 1 = 1$	$1 + 1 + 1 = 1 + \text{carry of 1}$

5.4 REPRESENTING SIGNED NUMBERS

In binary machines, the binary numbers are represented by a set of binary storage devices (usually flip-flops). Each device represents one bit. For example, a 6-bit FF register could store binary numbers ranging from 000000 to 111111 (0 to 63 in decimal). This represents the *magnitude* of the number. Since most digital computers and calculators handle negative as well as positive numbers, some means is required for representing the *sign* of the number ($+$ or $-$). This is usually done by adding another bit to the number called the *sign bit*. In general, the common convention which has been adopted is that a 0 in the sign bit represents a positive number and a 1 in the sign bit represents a negative number. This is illustrated in Figure 5.1. Register A contains the bits 0110100. The 0 in the leftmost bit (A_6) is the sign bit that represents $+$. The other six bits are the magnitude of the number 110100_2, which is equal to 52 in decimal. Thus, the number stored in the A register is $+52$. Similarly, the number stored in the B register is -31, since the sign bit is 1, representing $-$.

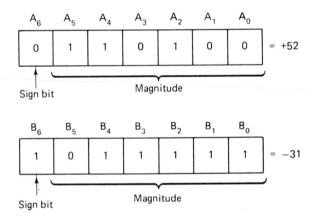

Figure 5.1 Representation of signed numbers.

The sign bit is used to indicate whether a stored binary number is positive or negative. For positive numbers, the rest of the bits are always used to represent the magnitude of the number in binary form. For negative numbers, however, three different forms in which to represent the magnitude are in common use: true-magnitude form, 1's-complement form, and 2's-complement form.

True-Magnitude Form

True-magnitude form is simply the representation shown in Figure 5.1, where the true magnitude of the number is given in binary form. The first bit is always the sign bit. As another example, 1111001 represents -57 in true-magnitude form ($111001_2 = 57$).

1's-Complement Form

The *1's-complement form* of any binary number is obtained simply by changing each 0 in the number to a 1 and each 1 in the number to a 0. In other words, change each bit to its complement. For example, the 1's complement of 101101 is 010010, and the 1's complement of 011010 is 100101.

When negative numbers are represented in 1's-complement form, the sign bit is made a 1 and the magnitude is converted from true binary form to its 1's complement. To illustrate, the number -57 would be represented as follows:

sign bits

$$-57 = 1\,111001 \qquad \text{(true-magnitude form)}$$
$$= 1\,000110 \qquad \text{(1's-complement form)}$$

Note that the sign bit is not complemented but is kept as a 1 to indicate a negative number.

Some additional examples of representing negative numbers in 1's-complement form are given below and should be verified by the reader:

$$-14 = 10001 \qquad -7.25 = 1000.10$$
$$-326 = 1010111001$$

2's-Complement Form

The *2's-complement form* of a binary number is formed simply by taking the 1's complement of the number and adding 1 to the least-significant bit position. The procedure is illustrated below for converting 111001 (decimal 57) to its 2's-complement form.

```
1 1 1 0 0 1    complement each bit to form 1's complement
↓ ↓ ↓ ↓ ↓ ↓
0 0 0 1 1 0

        +1    add 1 to LSB to form 2's complement
_____
0 0 0 1 1 1
```

Thus, -57 would be written as 1000111 in its 2's-complement representation. Again, the leftmost bit is the sign bit. The other 6 bits are the 2's-complement form of the magnitude. As another example, the 2's complement of -14 is written 10010.

The three forms of representing negative numbers are summarized in Figure 5.2 for -57. All three forms are currently used in digital systems. Some digital machines store negative numbers in true-magnitude form but convert them to 1's- or 2's-complement form before performing any arithmetic operations. Other

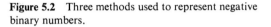

1	1	1	1	0	0	1		True magnitude
1	0	0	0	1	1	0		1's complement
1	0	0	0	1	1	1		2's complement

↳ Sign
   bits

Figure 5.2   Three methods used to represent negative binary numbers.

machines store the negative numbers in 1's- or 2's-complement form. In almost all modern digital machines, the negative numbers are in 1's- or 2's-complement form for the arithmetic operations. At present, the 2's-complement representation is the most frequently used.

Remember, in all three systems, true-magnitude, 1's complement and 2's complement, *positive* numbers are always in true binary form and a sign bit of 0. The differences are in the representation of negative numbers.

The 1's- and 2's-complement forms are used because, as we shall see, their use allows us to perform the operation of subtraction by actually using only the operation of addition. This is significant because it means that a digital machine can use the same circuitry to both add and subtract, thereby realizing a savings in hardware.

Converting from Complement Form to Binary

It is a relatively simple matter to take a number that is in its 1's- or 2's-complement form and convert it back to its true binary value. To go from 1's complement to true binary simply requires complementing each bit again. Similarly, to go from 2's complement to true binary simply requires complementing each bit and then adding 1 to the LSB. In both cases, the conversion back to binary is the same process that was used to produce the complement to begin with. The reader should verify this for a couple of binary numbers.

5.5 ADDITION IN THE 2'S-COMPLEMENT SYSTEM

The 1's-complement system and 2's-complement system are very similar. However, the 2's-complement system is generally in use because of several advantages it enjoys in circuit implementation.* We will now investigate how the operations of

*Some machines store their negative numbers in 1's-complement form but convert them to 2's-complement form while performing arithmetic operations by including the addition of the 1 to the LSB somewhere in the operation.

addition and subtraction are performed in digital machines that use the 2's-complement representation for negative numbers. In the various cases to be considered, it is important to note that the sign-bit of each number is operated on in the same manner as the magnitude portion.

Case I:
Two Positive Numbers

The addition of two positive numbers is straightforward. Consider the addition of +9 and +4:

$$
\begin{array}{llll}
+9 \longrightarrow & 0 \mid 1001 & \text{(augend)} \\
+4 \longrightarrow & 0 \mid 0100 & \text{(addend)} \\
\hline
& 0 \mid 1101 & \text{(sum} = +13) \\
& \uparrow & \\
& \text{sign bits}
\end{array}
$$

Note that the sign bits of the *augend* and *addend* are both 0 and the sign bit of the sum is 0, indicating that the sum is positive. Also note that the augend and addend are made to have the same number of bits. This must *always* be done in the 2's complement system.

Case II:
Positive Number and Smaller Negative Number

Consider the addition of +9 and −4. Remember that the −4 will be in its 2's-complement form. Thus, +4 (00100) must be converted to −4(11100)

$$
\begin{array}{llll}
& & \text{sign bits} \\
+9 \longrightarrow & 0 \mid 1001 & \text{(augend)} \\
-4 \longrightarrow & 1 \mid 1100 & \text{(addend)} \\
\hline
1 \mid 0 \mid 0101 \\
\uparrow \\
\text{this carry is disregarded, so the result is 00101 (sum} = +5)
\end{array}
$$

In this case the sign bit of the addend is 1. Note that the sign bits also participate in the addition process. In fact, a carry is generated in the last position of addition. *This carry is always disregarded*, so the final sum is 00101, which is equivalent to +5.

Case III:
Positive Number and Larger Negative Number

Consider the addition of -9 and $+4$:

$$
\begin{array}{r}
-9 \longrightarrow 10111 \\
+4 \longrightarrow 00100 \\
\hline
11011 \quad (\text{sum} = -5)
\end{array}
$$

↑——negative sign bit

The sum here has a sign bit of 1, indicating a negative number. Since the sum is negative, it is in 2's-complement form, so the last four bits, 1011, actually represent the 2's complement of the sum. To find the true magnitude of the sum, we must take the 2's complement of 1011; the result is 0101 (5). Thus, 11011 represents -5.

Case IV:
Two Negative Numbers

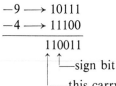

$$
\begin{array}{r}
-9 \longrightarrow 10111 \\
-4 \longrightarrow 11100 \\
\hline
110011
\end{array}
$$

↑ ↑——sign bit

↑——this carry is disregarded, so the result is 10011 (sum = -13)

This final result is again negative and in 2's-complement form with a sign bit of 1. Note that 0011 is the 2's complement of 1101 (13).

Case V:
Equal and Opposite Numbers

$$
\begin{array}{r}
-9 \longrightarrow 10111 \\
+9 \longrightarrow 01001 \\
\hline
0 \quad 100000
\end{array}
$$

↑——disregard, so the result is 00000 (sum = $+0$)

The result is obviously $+0$, as expected.

5.6 SUBTRACTION IN THE 2'S-COMPLEMENT SYSTEM

The subtraction operation using the 2's-complement system actually involves the operation of addition and is really no different than the various cases considered in Section 5.5. When subtracting one binary number (the *subtrahend*) from another binary number (the *minuend*), the procedure is as follows:

1. Take the 2's complement of the subtrahend, *including* the sign bit. If the subtrahend is a positive number, this will change it to a negative number in 2's-complement form. If the subtrahend is a negative number, this will change it to a positive number in true binary form. In other words, we are changing the sign of the subtrahend.

2. After taking the 2's complement of the subtrahend, it is *added* to the minuend. The minuend is kept in its original form. The result of this addition represents the required *difference*. The sign bit of this difference determines whether it is $+$ or $-$ and whether it is in true binary form or 2's-complement form.

Let us consider the case where $+4$ is to be subtracted from $+9$.

$$\text{minuend (9)} \longrightarrow 01001$$
$$\text{subtrahend } (+4) \longrightarrow 00100$$

Change the subtrahend to its 2's-complement form (11100). Now add this to the minuend:

$$
\begin{array}{r}
01001 \\
+\ 11100 \\
\hline
1\ \ 00101
\end{array}
$$

⌞—disregard, so the result is 00101 $= +5$

When the subtrahend is changed to its 2's complement, it actually becomes -4, so we are *adding* $+9$ and -4, which is the same as subtracting $+4$ from $+9$. This is the same as case II of Section 5.5. Any subtraction operation, then, actually becomes one of addition when the 2's-complement system is used. This feature of the 2's-complement system has made it the most widely used of the methods available, since it allows addition and subtraction to be performed by the same circuitry.

The reader should verify the results of using the above procedure for the following subtractions: (a) $+9 - (-4)$; (b) $-9 - (+4)$; (c) $-9 - (-4)$; (d) $+4 - (-4)$. Remember that when the result has a sign bit of 1, it is negative and in 2's-complement form.

Arithmetic Overflow

In each of the previous addition and subtraction examples, the numbers involved consisted of a sign bit and 4 magnitude bits. The answers also consisted of a sign bit and 4 magnitude bits. Any carry into the sixth bit position was disregarded. In all of the cases considered, the magnitude of the answer was small enough to fit into 4 bits. Let's look at the addition of $+9$ and $+8$.

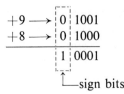

$$+9 \longrightarrow \boxed{0}\,1001$$
$$+8 \longrightarrow \boxed{0}\,1000$$
$$\boxed{1}\,0001$$

↑—sign bits

The answer has a negative sign bit, which is obviously incorrect. The answer should be $+17$, but the magnitude 17 requires more than 4 bits and therefore *overflows* into the sign bit position. This overflow condition always produces an incorrect result, and its occurrence is detected by examining the sign bit of the result and comparing it to the sign bits of the numbers being added. In a computer, a special circuit is used to detect any overflow condition and to signal that the answer is erroneous. We will encounter such a circuit in one of the problems following Chapter 6.

5.7 MULTIPLICATION OF BINARY NUMBERS

The multiplication of binary numbers is done in the same manner as the multiplication of decimal numbers. The process is actually simpler, since the multiplier digits are either 0 or 1, so we are always multiplying by 0 or 1 and no other digits. The following example illustrates:

$$
\begin{array}{l}
1001 \longleftarrow \text{multiplicand} = 9_{10}\\
1011 \longleftarrow \text{multiplier} = 11_{10}\\
\hline
1001 \left.\rule{0pt}{2.5em}\right\}\\
1001\\
0000 \quad\text{partial products}\\
1001\\
\hline
1100011 \}\text{final product} = 99_{10}
\end{array}
$$

In this example the multiplicand and multiplier are in true binary form and no sign bits are used. The steps followed in the process are exactly the same as in decimal multiplication. First, the LSB of the multiplier is examined; in our example it is a 1. This 1 multiplies the multiplicand to produce 1001, which is written down as the first partial product. Next, the second bit of the multiplier is examined. It is a 1, so 1001 is written for the second partial product. Note that this second partial product is *shifted* one place to the left relative to the first one. The third bit of the multiplier is 0, so 0000 is written as the third partial product; again, it is shifted one place to the left relative to the previous partial product. The fourth multiplier bit is 1, so the last partial product is 1001 shifted again one position to the left. The four partial products are then summed to produce the final product.

Most digital machines can only add two binary numbers at a time. For this reason, the partial products formed during multiplication cannot all be added together at the same time. Instead, they are added together two at a time; that is, the first is added to the second, their sum is added to the third, and so on. This process is now illustrated for the example above:

$$\text{Add} \begin{cases} 1001 & \longleftarrow \text{first partial product} \\ 1001 & \longleftarrow \text{second partial product shifted left} \end{cases}$$

$$\text{Add} \begin{cases} 11011 & \longleftarrow \text{sum of first two partial products} \\ 0000 & \longleftarrow \text{third partial product shifted left} \end{cases}$$

$$\text{Add} \begin{cases} 011011 & \longleftarrow \text{sum of first three partial products} \\ 1001 & \longleftarrow \text{fourth partial product shifted left} \end{cases}$$

$$\overline{1100011} \longleftarrow \text{sum of four partial products which equals final total product}$$

Multiplication in the 2's-Complement System

In machines that use the 2's-complement representation, multiplication is carried on in the manner described above provided that both the multiplicand and multiplier are put in true binary form. If the two numbers to be multiplied are positive, they are already in true binary form and are multiplied as they are. The resulting product is, of course, positive and is given a sign bit of 0. When the two numbers are negative, they will be in 2's-complement form. Each one is 2's-complemented to convert it to a positive number and then they are multiplied. The product is kept as a positive number and given a sign bit of 0.

When one of the numbers is positive and the other is negative, the negative number is first converted to a positive magnitude by taking its 2's complement. The product will be in true-magnitude form. However, the product has to be negative since the original numbers are of opposite sign. Thus, the product is then changed to 2's-complement form and given a sign bit of 1.

5.8 BINARY DIVISION

The process for dividing one binary number (the *dividend*) by another (the *divisor*) is the same as that which is followed for decimal numbers, that which we usually refer to as "long division." The actual process is simpler in binary because when we are checking to see how many times the divisor "goes into" the dividend, there are only two possibilities, 0 or 1. To illustrate, consider the following division examples:

$$
\begin{array}{r}
0011 \\
11 \,\overline{)\,1001} \\
\underline{011} \\
\overline{0011}
\end{array}
\qquad (9 \div 3 = 3)
$$

$$
\begin{array}{r}
0010.1 \\
100\overline{\smash{\big)}\,1010.0} \\
\underline{100} \\
100 \\
\underline{100} \\
0
\end{array}
\qquad (10 \div 4 = 2.5)
$$

In most modern digital machines the subtractions that are part of the division operation are usually carred out using 2's-complement subtraction, that is, 2's-complementing the subtrahend and then adding.

The division of signed numbers is handled in the same way as multiplication. Negative numbers are made positive by complementing and the division is then carried out. If the dividend and divisor are of opposite sign, the resulting quotient is changed to a negative number by 2's-complementing and is given a sign bit of 1. If the dividend and divisor are of the same sign, the quotient is left as a positive number and given a sign bit of 0.

5.9 THE OCTAL NUMBER SYSTEM

The octal number system is very important in digital computer work. The octal number system has a base of *eight*, meaning that it has eight possible digits: 0, 1, 2, 3, 4, 5, 6, and 7. Thus, each digit of an octal number can have any value from 0 to 7. The digit positions in an octal number have weights as follows:

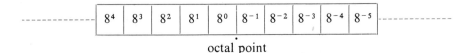

octal point

An octal number, then, can be easily converted to its decimal equivalent by multiplying each octal digit by its positional weight. For example,

$$
\begin{aligned}
372_8 &= 3 \times (8^2) + 7 \times (8^1) + 2 \times (8^0) \\
&= 3 \times 64 + 7 \times 8 + 2 \times 1 \\
&= 250_{10}
\end{aligned}
$$

Another example:

$$
\begin{aligned}
24.6_8 &= 2 \times (8^1) + 4 \times (8^0) + 6 \times (8^{-1}) \\
&= 20.75_{10}
\end{aligned}
$$

Conversion from Decimal to Octal

The methods for converting a decimal number to its octal equivalent are the same as those used to convert from decimal to binary (Section 5.2). To convert a decimal integer to octal, we progressively divide the decimal number by 8, writing down the remainders after each division. The remainders represent the digits of the octal number, the first remainder being the LSD. As an example, 266_{10} is converted to octal as follows:

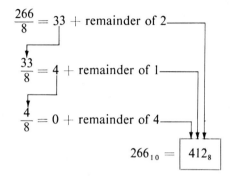

Decimal fractions are converted to octal by progressively *multiplying* by 8 and writing down the carries into the integer position. For example, $.38_{10}$ is converted to octal as follows:

$$0.38 \times 8 = 3.04 = 0.04 + \text{carry of } 3$$
$$0.04 \times 8 = 0.32 = 0.32 + \text{carry of } 0$$
$$0.32 \times 8 = 2.56 = 0.56 + \text{carry of } 2$$
etc.

$$.38_{10} = \boxed{.302_8}$$

Note that the first carry is the MSD of the fraction. More accuracy can be obtained by continuing the process to obtain more octal digits.

It is useful when converting a relatively large decimal number to binary to first convert it to octal. The octal number can then be converted to binary. This method is usually a lot quicker than direct decimal-to-binary conversion because of the simplicity of the octal-to-binary conversion.

Conversion Between Octal and Binary

The primary advantage of the octal number system is the ease with which conversion can be made between binary and octal numbers. The conversion from octal to binary is performed by converting *each* octal digit to its 3-bit binary equivalent. The eight possible digits are converted as indicated in Table 5.1.

Table 5.1

Octal Digit	0	1	2	3	4	5	6	7
Binary Equivalent	000	001	010	011	100	101	110	111

Using these conversions, any octal number is converted to binary by individually converting each digit. For example, we can convert 472_8 to binary as follows:

$$
\begin{array}{ccc}
4 & 7 & 2 \\
\downarrow & \downarrow & \downarrow \\
100 & 111 & 010
\end{array}
$$

Hence, octal 472 is equivalent to binary 100111010. As another example, consider converting 54.31_8 to binary:

$$
\begin{array}{ccccc}
5 & 4 & . & 3 & 1 \\
\downarrow & \downarrow & \downarrow & \downarrow & \downarrow \\
101 & 100 & . & 011 & 001
\end{array}
$$

Thus, $54.31_8 = 101100.011001_2$.

Converting from binary to octal is simply the reverse of the foregoing process. The binary digits are grouped into groups of three on each side of the binary point with zeros added on either side if needed to complete a group of three. Then each group of three bits is converted to its octal equivalent (Table 5.1). To illustrate, consider the conversion of 11010.1011_2 to octal:

$$
\begin{array}{ccccc}
011 & 010 & . & 101 & 100 \\
\downarrow & \downarrow & . & \downarrow & \downarrow \\
3 & 2 & . & 5 & 4
\end{array}
$$

Note that 0s were added on each end to complete the groups of three. Thus, the desired octal conversion is 32.54_8. Here are several more examples:

$$10110_2 = 26_8$$
$$10011.011 = 23.3_8$$
$$111011.1111 = 73.74_8$$

Counting in Octal

The largest octal digit is 7, so that when counting in octal, a digit is incremented upward from 0 to 7. Once it reaches 7, it recycles to 0 on the next count and causes the next-higher digit to be incremented. This is illustrated in the following sequences of octal counting: 64, 65, 66, 67, 70; 275, 276, 277, 300.

With N octal digits, we can count from zero up to $8^N - 1$, for a total of 8^N

different counts. For example, with three octal digits we can count from 000_8 to 777_8, which is a total of $8^3 = 512_{10}$ different octal numbers.

Usefulness of Octal System

The ease with which conversions can be made between octal and binary make the octal system attractive as a "shorthand" means of expressing large binary numbers. In computer work, binary numbers with up to 36 bits are not uncommon. These binary numbers, as we shall see, do not always represent a numerical quantity but are often some type of code which conveys nonnumerical information. In computers, binary numbers might represent (1) actual numerical data; (2) numbers corresponding to a location (address) in memory; (3) an instruction code; (4) a code representing alphabetic and other nonnumerical characters; or (5) a group of bits representing the status of devices internal or external to the computer.

When dealing with a large quantity of binary numbers of many bits, it is convenient and more efficient for us to write the numbers in octal rather than binary. Keep in mind, however, that the digital circuits and systems work strictly in binary; we are using octal only as a convenience for the operators of the system.

5.10 THE HEXADECIMAL NUMBER SYSTEM

The hexadecimal system uses base 16. Thus, it has 16 possible digit symbols. It uses the digits 0 through 9 plus the letters A, B, C, D, E, and F as the 16 digit symbols. Table 5.2 shows the relationships among hexadecimal, decimal, and binary. Note that each hexadecimal digit represents a group of four binary digits.

Table 5.2

Hexadecimal	Decimal	Binary
0	0	0000
1	1	0001
2	2	0010
3	3	0011
4	4	0100
5	5	0101
6	6	0110
7	7	0111
8	8	1000
9	9	1001
A	10	1010
B	11	1011
C	12	1100
D	13	1101
E	14	1110
F	15	1111

Many computers utilize the hexadecimal system, rather than octal, to represent large binary numbers. The conversions between hexadecimal and binary are done in exactly the same manner as octal and binary except that groups of *four* bits are used. The following examples illustrate:

$$1\ 1\ 1\ 0\ 1\ 0\ 0\ 1\ 1\ 0_2 = \underbrace{0\ 0\ 1\ 1}_{3}\ \underbrace{1\ 0\ 1\ 0}_{A}\ \underbrace{0\ 1\ 1\ 0}_{6}$$

$$= 3\ A\ 6_{16}$$

$$9\ F\ 2_{16} = \quad \underset{\downarrow}{9} \qquad \underset{\downarrow}{F} \qquad \underset{\downarrow}{2}$$

$$1\ 0\ 0\ 1\ \ \ 1\ 1\ 1\ 1\ \ \ 0\ 0\ 1\ 0$$

$$= 100111110010_2$$

Counting in Hexadecimal

When counting in *hex* (abbreviation for hexadecimal) each digit can be incremented from 0 up to F. Once it reaches F, the next count causes it to recycle to 0 and the next-higher digit is incremented. This is illustrated in the following hex counting sequences: 38, 39, 3A, 3B, 3C, 3D, 3E, 3F, 40; 6B8, 6B9, 6BA, 6BB, 6BC, 6BD, 6BE, 6BF, 6C0.

Hex-to-Decimal Conversion

Occasionally, it is necessary to convert from a hexadecimal number to its equivalent decimal number. This is easily done if it is remembered that the various digit positions in a hex number have weights that are powers of 16. The conversion process is as follows:

$$356_{16} = 3 \times 16^2 + 5 \times 16^1 + 6 \times 16^0$$
$$= 768 + 80 + 6$$
$$= 854_{10}$$
$$2AF_{16} = 2 \times 16^2 + 10 \times 16^1 + 15 \times 16^0$$
$$= 512 + 160 + 15$$
$$= 687_{10}$$

Note that in the second example the value 10 was substituted for A and the value 15 for F in the conversion to decimal.

5.11   THE BCD CODE

When numbers, letters, or words are represented by a special group of symbols, this is called *encoding*, and the group of symbols is called a *code*. Probably one of the most familiar codes is the Morse Code, where series of dots and dashes represent letters of the alphabet.

We have seen that any decimal number can be represented by an equivalent binary number. The group of 0s and 1s in the binary number can be thought of as a code representing the decimal number. When a decimal number is represented by its equivalent binary number, we call it *straight binary coding*.

Digital systems all use some form of binary numbers for their internal operation, but the external world is decimal in nature. This means that conversions between the decimal and binary systems are being performed often. We have seen that the conversions between decimal and binary can become long and complicated for large numbers. For this reason, another means of encoding decimal numbers, which combines some features of both the decimal and binary system, is sometimes used.

Binary-Coded-Decimal Code

If *each* digit of a decimal number is represented by its binary equivalent, this produces a code called *binary-coded-decimal* (hereafter abbreviated BCD). Since a decimal digit can be as large as 9, 4 bits are required to code each digit (the binary code for 9 is 1001).

To illustrate the BCD code, take a decimal number such as 874. Each digit is changed to its binary equivalent as follows:

$$
\begin{array}{ccc}
8 & 7 & 4 \\
\downarrow & \downarrow & \downarrow \\
1000 & 0111 & 0100
\end{array}
$$

As another example, let us change 94.3 to its BCD-code representation:

$$
\begin{array}{ccccc}
9 & 4 & . & 3 \\
\downarrow & \downarrow & & \downarrow \\
1001 & 0100 & . & 0011 \ .
\end{array}
$$

Once again, each decimal digit is changed to its straight binary equivalent. Note that 4 bits are always used for each digit.

The BCD code, then, represents each digit of the decimal number by a 4-bit binary number. Clearly, only the 4-bit binary numbers from 0000 through 1001 are used. The BCD code does not use the numbers 1010, 1011, 1100, 1101, 1110, and 1111. In other words, only 10 of the 16 possible 4-bit binary code groups are used. If any of these "forbidden" 4-bit numbers ever occurs in a machine using the BCD code, it is usually an indication that an error has occurred.

EXAMPLE 5.1   Convert the BCD number 0110100000111001 to its decimal equivalent.

Solution:

$$
\begin{array}{cccc}
0110 & 1000 & 0011 & 1001 \\
6 & 8 & 3 & 9
\end{array}
$$

EXAMPLE 5.2   Convert the BCD number 011111000001 to its decimal equivalent.

Solution:

$$0111 \quad 1100 \quad 0001$$

$$7 \qquad \quad \; \; 1$$

forbidden code group indicates error in BCD number

Comparison of BCD and Binary

It is important to realize that BCD is not another number system like binary, octal, decimal, and hexadecimal. It is, in fact, the decimal system with each digit encoded in its binary equivalent. It is also important to understand that a BCD number is *not* the same as a straight binary number. A straight binary code takes the *complete* decimal number and represents it in binary; the BCD code converts *each* decimal *digit* to binary individually. To illustrate, take the number 137 and compare its straight binary and BCD codes:

$$137_{10} = 10001001_{2} \qquad \text{(binary)}$$

$$137_{10} = 0001 \quad 0011 \quad 0111 \qquad \text{(BCD)}$$

The BCD code requires 12 bits while the straight binary code requires only 8 bits to represent 137. It is always true that the BCD code for a given decimal number requires more bits than straight binary. This is because BCD does not use all possible 4-bit groups, as pointed out earlier and is therefore somewhat inefficient.

The main advantage of the BCD code is the relative ease of converting to and from decimal. Only the 4-bit code groups for the decimal digits 0–9 need be remembered. This ease of conversion is especially important from a hardware standpoint because in a digital system it is the logic circuits that perform the conversions to and from decimal.

BCD is used in digital machines whenever decimal information is either applied as inputs or displayed as outputs. Digital voltmeters, frequency counters, and digital clocks all use BCD because they display output information in decimal. Electronic calculators use BCD because the input numbers are entered in decimal via the keyboard and the output numbers are displayed in decimal.

BCD is not often used in modern high-speed digital computers for two good reasons. First, as was already pointed out, the BCD code for a given decimal number requires more bits than the straight binary code and is therefore less efficient. This is important in digital computers because the number of places in memory where these bits can be stored is limited. Second, the arithmetic processes for numbers represented in BCD code are more complicated than straight binary and thus require more complex circuitry. The more complex circuitry contributes to a decrease in the speed at which arithmetic operations take place. Calculators that use BCD are therefore considerably slower in their operation than computers.

5.12 BCD ADDITION

The addition of decimal numbers that are in BCD form can be best understood by considering the two cases that can occur when two decimal digits are added.

Sum Equals Nine or Less

Consider adding 5 and 4 using BCD to represent each digit:

$$
\begin{array}{rl}
5 & \quad 0101 \longleftarrow \text{BCD for 5} \\
+\,4 & \quad +\,0100 \longleftarrow \text{BCD for 4} \\
\hline
9 & \quad \overline{1001} \longleftarrow \text{BCD for 9}
\end{array}
$$

The addition is carried out as in normal binary addition and the sum is 1001, which is the BCD code for 9. As another example, take 45 added to 33:

$$
\begin{array}{rll}
45 & \quad 0100 & 0101 \longleftarrow \text{BCD for 45} \\
+\,33 & \quad +\,0011 & 0011 \longleftarrow \text{BCD for 33} \\
\hline
78 & \quad 0111 & 1000 \longleftarrow \text{BCD for 78}
\end{array}
$$

In this example the 4-bit codes for 5 and 3 are added in binary to produce 1000, which is BCD for 8. Similarly, adding the second-decimal-digit positions produces 0111, which is BCD for 7. The total is 0111 1000, which is the BCD code for 78.

In the examples above, none of the sums of the pairs of decimal digits exceeded 9; therefore, *no decimal carries were produced*. For these cases the BCD addition process is straightforward and is actually the same as binary addition.

Sum Greater than Nine

Consider the addition of 6 and 7 in BCD:

$$
\begin{array}{rl}
6 & \quad 0110 \longleftarrow \text{BCD for 6} \\
+\,7 & \quad +\,0111 \longleftarrow \text{BCD for 7} \\
\hline
+13 & \quad \overline{1101} \longleftarrow \text{invalid code group for BCD}
\end{array}
$$

The sum 1101 does not exist in the BCD code; it is one of the six forbidden or invalid 4-bit code groups. This has occurred because the sum of the two digits exceeds 9. Whenever this occurs the sum has to be corrected by the addition of six (0110) to take into account the skipping of the six invalid code groups:

$$
\begin{array}{rl}
6 & \quad 0110 \longleftarrow \text{BCD for 6} \\
+\,7 & \quad +\,0111 \longleftarrow \text{BCD for 7} \\
\hline
13 & \quad 1101 \longleftarrow \text{invalid sum} \\
& \quad +\,0110 \longleftarrow \text{add 6 for correction} \\
\hline
& \quad 0001 \quad 0011 \longleftarrow \text{BCD for 13} \\
& \qquad \underbrace{\;\;}_{1} \quad \underbrace{\;\;}_{3}
\end{array}
$$

As shown above, 0110 is added to the invalid sum and produces the correct BCD result. Note that a carry is produced into the second decimal position. This addition of 0110 has to be performed whenever the sum of the two decimal digits is greater than 9.

As another example, take 47 plus 35 in BCD:

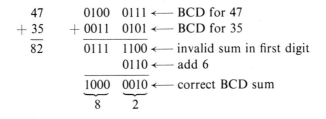

The addition of the 4-bit codes for the 7 and 5 digits results in an invalid sum and is corrected by adding 0110. Note that this generates a carry of 1, which is carried over to be added to the BCD sum of the second-position digits.

To summarize the BCD addition procedure:

1. Add, using ordinary binary addition, the BCD code groups for each digit position.

2. For those positions where the sum is 9 or less, no correction is needed. The sum is in proper BCD form.

3. When the sum of two digits is greater than 9, a correction of 0110 should be added to that sum to get the proper BCD result. This will always produce carry into the next decimal position.

The procedure for BCD addition is clearly more complicated than straight binary addition. This is also true of the other BCD arithmetic operations. Readers should perform the addition of 275 + 241 to check their understanding of the BCD addition procedure.

5.13 EXCESS-3 CODE

The *excess-3 code* is related to the BCD code and is sometimes used instead of BCD because it possesses advantages in certain arithmetic operations. The excess-3 code for a decimal number is performed in the same manner as BCD except that *3* is added to *each* decimal digit before encoding it in binary. For example, to encode the decimal number 4 into excess-3 code, we must first add 3 to obtain 7. Then the 7 is encoded in its equivalent 4-bit binary code 0111.

As another example, let us convert 46 into its excess-3-code representation.

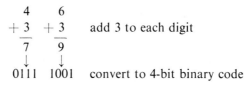

$$
\begin{array}{cc}
4 & 6 \\
+\,3 & +\,3 \\
\hline
7 & 9 \\
\downarrow & \downarrow \\
0111 & 1001
\end{array}
$$

add 3 to each digit

convert to 4-bit binary code

Table 5.3 lists the BCD and excess-3-code representations for the decimal digits. Note that both codes use only 10 of the 16 possible 4-bit code groups. The excess-3 code, however, does not use the same code groups. For excess-3, the invalid code groups are 0000, 0001, 0010, 1101, 1110, and 1111.

Table 5.3

Decimal	BCD	Excess-3
0	0000	0011
1	0001	0100
2	0010	0101
3	0011	0110
4	0100	0111
5	0101	1000
6	0110	1001
7	0111	1010
8	1000	1011
9	1001	1100

5.14 THE GRAY CODE

The *Gray code* belongs to a class of codes called *minimum-change codes*, in which only *one* bit in the code group changes when going from one step to the next. The Gray code is an *unweighted* code, meaning that the bit positions in the code groups do not have any specific weight assigned to them. Because of this, the Gray code is not suited for arithmetic operations but finds application in input/output devices and some types of analog-to-digital convertors.

Table 5.4 shows the Gray-code representation for the decimal numbers 0 through 15, together with the straight binary code. If we examine the Gray-code groups for each decimal number, it can be seen that in going from *any* one decimal number to the next, only *one* bit of the Gray code changes. For example, in going from 3 to 4, the Gray code changes from 0010 to 0110, with only the second bit from the left changing. Going from 14 to 15 the Gray-code bits change from 1001 to 1000, with only the last bit changing. This is the principal characteristic of the Gray code. Compare this with the binary code, where anywhere from one to all of the bits change in going from one step to the next.

The Gray code is often used in situations where other codes, such as binary, might produce erroneous or ambiguous results during those transitions in which

Table 5.4

Decimal	Binary Code	Gray Code
0	0000	0000
1	0001	0001
2	0010	0011
3	0011	0010
4	0100	0110
5	0101	0111
6	0110	0101
7	0111	0100
8	1000	1100
9	1001	1101
10	1010	1111
11	1011	1110
12	1100	1010
13	1101	1011
14	1110	1001
15	1111	1000

more than one bit of the code is changing. For instance, using binary code and going from 0111 to 1000 requires that all four bits change simultaneously. Depending on the device or circuit that is generating the bits, there may be a significant difference in the transition times of the different bits. If so, the transition from 0111 to 1000 could produce one or more intermediate states. For example, if the most-significant bit changes faster than the rest, the following transitions will occur:

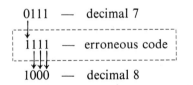

$$0111 \quad - \quad \text{decimal 7}$$

$$1111 \quad - \quad \text{erroneous code}$$

$$1000 \quad - \quad \text{decimal 8}$$

The occurrence of 1111 is only momentary but it could conceivably produce erroneous operation of the elements that are being controlled by the bits. Obviously, using the Gray code would eliminate this problem, since only one bit change occurs per transition and no "race" between bits can occur.

Converting from Binary to Gray Code

Any binary number can be converted to its Gray-code representation as follows:

1. The first bit of the Gray code is the same as the first bit of the binary number.

2. The second bit of the Gray code equals the exclusive-OR of the first and

second bits of the binary number; that is, it will be 1 if these binary-code bits are *different*, 0 if they are the *same*.

3. The third Gray-code bit equals the exclusive-OR of the second and third bits of the binary number, and so on.

To illustrate, let us convert binary 10110 to Gray code:

$$\begin{array}{ccccc} 1 & 0 & 1 & 1 & 0 \\ & & & & \\ 1 & 1 & 1 & 0 & 1 \end{array} \quad \begin{array}{l} \text{binary} \\ \\ \text{Gray} \end{array}$$

The first bit of the Gray code is the same as the first bit of the binary code. The first and second bits of the binary code are *different*, giving a 1 for the second Gray bit. The second and third bits of the binary number are *different*, giving a 1 for the third Gray bit. The third and fourth bits of the binary number are the *same*, so the fourth Gray bit is 0. Finally, the fourth and fifth binary bits are *different*, giving a fifth Gray bit of 1.

Another example is as follows:

$$\begin{array}{cccccccc} 1 & 0 & 0 & 1 & 1 & 0 & 0 & 1 \\ & & & & & & & \\ 1 & 1 & 0 & 1 & 0 & 1 & 0 & 1 \end{array} \quad \begin{array}{l} \text{binary} \\ \\ \text{Gray} \end{array}$$

Converting from Gray to Binary

To convert from Gray to binary requires the opposite procedure to that given above.

1. The first binary bit is the same as the first Gray-code bit.

2. If the second Gray bit is 0, the second binary bit is the same as the first; if the second Gray bit is 1, the second binary bit is the inverse of the first binary bit.

3. Step 2 is repeated for each successive bit.

To illustrate, let us convert 1101 from Gray to binary:

$$\begin{array}{cccc} 1 & 1 & 0 & 1 \\ \downarrow & \downarrow & \downarrow & \downarrow \\ 1 \longrightarrow & 0 \longrightarrow & 0 \longrightarrow & 1 \end{array} \quad \begin{array}{l} \text{Gray} \\ \\ \text{binary} \end{array}$$

The first Gray bit is a 1, so the first binary bit is written as a 1. The second Gray bit is a 1, so the second binary bit is made a 0 (inverse of the first binary bit). The third Gray bit is a 0, so the third binary bit is made a 0 (same as the second binary bit). The fourth Gray bit is 1, making the fourth binary bit a 1 (inverse of the third binary bit).

This process can be viewed in another way. Each binary bit (except the first) can be obtained by taking the exclusive-OR of the corresponding Gray-code bit and the previous binary bit. The reader should verify that this process gives the proper result for the following conversions:

Gray	Binary
110101 $\longrightarrow$	100110
010110 $\longrightarrow$	011011

5.15   ALPHANUMERIC CODES

A computer must be capable of handling nonnumeric information if it is to be very useful. In other words, a computer must be able to recognize codes that represent numbers, letters, and special characters. These codes are classified as *alphanumeric codes*. A complete and adequate set of necessary characters includes: (1) 26 lower-case letters; (2) 26 uppercase letters; (3) 10 numeric digits; and (4) about 25 special characters, such as $+$, $/$, $\#$, $\%$, and so on.

This totals up to 87 characters. To represent 87 characters with some type of binary code would require at least 7 bits. With 7 bits, there are $2^7 = 128$ possible binary numbers; 87 of these arrangements of 0 and 1 bits serve as the code groups representing the 87 different characters. For example, the code group 1010101 might represent the letter U.

The most common alphanumeric code is known as the American Standard Code for Information Interchange (ASCII) and is used by most minicomputer and microcomputer manufacturers. Table 5.5 shows a partial listing of the 7-bit ASCII code. For each character, the octal and hex equivalents are also shown. The ASCII code is used in the transmission of alphanumeric information between a computer and external input/output devices like a teletypewriter (TTY) or cathode-ray tube (CRT).

5.16   PARITY METHOD FOR ERROR DETECTION

The transmission of binary data from one location to another is commonplace in all digital systems. Here are several examples of this:

1. Binary-data output from a computer being recorded on magnetic tape.

2. Transmission of binary data over telephone lines, such as between a computer and a remote console.

3. A number is taken from the computer memory and placed in the arithmetic unit, where it is added to another number. The sum is then transferred back into memory.

Table 5-5 Partial Listing of ASCII Code

Character	7-Bit ASCII	Octal	Hex
A	100 0001	101	41
B	100 0010	102	42
C	100 0011	103	43
D	100 0100	104	44
E	100 0101	105	45
F	100 0110	106	46
G	100 0111	107	47
H	100 1000	110	48
I	100 1001	111	49
J	100 1010	112	4A
K	100 1011	113	4B
L	100 1100	114	4C
M	100 1101	115	4D
N	100 1110	116	4E
O	100 1111	117	4F
P	101 0000	120	50
Q	101 0001	121	51
R	101 0010	122	52
S	101 0011	123	53
T	101 0100	124	54
U	101 0101	125	55
V	101 0110	126	56
W	101 0111	127	57
X	101 1000	130	58
Y	101 1001	131	59
Z	101 1010	132	5A
0	011 0000	060	30
1	011 0001	061	31
2	011 0010	062	32
3	011 0011	063	33
4	011 0100	064	34
5	011 0101	065	35
6	011 0110	066	36
7	011 0111	067	37
8	011 1000	070	38
9	011 1001	071	39
blank	010 0000	040	20
.	010 1110	056	2E
(	010 1000	050	28
+	010 1011	053	2B
$	010 0100	044	24
*	010 1010	052	2A
)	010 1001	051	29
−	010 1101	055	2D
/	010 1111	057	2F
,	010 1100	054	2C
=	011 1101	075	3D

4. Punched IBM cards are fed to a card reader, which transfers the information from the cards into the computer.

The process of transferring data is subject to error, although modern equipment has been designed to reduce the probability of error. However, even relatively infrequent errors can cause useless results, so it is desirable to detect them whenever possible. One of the most widely used schemes for error detection is the *parity method.*

The Parity Bit

A *parity bit* is an extra bit that is attached to a code group which is being transferred from one location to another. The parity bit is made either 0 or 1, depending on the number of 1s that are contained in the code group. There are two different methods that are used.

In the *even-parity* method, the value of the parity bit is chosen so that the total number of 1s in the code group (including the parity bit) is an *even* number. For example, suppose that the code group is 1000011. This is the ASCII character C. The code group has *three* 1s. Therefore, we will add a parity bit of 1 to make the total number of 1s an even number. The *new* code group, including the parity bit, thus becomes

$$\boxed{1}\ 1\ 0\ 0\ 0\ 0\ 1\ 1$$
$$\llcorner\!\!-\!\!-\!\!-\!\!-\!\!-\!\!-\!\!-\!\!-\text{added parity bit*}$$

If the code group contains an even number of 1s to begin with, the parity bit is given a value of 0. For example, if the code group is 1000001 (the ASCII code for "A"), the assigned parity bit would be 0, so the new code, including the parity bit, would be 01000001.

The *odd-parity* method is used in exactly the same way except that the parity bit is chosen so the total number of 1s (including the parity bit) is an *odd* number. For example, for the code group 1000001, the assigned parity bit would be a 1. For the code group 1000011, the bit would be a 0.

Regardless of whether even parity or odd parity is used, the parity bit becomes an actual part of the code word. For example, adding a parity bit to the 7-bit ASCII code produces an 8-bit code.

The parity bit is used to ensure that during the transmission of a character code from one place to another (e.g., TTY to computer), any *single* errors can be detected. For example, if odd parity is being used and the recipient of the transmitted character detects an *even* number of 1 bits, clearly the character code must be in error. Figure 5.3 shows how the parity method is used. The code-group bits

---

*The parity bit can be placed at either end of the code word. Of course, the receiver of the data must know which is the parity bit and which is data.

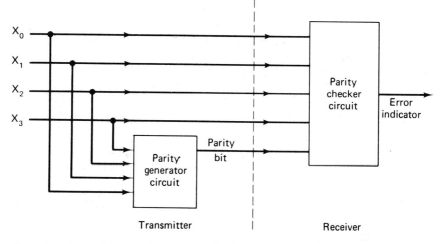

Figure 5.3   General diagram for parity method.

are represented by X_0, X_1, X_2, and X_3. These bits might be coming from the outputs of a FF register. They are fed to a *parity generator circuit*, which is a logic circuit that examines the input bits and produces an output parity bit of correct value. The parity bit is transmitted along with the X bits.

When the transmitted bits reach their destination, they are fed to a *parity-checker circuit*, which is a logic circuit that examines all the bits to determine whether the correct parity is present. In an even-parity system, the parity checker will generate a low error output if the number of input 1s is an even number, a high error output (indicating an error) if the number of 1s is odd. In an odd-parity system this would be reversed.

If an error occurs in *one* of the transmitted bits, the parity-checker circuit will detect it. For example, let us assume that the X code-group bits are 0110 and that we are using an odd-parity system. The parity generator circuit will thus generate a 1 for a parity bit, so 01101 will be transmitted. If these bits reach the parity checker unchanged, the parity checker will produce a 0 output (no error). However, if *one* of the bits changes before it reaches the parity checker (such as 00101 instead of 01101), the parity checker will go high, indicating that an error has occurred in the transmission. The error output can be used to sound an alarm, to stop the system operation, or to activate an error indicator.

It should be apparent that this parity method can detect *single* errors but cannot detect *double* errors. This is because a double error will not change the parity of the group of bits, so the parity checker will indicate no error. Also, this parity method does not pinpoint the error; that is, it does not determine which bit is in error. More sophisticated methods can be used to detect double errors and to pinpoint the bit or bits which are in error so that corrections can be made.

The parity method described above is useful for systems in which the probabilities of errors are very low; that is, single errors occur only rarely and double errors occur almost never. Some of the logic circuits used in a typical parity-checking system are covered in Problems 5.34–5.37.

QUESTIONS AND PROBLEMS

5.1   Convert the following binary numbers to equivalent decimal numbers:
(a)   10110   (b)   10001101   (c)   10111.1011   (d)   0.011011   (e)   110111.101
(f)   100100001001

5.2   Convert the following decimal numbers to their binary equivalents:
(a)   37   (b)   14   (c)   189   (d)   72.45   (e)   0.4475   (f)   4097.188   (g)   52
(h)   205

5.3   Add the following groups of binary numbers using binary addition:
(a)   1010 + 1011   (b)   1111 + 0011   (c)   1011.1101 + 11.1
(d)   0.1011 + 0.1111   (e)   10011011 + 10011101

5.4   Represent each of the following numbers in true-magnitude form using a sign bit:
(a)   +32   (b)   −14   (c)   +63.75   (d)   −19.625

5.5   Repeat Problem 5.4 using the 1's-complement system.

5.6   Repeat Problem 5.4 using the 2's-complement system.

5.7   The following numbers are represented in a true-magnitude form with a sign bit. Determine their decimal equivalent.
(a)   10110   (b)   0110.1001   (c)   11101101   (d)   011011.11

5.8   Determine the decimal equivalents of the numbers in Problem 5.7 assuming that they are represented in the 1's-complement system.

5.9   Determine the decimal equivalents of the numbers in Problem 5.7 assuming they are represented in 2's-complement form.

5.10   How many FFs are required to represent numbers in the range from −127 to +127 using 2's-complement representation?

5.11   Perform the following operations using the 2's-complement form to represent negative numbers.
(a)   Add +9 to +6.
(b)   Add −7 to +4.
(c)   Add +9 to −14.
(d)   Add −3.5 to −2.625.
(e)   Add +11.0 to −9.5.
(f)   Add −18 to −13.
(g)   Subtract +6 from +7.
(h)   Subtract +11 from −3.
(i)   Subtract +19 from +19.
(j)   Subtract 14.125 from 10.500.

5.12   The following numbers (denoted *A*, *B*, and *C*) are stored in a computer that uses 2's-complement form for negative numbers. The leftmost bit of each number is the sign bit.

$$A = 01010 \qquad B = 11100 \qquad C = 00101$$

Perform the following operations on these numbers:
(a) $A + B$ (b) $A - B$ (c) $A - C$ (d) $B - C$ (e) $C - B$ (f) $A \times B$
(g) $A \times C$ (h) $B \times C$ (i) $B \times B$

5.13 Multiply the following pairs of binary numbers:
(a) 111×101 (b) 1011×1011 (c) 101.101×110.010
(d) $.1101 \times .1011$

5.14 Perform the following divisions:
(a) $1100 \div 100$ (b) $111111 \div 1001$ (c) $10111 \div 100$
(d) $10110.1101 \div 1.1$

5.15 Convert the following numbers from octal to decimal:
(a) 743_8 (b) 36.4_8 (c) 124.25_8

5.16 Convert the following numbers from decimal to octal and then to binary.
(a) 59 (b) 372 (c) 0.58 (d) 64.125

5.17 Convert the numbers of Problem 5.15 from octal to binary.

5.18 Convert the following binary numbers to octal and then to hexadecimal.
(a) 101100110011 (b) 1011101.1011 (c) 10011000010110
(d) 10110.01101101

5.19 Convert the following hex numbers to binary:
(a) 2AC (b) 7B4

5.20 Encode the following decimal numbers in BCD code:
(a) 47 (b) 924.62 (c) 187 (d) 6.379

5.21 The following numbers are in BCD code. Convert them to their decimal equivalents:
(a) 10010110.01010001 (b) 1000011101010010.10010101
(c) 0010100001111001

5.22 How many bits are required to represent the decimal numbers 0–999 in straight binary code? In BCD code?

5.23 Add the following decimal numbers after converting each to its BCD code:
(a) $74 + 23$ (b) $58 + 37$ (c) $147 + 380$ (d) $385 + 118$

5.24 Convert the following numbers to excess-3 code:
(a) 82 (b) 49.71

5.25 Convert 101101101 from binary to Gray code.

5.26 Convert 1000110101 from binary to Gray code.

5.27 Convert 1101100101 from Gray code to binary.

5.28 Convert 10111101101 from Gray code to binary.

5.29 What is the principal characteristic of the Gray code and when is it useful?

5.30 Use the ASCII code of Table 5.5 to represent the statement "$X = 25/Y$."

5.31 The following code groups are being transmitted. Attach an *even*-parity bit to each group.
(a) 10110110 (b) 00101000 (c) 11110111

5.32 Convert the following decimal numbers to BCD code and then attach an *odd*-parity bit.
(a) 74 (b) 38 (c) 165 (d) 9201

5.33   In a certain digital system, the decimal numbers from 000 through 999 are represented in BCD code. An *odd*-parity bit is also included at the end of each code group. Examine each of the code groups below and assume that each one has just been transferred from one location to another. Some of the groups contain errors. Assume that *no more than* two errors have occurred for each group. Determine which of the code groups has a single error and which of them *definitely* has a double error. (*Hint:* Remember that this is a BCD code.)

    (a)   1001010110000

                └─parity bit

    (b)   0100011101100

    (c)   0111110000011

    (d)   1000011000101

5.34   Figure 5.4 shows a *parity-generator circuit*, which takes inputs X_0–X_3 and generates an even-parity-bit output. Verify its operation.

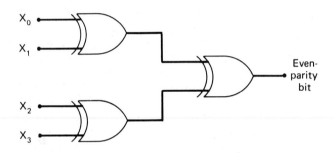

Figure 5.4

5.35   Change the circuit of Figure 5.4 so that it generates an odd-parity bit.

5.36   Modify the circuit of Figure 5.4 so that it can handle a 5-bit code X_0–X_4 and generates an even-parity bit.

5.37   Design a parity-checker circuit that will take X_0–X_3 and the even-parity bit from Figure 5.4 and will generate a HIGH error output if the parity is not correct.

5.38   A four-FF register is storing the BCD code for a decimal digit. Design a *BCD error-detector circuit* which examines the four FF outputs to determine if any illegal BCD code groups occur. Simplify the circuit using the techniques learned previously.

5.39   List the octal numbers in sequence from 165_8 to 185_8.

5.40   List the hexadecimal numbers in sequence from $2E8_{16}$ to 300_{16}.

ARITHMETIC CIRCUITS

One essential function of most computers and calculators is the performance of arithmetic operations. These operations are all performed in the arithmetic unit of a computer, where logic gates and flip-flops are combined so that they can add, subtract, multiply, and divide binary numbers. These circuits perform arithmetic operations at speeds that are not humanly possible. Typically, an addition operation will take less than 1 μs.

In this chapter we will study some of the basic arithmetic circuits that are used to perform the arithmetic operations discussed in Chapter 5. In some cases we will go through the actual design process, even though the circuits may be commercially available in integrated-circuit form, to provide more practice in the use of the techniques learned in Chapter 3.

6.1 THE ARITHMETIC UNIT

All arithmetic operations take place in the *arithmetic unit* of a computer. Figure 6.1 is a block diagram showing the major elements included in a typical arithmetic unit. The main purpose of the arithmetic unit is to accept binary data that are stored in the memory and to execute arithmetic operations on these data according to instructions from the control unit.

The arithmetic unit contains at least two flip-flop registers: the *B register* and the *accumulator register*. It also contains combinatorial logic, which performs the arithmetic operations on the binary numbers that are stored in the *B* register and the accumulator. A typical sequence of operations may occur as follows:

1. The control unit receives an instruction (from the memory unit) specifying

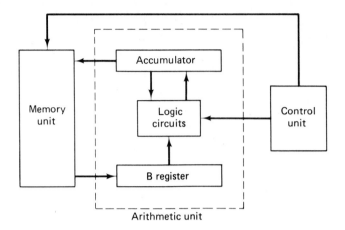

Figure 6.1   Functional parts of arithmetic unit.

that a number stored in a particular memory location (address) is to be added to the number presently stored in the accumulator register.

2.   The number to be added is transferred from memory to the *B* register.

3.   The number in the *B* register and the number in the accumulator register are added together in the logic circuits (upon command from control unit). The resulting sum is then sent to the accumulator to be stored.

4.   The new number in the accumulator can remain there so that another number can be added to it or, if the particular arithmetic process is finished, it can be transferred to memory for storage.

These steps should make it apparent how the accumulator register derives its name. This register "accumulates" the sums that occur when performing successive additions between new numbers acquired from memory and the previously accumulated sum. In fact, for any arithmetic problem containing several steps, the accumulator always contains the results of the intermediate steps as they are completed as well as the final result when the problem is finished.

6.2   THE PARALLEL BINARY ADDER

Computers and calculators perform the addition operation on two binary numbers at a time, where each binary number can have several binary digits. Figure 6.2 illustrates the addition of two 5-bit numbers. The *augend* is stored in the accumulator register; that is, the accumulator contains five FFs, storing the values 10101 in successive FFs. Similarly, the *addend*, the number that is to be added to the augend, is stored in the *B* register (in this case, 00111).

The addition process starts by adding the least significant bits (LSBs) of the

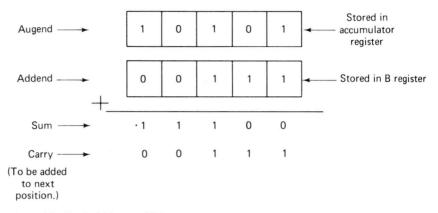

Figure 6.2   Typical binary addition process.

augend and addend. Thus, $1 + 1 = 10$,* which means that the *sum* for that position is 0 with a *carry* of 1.

This carry has to be added to the next position along with the augend and addend bits in that position. Thus, in the second position, $1 + 0 + 1 = 10$, which is again a sum of 0 and a carry of 1. This carry is added to the next position together with the augend and addend bits in that position, and so on for the remaining positions.

At each step in this addition process we are performing the addition of 3 bits; the augend bit, the addend bit, and a carry bit from the previous position. The result of the addition of these 3 bits produces 2 bits: a *sum* bit, and a *carry* bit which is to be added to the next position. It should be clear that the same process is followed for each bit position. As such, if we can design a logic circuit that can duplicate this process, then we simply have to use identical circuits for each of the bit positions. This is illustrated in Figure 6.3.

In this diagram variables A_4, A_3, A_2, A_1, and A_0 represent the bits of the augend that are stored in the accumulator (which is also called the A register). Variables B_4, B_3, B_2, B_1, and B_0 represent the bits of the addend stored in the B register. Variables C_4, C_3, C_2, C_1, and C_0 represent the carry bits into the corresponding positions. Variables S_4, S_3, S_2, S_1, S_0 are the sum bits for each position. Corresponding bits of the augend and addend are fed to a logic circuit called a *full adder*, along with a carry bit from the previous position. For example, bits A_1 and B_1 are fed into full adder 1 along with C_1, which is the carry bit produced by the addition of the A_0 and B_0 bits. Bits A_0 and B_0 are fed into full adder 0 along with C_0. Since A_0 and B_0 are the LSBs of the augend and addend, it appears that C_0 would always have to be 0, since there can be no carry into that position. However, we shall see that there will be situations when C_0 can also be 1.

The full-adder circuit used in each position has three inputs: an A bit, a B bit,

---

*Remember: now we are talking about *arithmetic* addition, not OR addition.

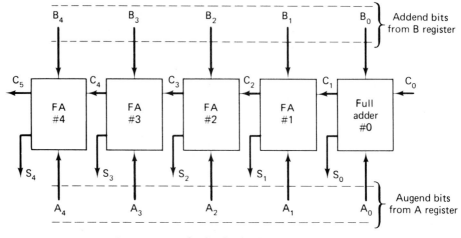

Sum appears at S_4, S_3, S_2, S_1, S_0 outputs.

Figure 6.3   Block diagram of 5-bit parallel adder circuit using full adders.

and a C bit, and it produces two outputs: a sum bit and a carry bit. For example, full adder 0 has inputs A_0, B_0, and C_0, and it produces outputs S_0 and C_1. Full adder 1 has A_1, B_1, and C_1 as inputs and S_1 and C_2 as outputs; and so on. This arrangement is repeated for as many positions as there are in the augend and addend. Although this illustration is for 5-bit numbers, in modern computers and calculators the numbers usually range from 8 to 36 bits.

The arrangement in Figure 6.3 is called a *parallel adder* because all the bits of the augend and addend are present and are fed into the adder circuits *simultaneously.* This means that the additions in each position are taking place at the same time. This is different from how we add on paper, taking each position one at a time starting with the LSB. Clearly, parallel addition is extremely fast. More will be said about this later.

6.3 DESIGN OF A FULL ADDER

Now that we know the function of the full adder we can proceed to design a logic circuit that will perform this function following the steps outlined in Chapter 3. First, we must construct a truth table showing the various input and output values for all possible cases. Figure 6.4 shows the truth table having three inputs A, B, and C_{IN}, and two outputs, S and C_{OUT}. There are eight possible cases for the three inputs, and for each case the desired output values are listed. For example, consider the case $A = 1$, $B = 0$, and $C_{IN} = 1$. The full adder (hereafter abbreviated FA) must add these bits to produce a sum (S) of 0 and a carry (C_{OUT}) of 1. The reader should check the other cases to be sure they are understood.

Augend bit input	Addend bit input	Carry bit input	Sum bit output	Carry bit output
A	B	C_{IN}	S	C_{OUT}
0	0	0	0	0
0	0	1	1	0
0	1	0	1	0
0	1	1	0	1
1	0	0	1	0
1	0	1	0	1
1	1	0	0	1
1	1	1	1	1

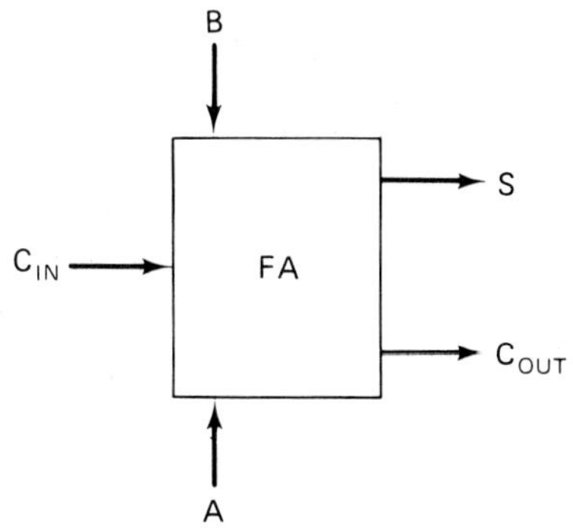

Figure 6.4   Truth table for full-adder circuit.

Since there are two outputs, we will design the circuitry for each output individually, starting with the S output. The truth table shows that there are four cases where S is to be a 1. Using the sum-of-products method (Section 3.2) we can write the expression for S as

$$S = \bar{A}\bar{B}C_{IN} + \bar{A}B\bar{C}_{IN} + A\bar{B}\bar{C}_{IN} + ABC_{IN} \tag{6.1}$$

We can now try to simplify this expression by factoring. Unfortunately, none of the terms in the expression has two variables in common with any of the other terms. However, $\bar{A}$ can be factored from the first two terms and A can be factored from the last two terms:

$$S = \bar{A}(\bar{B}C_{IN} + B\bar{C}_{IN}) + A(\bar{B}\bar{C}_{IN} + BC_{IN})$$

The first term in parentheses should be recognized as the exclusive-OR combination of B and C_{IN}, which can be written as $B \oplus C_{IN}$. The second term in parentheses should be recognized as the exclusive-NOR of B and C_{IN}, which can be written as $\overline{B \oplus C_{IN}}$. Thus, the expression for S becomes

$$S = \bar{A}(B \oplus C_{IN}) + A(\overline{B \oplus C_{IN}})$$

If we let $X = B \oplus C_{IN}$, this can be written as

$$S = \bar{A} \cdot X + A \cdot \bar{X} = A \oplus X$$

which is simply the EX-OR of A and X. Thus, we have finally,

$$S = A \oplus (B \oplus C_{\text{IN}}) \tag{6.2}$$

Consider now the output C_{OUT} in the truth table of Figure 6.4. We can write the sum-of-products expression for C_{OUT} as follows:

$$C_{\text{OUT}} = \bar{A}BC_{\text{IN}} + A\bar{B}C_{\text{IN}} + AB\bar{C}_{\text{IN}} + ABC_{\text{IN}}$$

This expression can be simplified by factoring. We will employ the trick introduced in Chapter 2 (Example 2.10) whereby we will use the ABC_{IN} term *three* times since it has common factors with each of the other terms. Hence,

$$C_{\text{OUT}} = BC_{\text{IN}}(\bar{A} + A) + AC_{\text{IN}}(\bar{B} + B) + AB(\bar{C}_{\text{IN}} + C_{\text{IN}})$$
$$= BC_{\text{IN}} + AC_{\text{IN}} + AB \tag{6.3}$$

This expression cannot be simplified further.

Expressions (6.2) and (6.3) can be implemented as shown in Figure 6.5. There are several other implementations that can be used to produce the same expressions for S and C_{OUT}, none of which has any particular advantage over those shown. The complete circuit with inputs A, B, and C_{IN} and outputs S and C_{OUT}

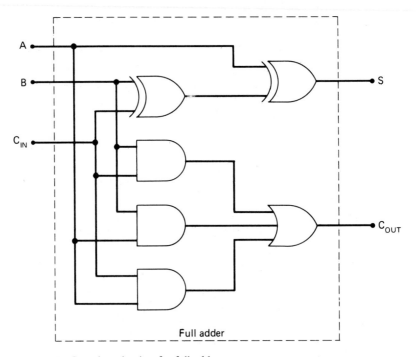

Figure 6.5 Complete circuitry for full adders.

represents the full adder. Each of the FAs in Figure 6.3 contains this same circuitry (or its equivalent).

The Half Adder

The FA operates on three inputs to produce a sum and carry output. In some cases a circuit is needed that will add only 2 input bits, to produce a sum and carry output. An example would be the addition of the LSB position of two binary numbers where there is no carry input to be added. A special logic circuit can be designed to take *two* input bits, A and B, and to produce sum (S) and carry (C_{OUT}) outputs. This circuit is called a *half adder* (HA). Its operation is similar to a FA except that it operates on only 2 bits. We shall leave the design of the HA as an exercise at the end of the chapter since it is relatively simple. However, we will use the HA symbol whenever it is appropriate in the discussions to follow.

6.4 COMPLETE PARALLEL ADDER WITH REGISTERS

The parallel adder of Figure 6.3 shows how two binary numbers can be added together to produce a binary number equal to their sum. As stated earlier, in a computer the two numbers to be added are stored in FF registers. Figure 6.6 shows the complete schematic for the addition of two 4-bit numbers. The augend bits A_3–A_0 are stored in the accumulator (A register); the addend bits B_3–B_0 are stored in the B register. The implementation of adders for more than 4 bits is accomplished by adding more FFs to the registers and including a FA for each new position.

The contents of registers A and B are added in the four FA circuits and their sum is produced at the S_3–S_0 outputs. These sum outputs are then transferred into the A register through the D inputs of the FFs when an add command pulse is applied. The sum transferred into the A register will destroy the previously stored augend. Note that C_4 is the output carry from the last stage and can be used as the carry input to a fifth stage or as an *overflow* bit to indicate that the sum exceeds 1111.

It will be helpful to go through the complete process of adding two binary numbers. Let us add 1001 to 0101. To begin, we will assume that both registers have been cleared to the 0000 state. Also, we will make $C_0 = 0$ since there is no carry into the first position. The first number to be added (1001) is transferred into the B register. (In a computer this number usually comes from the memory unit.) At this point the B register holds 1001 and the A register holds 0000. These inputs to the FA circuits produce sum outputs 1001; that is, $S_3 = 1, S_2 = 0, S_1 = 0$, and $S_0 = 1$. The levels present on these sum output lines will be transferred to the A register on the positive transition of the add pulse. Thus, after the add pulse

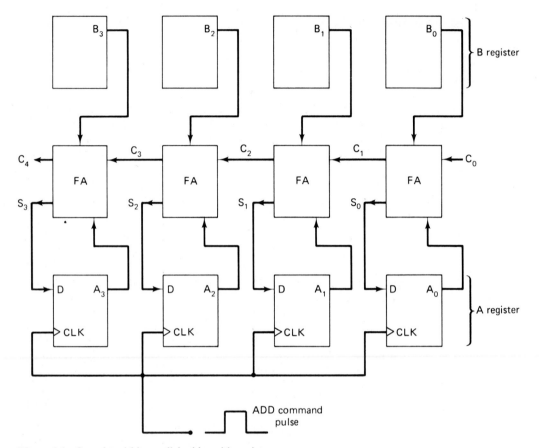

Figure 6.6 Complete 4-bit parallel adder with registers.

occurs, the *A* register will contain 1001. So far, the 1001 that was in the *B* register has been added to the 0000 that was in the *A* register to produce a sum of 1001, which has been transferred to the *A* register.

The second number to be added (0101) is now transferred into the *B* register, so the inputs to the FA circuits are 1001 (from the *A* register) and 0101 (from the *B* register). This produces sum outputs 1110; that is, $S_3 = S_2 = S_1 = 1$, and $S_0 = 0$. These levels are then transferred to the *A* register on the next add pulse, so the final result 1110 is now in the *A* register. In a computer this result would probably be transferred into the main memory for storage while the adder circuit was utilized for further arithmetic operations.

6.5   CARRY PROPAGATION

The parallel adder of Figure 6.6 performs additions at a relatively high speed since it adds the bits from each position simultaneously. However, its speed is limited by an effect called *carry propagation* or *carry ripple*, which can best be explained by considering the following addition:

Addition of the LSB position produces a carry into the second position. This carry, when added to the bits of the second position, produces a carry into the third position. This latter carry, when added to the bits of the third position, produces a carry into the last position. The key thing to notice in this example is that the sum bit generated in the *last* position (MSB) depended on the carry that was generated by the addition in the *first* position.

Looking at this from the viewpoint of the circuit of Figure 6.6, S_3 out of the last full adder depends on C_1 out of the first full adder. But the C_1 signal must pass through three FAs before it produces S_3. What this means is that the S_3 output will not reach its correct value until C_1 has propagated through the intermediate FAs. This represents a time delay that depends on the propagation delay produced in a FA. For example, if each FA is considered to have a propagation delay of 40 ns, then S_3 will not reach its correct level until 120 ns after C_1 is generated. This means that the add command pulse cannot be applied until 160 ns after the augend and addend numbers are present in the FF registers (the extra 40 ns is due to the delay of the LSB FA, which generates C_1).

Obviously, the situation becomes much worse if we extend the adder circuitry to add a greater number of bits. If the adder were handling 32-bit numbers, which is typical for a large computer, the carry propagation delay could be 1280 ns = 1.28 μs. The add pulse could not be applied until at least 1.28 μs after the numbers were present in the registers.

This magnitude of delay is prohibitive for high-speed computers. Fortunately, logic designers have come up with several ingenious schemes for reducing this delay. One of the schemes, called *look-ahead carry*, utilizes logic gates to look at the lower-order bits of the augend and addend to see if a higher-order carry is to be generated. For example, it is possible to build a logic circuit with B_2, B_1, B_0, A_2, A_1, and A_0 as inputs and C_3 as an output. This logic circuit would have a shorter delay than is obtained by the carry propagation through the FAs. This scheme requires a large amount of extra circuitry but is necessary to produce high-speed adders. The extra circuitry is not a significant consideration with the present use of integrated circuits. Many high-speed adders available in integrated-circuit form utilize the look-ahead carry or a similar technique for reducing overall propagation delay times. Problem 6.7 will investigate the look-ahead carry in more detail.

6.6 INTEGRATED-CIRCUIT PARALLEL ADDERS

Integrated-circuit manufacturers have made available multibit parallel adders which are fabricated as a single IC package. One of the most common of these has the capability of adding two 4-bit numbers. Figure 6.7(a) shows the block diagram

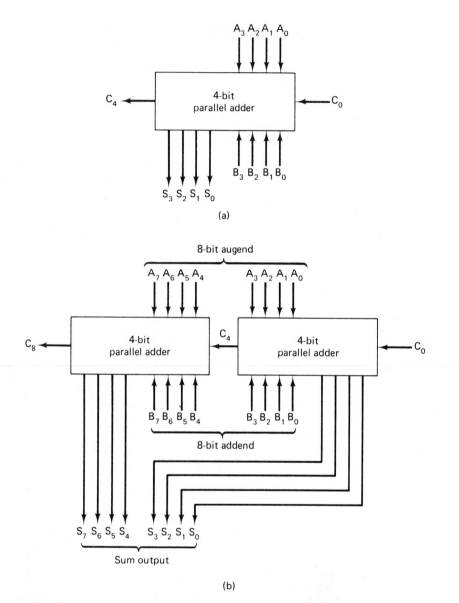

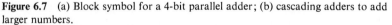

Figure 6.7   (a) Block symbol for a 4-bit parallel adder; (b) cascading adders to add larger numbers.

for such an adder. The adder inputs are two 4-bit numbers, $A_3A_2A_1A_0$ and $B_3B_2B_1B_0$, and the carry into the first position, C_0. Its outputs are the sum bits, $S_3S_2S_1S_0$, and the carry out of the last position, C_4. The adder block consists of the full adders and their interconnections, such as those shown in Figure 6.6. If it is a high-speed adder it will also contain extra logic gates for reducing carry propagation time, such as the look-ahead carry idea mentioned previously. In all the work to follow, we will use this block symbol to represent the 4-bit parallel adder, for clarity.

Cascading Parallel Adders

Two or more parallel adder blocks may be connected in cascade to accommodate the addition of larger binary numbers. To illustrate, Figure 6.7(b) shows how two 4-bit parallel adders can be connected to add two 8-bit numbers. The adder on the right adds the four least-significant bits of the numbers. The C_4 output of this adder is connected as the input carry to the first position of the second adder, which adds the four most-significant bits of the numbers. The eight sum outputs represent the resultant sum of the two 8-bit numbers. C_8 is the carry out of the last position (MSB) of the second adder. C_8 can be used as an overflow bit or as a carry into another adder stage if larger binary numbers are to be handled.

6.7   THE 2'S-COMPLEMENT SYSTEM

Most modern computers use the 2's-complement system to represent negative numbers and to perform subtraction. The operations of addition and subtraction of signed numbers can be performed using only the addition operation if we use the 2's-complement form to represent negative numbers. (It may be helpful for purposes of the following discussion to review Sections 5.4–5.6.)

Addition

Positive and negative numbers, including the sign bits, can be added together in the basic parallel adder circuit when the negative numbers are in 2's-complement form. This is illustrated in Figure 6.8 for the addition of -3 and $+6$. The -3 is represented in its 2's-complement form as 1101, where the first 1 is the sign bit; the $+6$ is represented as 0110, with the first zero as the sign bit. These numbers are stored in their corresponding registers. The 4-bit parallel adder produces sum outputs of 0011, which represents $+3$. The C_4 output is 1 but is disregarded in the 2's-complement method.

Subtraction

When the 2's-complement system is used, the number to be subtracted (the subtrahend) is 2's-complemented and then *added* to the minuend (the number the subtrahend is being subtracted from). For example, we can assume that the minuend is already stored in the accumulator (*A* register). The subtrahend is then placed in the *B* register (in a computer it would be transferred here from memory)

From A register

A_3 A_2 A_1 A_0

2's-complement
representation of -3 (augend)

| 1 | 1 | 0 | 1 |

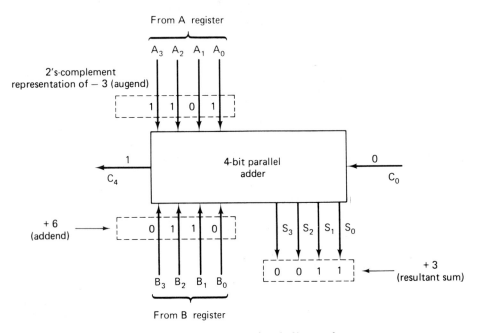

1   4-bit parallel   0
C_4 adder C_0

+6
(addend)

| 0 | 1 | 1 | 0 | S_3 S_2 S_1 S_0

B_3 B_2 B_1 B_0 | 0 | 0 | 1 | 1 | +3
(resultant sum)

From B register

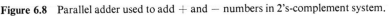

Figure 6.8   Parallel adder used to add $+$ and $-$ numbers in 2's-complement system.

and is changed to its 2's-complement form before it is added to the number in the *A* register. The sum outputs of the adder circuit now represent the *difference* between the minuend and subtrahend.

The parallel-adder circuit that we have been discussing can be adapted to perform the subtraction described above if we provide a means for taking the 2's complement of the *B* register number. The 2's complement of a binary number is obtained by complementing (inverting) each bit and then adding 1 to the LSB. Figure 6.9 shows how this can be accomplished. The *inverted* outputs of the *B* register are used rather than the normal outputs; that is, $\bar{B}_0$, $\bar{B}_1$, $\bar{B}_2$, and $\bar{B}_3$ are fed to the adder inputs (remember, B_3 is the sign bit). This takes care of complementing each bit of the *B* number. Also, C_0 is made a logical 1, so it adds an extra 1 into the LSB of the adder; this accomplishes the same effect as adding 1 to the LSB of the *B* register for forming the 2's complement.

The S_3–S_0 outputs represent the results of the subtraction operation. Of course, S_3 is the sign bit of the result and indicates whether the result is $+$ or $-$. The carry output C_4 is again disregarded.

To help clarify this operation, study the following steps for subtracting $+6$ from $+4$:

1. $+4$ is stored in the *A* register as 0100.
2. $+6$ is stored in the *B* register as 0110.

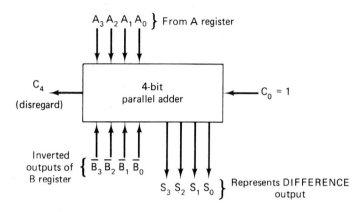

Figure 6.9 Parallel adder used to perform subtraction using 2's-complement system.

3. The inverted outputs of the *B*-register FFs are fed to the adder, that is, 1001.

4. The 1001 is added to 0100 by the parallel adder along with a 1 added to the LSB position by making $C_0 = 1$. This produces sum output bits 1110 and a $C_4 = 1$, which is disregarded. This 1110 represents the required difference. Since the sign-bit = 1, it is a negative result and is in 2's-complement form. Verify that it represents −2.

Combined Addition and Subtraction

It should now be clear that the basic parallel adder circuit can be used to perform addition or subtraction depending on whether the *B* number is left unchanged or is 2's-complemented. A complete circuit that can perform *both* addition and subtraction in the 2's-complement system is shown in Figure 6.10.

This adder/subtractor circuit is controlled by the two control signals ADD and SUB. When the ADD level is HIGH, the circuit performs addition of the numbers stored in the *A* and *B* registers. When the SUB level is HIGH, the circuit subtracts the *B*-register number from the *A*-register number. The operation is described as follows:

1. Assume that ADD = 1 and SUB = 0. The SUB = 0 *disables* (inhibits) AND gates 2, 4, 6, and 8, holding their outputs at 0. The ADD = 1 *enables* AND gates 1, 3, 5, and 7, allowing their outputs to pass the B_0, B_1, B_2, and B_3 levels, respectively.

2. The B_0–B_3 levels pass through the OR gates into the 4-bit parallel adder to be added to the A_0–A_3 bits. The *sum* appears at the S_0–S_3 outputs.

3. Note that SUB = 0 causes $C_0 = 0$ into the adder.

4. Now assume that ADD = 0 and SUB = 1. The ADD = 0 disables AND

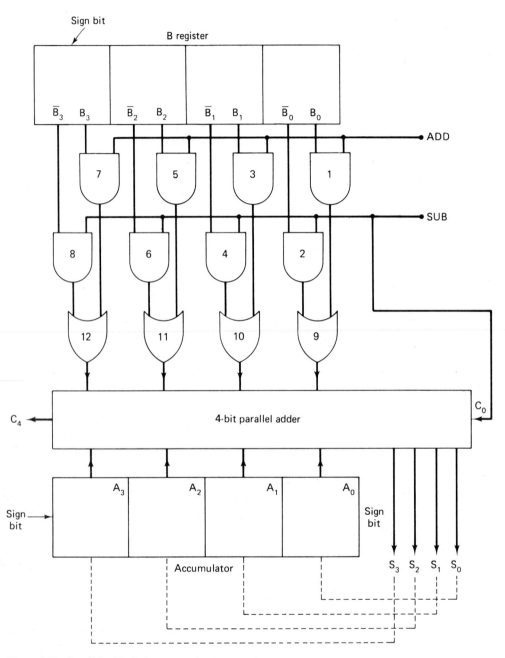

Figure 6.10 Parallel adder/subtractor using 2's-complement system.

gates 1, 3, 5, and 7. The SUB $= 1$ enables AND gates 2, 4, 6, and 8, so their outputs pass the $\bar{B}_0$, $\bar{B}_1$, $\bar{B}_2$, and $\bar{B}_3$ levels, respectively.

5. The $\bar{B}_0$–$\bar{B}_3$ levels pass through the OR gates into the adder to be added to the A_0–A_3 bits. Note also that C_0 is now 1. Thus, the B-register number has essentially been 2's-complemented.

6. The *difference* appears at the S_0–S_3 outputs.

The configuration of Figure 6.10 is one of the most widely used in computers because it provides a relatively simple method for adding and subtracting positive or negative numbers. In most computers the outputs present at the S output lines are usually transferred into the A register (accumulator) such as was done in Figure 6.6, so the results of the addition or subtraction always end up stored in the A register.

6.8 THE SERIAL ADDER

The parallel adder that we have been studying performs the addition of two binary numbers at a relatively fast rate since all the bit positions are operated on simultaneously. Its speed is limited by the carry propagation time, which, as was mentioned earlier, is minimized by using carry look-ahead techniques. The major disadvantage of the parallel addition process is that it requires a large amount of logic circuitry, which increases in direct proportion to the number of bits in the numbers being added. In *serial* addition, the addition process is carried out in a manner more closely related to the way we perform addition on paper, that is, one position at a time. This results in much simpler circuitry than parallel addition but results in a much slower speed of operation.

Figure 6.11 shows the diagram of a 4-bit serial adder. The A and B registers once again are used to store the numbers to be added. However, in the serial adder these registers are *shift registers* whose binary values shift from left to right upon application of each clock pulse.* The A_0 and B_0 outputs (LSBs) are fed into a *single* full-adder circuit along with the Q output of the carry FF. The carry FF is a separate FF used to store the carry output of the FA so that it can be added to the next significant position of the numbers in the registers.

Notice that SUM output of the FA is fed to the D input of the MSB of the A register (in this case, A_3), so the SUM value is transferred into A_3 on the occurrence of the clock pulse. Also note that the output of B_0 is connected to the D input of B_3, so B_3 takes on the value of B_0 when a clock pulse occurs. In this way the B-register contents will be unchanged after all the shifting operations are completed.

The operation of this serial adder can be easily understood by following through a complete example. We will assume that the augend 0111 is in the A

---

*Shift registers were first discussed in Section 4.11.

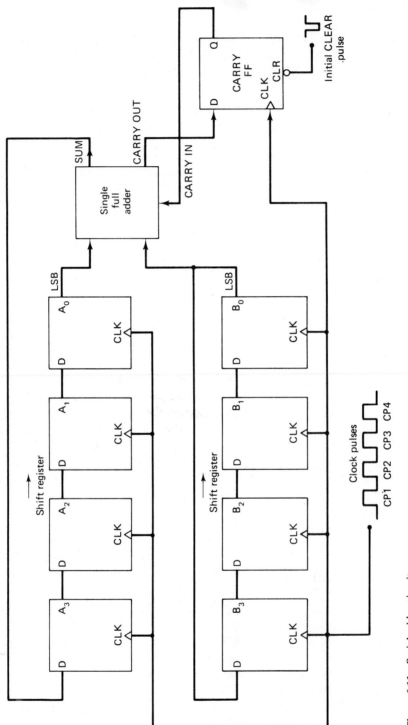

Figure 6.11 Serial adder circuit.

register and the addend 0010 is in the *B* register. Also, we will assume that the carry FF has been initially cleared to the 0 state, so CARRY-IN = 0. Refer to the first entry in Figure 6.12, which shows the various logic levels before the first clock pulse (CP) is applied.

	A_3 A_2 A_1 A_0	B_3 B_2 B_1 B_0	CARRY-IN (output Q)	SUM Output	CARRY Output
Before first clock pulse	0 1 1 1	0 0 1 0	0	1	0
After CP1	1 0 1 1	0 0 0 1	0	0	1
After CP2	0 1 0 1	1 0 0 0	1	0	1
After CP3	0 0 1 0	0 1 0 0	1	1	0
After CP4	1 0 0 1	0 0 1 0	0	1	0
	Final sum				

Figure 6.12 Example showing sequence of states for serial adder.

First Clock Pulse

Since $A_0 = 1$, $B_0 = 0$, and CARRY-IN = 0, the FA outputs will be SUM = 1 and CARRY-OUT = 0. These levels are present at the FA outputs before the first CP occurs. When the first CP occurs, the values in the *A* register shift from left to right one bit, as do the values in the *B* register. In addition, the SUM level is transferred into A_3, the B_0 level is transferred into B_3, and the CARRY-OUT level is transferred to the carry *FF*, whose output becomes CARRY-IN = 0. The results of these transfers are listed in the second entry in Figure 6.12.

Second Clock Pulse

At this point, A_0 and B_0 contain the bits (1 and 1) that were in the second position of the original augend and addend. These are fed to the FA along with CARRY-IN (now 0) from the carry FF. Thus, the FA outputs are now SUM = 0 and CARRY-OUT = 1. When the second CP occurs, the *A* and *B* registers again shift right, SUM = 0 is transferred to A_3, and CARRY-OUT = 1 is transferred to the carry FF. The conditions at the end of the second CP are shown in Figure 6.12.

Third Clock Pulse

The values of A_0 and B_0 are now 1 and 0, respectively, and CARRY-IN is 1. These values produce SUM = 0 and CARRY-OUT = 1 at the FA outputs. On the occurrence of the third CP, the *A* and *B* registers shift right, SUM = 0 goes to

A_3, and CARRY-OUT $= 1$ goes to the carry FF. Figure 6.12 shows the conditions at the end of the third CP.

Fourth Clock Pulse

A_0 and B_0 are now both 0 and CARRY-IN $= 1$, so the FA produces SUM $= 1$ and CARRY-OUT $= 0$. The fourth CP transfers SUM $= 1$ into A_3 and initiates all the other transfers. At the completion of this fourth CP the A register holds the number 1001, which is, of course, the sum of the original augend and addend. In addition, the B register holds the original addend 0010. The addition process is now complete.

6.9 SERIAL SUBTRACTION USING 2's COMPLEMENT

The serial adder of Figure 6.11 can be easily converted to a serial subtractor using the 2's-complement system. For subtraction, the subtrahend is stored in the B register and must be 2's-complemented before it is added to the minuend stored in the A register. One simple way to accomplish this is to complement the B register and to make the initial CARRY-IN $= 1$ instead of 0 prior to the first CP. This can be easily accomplished by feeding the inverted output $\bar{B}_0$ into the FA instead of B_0 and initially *setting* the carry FF to 1 instead of clearing it. The remaining circuitry is the same as for addition. The reader should verify the subtractor operation by following through an example.

A combination serial adder/subtractor can be constructed in a manner similar to its parallel counterpart. This will be left as an exercise at the end of the chapter.

6.10 SERIAL VERSUS PARALLEL

The relative advantages and disadvantages of parallel arithmetic circuits and serial arithmetic circuits should now be clear. Stated simply: *Parallel is generally faster than serial but serial requires fewer components.*

For the parallel adder we saw that all the outputs of the A and B registers were available for addition at the same time. In the serial adder only the rightmost FFs of these registers can be added at one time. In the parallel adder all the bit positions are processed simultaneously; in the serial adder the bit positions are processed one at a time. As such, the parallel adder produces the desired output only after a short delay—the time it takes the logic levels to propagate through the FA circuits. Using fast logic circuits and look-ahead carry techniques, a parallel adder can add two 36-bit numbers in about 100 ns.

A serial adder would require 36 clock pulses to add two 36-bit numbers. Using fast logic circuits and FFs, a clock frequency of about 20 MHz could be used (50 ns per pulse). Thus, it could perform this addition in $36 \times 50 = 1800$ ns $= 1.8$ μs, which is much slower than the parallel adder.

On the other hand, the parallel adder requires 36 full adders to add two 36-bit numbers, whereas the serial adder requires only one full adder for any number of bits. In this respect the serial adder is obviously superior.

Whenever high speed is the prime consideration, such as in modern digital computers, parallel arithmetic circuits are used, despite their complexity. Where speed is secondary to the minimization of components, serial arithmetic circuits are used. Electronic calculators, for example, do not require very high speed operation, but they do require minimizing the circuitry as much as possible, especially if they are battery-operated. For this reason, calculators almost universally use serial arithmetic circuits.

Of course, there are situations where serial arithmetic circuits that use very fast FFs and logic will operate at speeds comparable to parallel arithmetic circuits which use slower logic.

6.11   BCD ADDER

As noted earlier, some digital systems use BCD numbers instead of straight binary. Recall that in the BCD code each decimal digit is represented by its equivalent 4-bit binary code. For example, 478 becomes

$$
\begin{array}{cccl}
4 & 7 & 8 & \text{decimal} \\
\downarrow & \downarrow & \downarrow & \\
0100 & 0111 & 1000 & \text{BCD}
\end{array}
$$

The BCD addition process was discussed in Section 5.12 and is now summarized:

1. Add the BCD code groups for each decimal digit position; use ordinary binary addition.

2. For those positions where the sum is 9 or less, the sum is in proper BCD form and no correction is needed.

3. When the sum of two digits is greater than 9, a correction of 0110 should be added to that sum to produce the proper BCD result. This will produce a carry to be added to the next decimal position.

A BCD adder circuit must be able to operate in accordance with the above steps. In other words, the circuit must be able to do the following:

1. Add two 4-bit BCD code groups, using straight binary addition.

2. Determine if the sum of this addition is greater than 1001 (decimal 9); if it is, add 0110 (6) to this sum and generate a carry to the next decimal position.

The first requirement is easily met by using a 4-bit binary parallel adder such as the one discussed earlier (Figure 6.7). For example, if the two BCD code groups represented by $A_3A_2A_1A_0$ and $B_3B_2B_1B_0$, respectively, are applied to a 4-bit parallel adder, the adder will perform the following operation:

$$
\begin{array}{ll}
A_3A_2A_1A_0 & \longleftarrow \text{BCD code group} \\
+\ B_3B_2B_1B_0 & \longleftarrow \text{BCD code group} \\
\hline
S_4{}^*S_3S_2S_1S_0 & \longleftarrow \text{straight binary sum}
\end{array}
$$

The sum output number $S_4S_3S_2S_1S_0$ can range anywhere from 00000 (when the A number and the B number are both 0000) to 10010 (when both numbers are 1001). In accordance with the second requirement for the BCD adder, it is necessary to detect whenever the $S_4S_3S_2S_1S_0$ sum is greater than 01001 (decimal 9). Let us define X as a logic variable that will be *high* only when the $S_4S_3S_2S_1S_0$ number is greater than 01001. It can be reasoned that X will be high for either of the following conditions:

1. Whenever $S_4 = 1$ (sums greater than 15).

2. Whenever $S_3 = 1$ and either S_2 or S_1 or both are 1 (sums 10–15).

This can be expressed as

$$X = S_4 + S_3(S_2 + S_1)$$

Whenever $X = 1$, it is necessary to add the correction 0110 to the sum bits and to generate a carry. Figure 6.13 shows the complete circuitry for a BCD adder, including the logic-circuit implementation for X.

Examination of Figure 6.13 should show that the circuit satisfies the requirements for a BCD adder. The two BCD code groups A_3–A_0 and B_3–B_0 are added together in the parallel *binary* adder, producing the sum bits $S_4S_3S_2S_1S_0$. The gate network produces the X output, which goes HIGH if the sum from the parallel adder exceeds 01001. When $X = 1$, the bottom group of adders will add 0110 to this sum, producing the final BCD sum output bits represented by $\Sigma_3\Sigma_2\Sigma_1\Sigma_0$ (the Greek letter Σ, sigma, is a common mathematical symbol for *sum*). X is also the carry output that is produced when the sum is greater than 01001. Of course, when $X = 0$, there is no carry and no addition of 0110. In such cases, $\Sigma_3\Sigma_2\Sigma_1\Sigma_0 = S_3S_2S_0S_0$.

To help in the understanding of the BCD adder, the reader should try several cases by following them though the circuit. The following cases would be particularly instructive:

	A_3	A_2	A_1	A_0	B_3	B_2	B_1	B_0
(a)	0	1	0	1	0	0	1	1
(b)	0	1	1	1	0	1	1	0

$^*S_4$ is actually C_4, the carry generated in the fourth position.

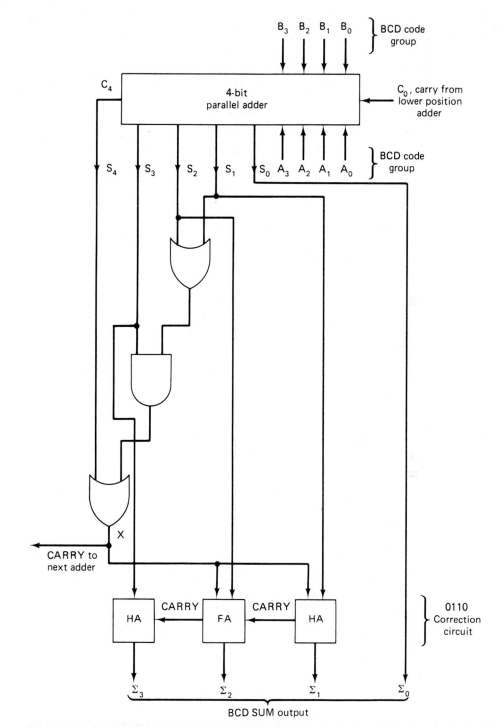

Figure 6.13 BCD adder.

Cascading BCD Adders

The circuit of Figure 6.13 is used for adding two decimal digits that have been encoded in BCD code. When decimal numbers with several digits are to be added together, it is necessary to use a separate BCD adder for each digit position. Figure 6.14 illustrates for the addition of two three-digit decimal numbers. The A register contains 12 bits, which are the three BCD code groups for one of the three-digit decimal numbers; similarly, the B register contains the BCD representation of the other three-digit decimal number. The A_3-A_0 and B_3-B_0 code groups representing the least-significant digits are fed to the first BCD adder. The BCD adder block is assumed to contain the circuitry of Figure 6.13. This first BCD adder produces sum outputs $\Sigma_3\Sigma_2\Sigma_1\Sigma_0$, which is the BCD code for the least-significant digit of the sum. It also produces a carry output that is sent to the second BCD adder, which is adding A_7-A_4 and B_7-B_4, the BCD code groups for the second-decimal-digit position. The second BCD adder produces $\Sigma_7\Sigma_6\Sigma_5\Sigma_4$, the BCD code for the second digit of the sum, and so on. This arrangement can, of course, be extended to decimal numbers of any size by simply adding more FFs to the registers and including a BCD adder for each digit position.

6.12 BINARY MULTIPLIERS

The multiplication of two binary numbers is done with paper and pencil by performing successive additions and shifting. To illustrate:

```
        1011        multiplicand (11)
     ×  1101        multiplier (13)
     ───────
        1011
       0000
      1011
      1011
     ──────────
     10001111      product (143)
```

This process consists of examining the successive bits of the multiplier, beginning with the LSB. If the multiplier bit is a 1, the multiplicand is copied down; if it is a 0, zeros are written down. The numbers written down in successive lines are shifted one position to the left relative to the previous line. When all the multiplier bits have been examined, the various lines are *added* to produce the final product.

In digital machines this process is modified somewhat because the binary adder is designed to add only two binary numbers at a time. Instead of adding all the lines at the end, they are added two at a time and their sum is accumulated in a register (the accumulator register). In addition, when the multiplier bit is 0,

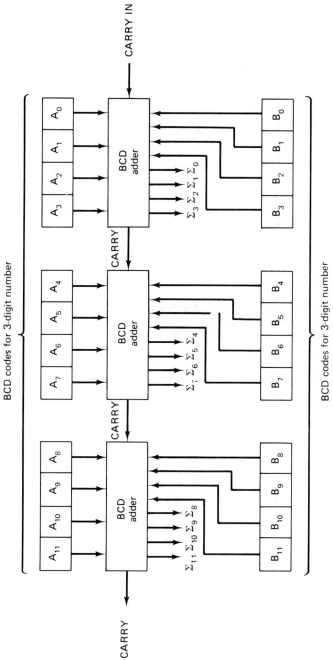

Figure 6.14 Cascading BCD adders to add two three-digit decimal numbers.

175

there is no need to write down and add zeros since it does not affect the final result. The previous example is redone here showing the modified process.

multiplicand: 1011
multiplier: 1101

1011	1st multiplier bit $= 1$; write down multiplicand shift multiplicand one position to the left (10110)
1011	2nd multiplier bit $= 0$; write down previous result shift new multiplicand to the left again (101100)
$+101100$	3rd multiplier bit $= 1$; write down new multiplicand (101100)
110111	add shift new multiplicand to the left (1011000)
$+1011000$	4th multiplier bit $= 1$; write down new multiplicand (1011000)
10001111	add to obtain final product

This multiplication process can be implemented as shown in Figure 6.15. It contains three registers. The X register is used to store the multiplier bits. The B register is used to store the multiplicand bits. These two registers are connected as shift registers (B shifts right and X shifts left on each clock pulse). The A register is the accumulator register, which is used to accumulate the partial products. Note that the adder outputs S_0–S_7 are connected to the D inputs of the A register so that the sum is transferred to the A register when a clock pulse gets through the AND gate. Also note that the B and X registers will shift on the negative-going edge of the clock pulses while the A register responds to the positive clock transitions.

The circuit operation can best be described with the aid of Figure 6.16, which shows the contents of all the registers and the adder outputs at each point during the operation. The step-by-step process takes place as follows:

1. Initially the A register is at 00000000. The B register contains the multiplicand 00001011. The X register contains the multiplier 1101. The adder outputs represent the sum of A and B, that is, 00001011. These conditions are shown in Figure 6.16(a).

2. The first clock pulse occurs. Since the LSB of the multiplier (X_0) is a 1, this clock pulse gets through the AND gate and its positive edge transfers the adder outputs into the A register (adds multiplicand to accumulator). The negative edge of this clock pulse causes the B register to shift left and the X register to shift right. This situation is shown in Figure 6.16(b).

3. The second pulse occurs. Since X_0 is now at 0, this pulse will not affect the A register. Thus, the accumulator contents does not change. The

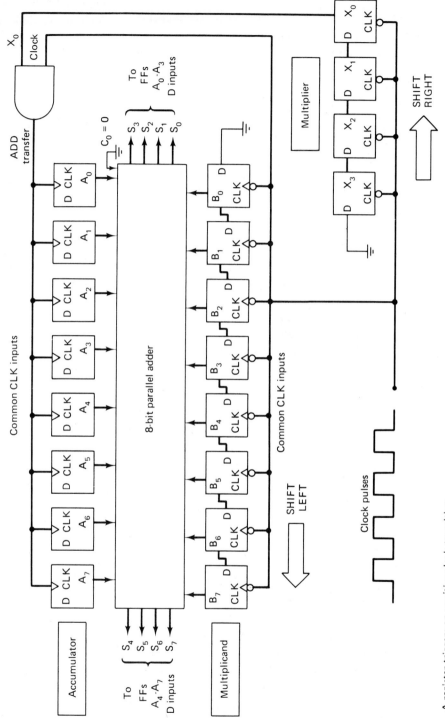

Figure 6.15 Binary multiplier circuit.

A register triggers on *positive* clock transition.
B and X registers trigger on *negative* edge.

177

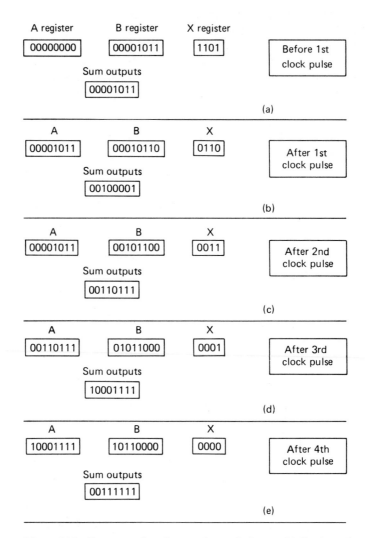

Figure 6.16 Contents of various registers during multiplication of 1011 × 1101.

negative edge of this pulse shifts the *B* and *X* registers. This situation is shown in Figure 6.16(c).

4. The third clock pulse occurs. X_0 is now a 1, so the adder outputs are transferred to the *A* register. Then, the *B* and *X* registers are shifted on the negative edge of the clock pulse.

5. The fourth clock pulse occurs. X_0 is at 1, so the adder outputs transfer to the *A* register. *B* and *X* both shift. After this fourth pulse the *A* register contains the final product, 10001111.

The reader is encouraged to try other examples with the circuit until its operation is fully understood.

The multiplier circuit of Figure 6.15 is only one of several types in use today. It is not the intent here to show all the possible variations but rather to show the multiplier circuit as yet another application of the digital principles we have learned so far. Anyone interested in a more intense treatment of multiplier circuits and other arithmetic circuits, including dividers, should consult textbooks devoted to digital computer design.

QUESTIONS AND PROBLEMS

6.1　What is the function of a full adder? Write down the truth table showing all the possible input and output conditions.

6.2　Convert the FA circuit of Figure 6.5 to all NAND gates.

6.3　What is the function of a half adder? How is it different from a full adder?

6.4　Write the truth table for a half adder (inputs *A* and *B*; outputs *sum* and *carry*). From the truth table design a logic circuit that will act as a half adder.

6.5　A full adder can be implemented in many different ways. Figure 6.17 shows how one may be constructed from two half adders. Construct a truth table for this arrangement and verify that it operates as a FA.

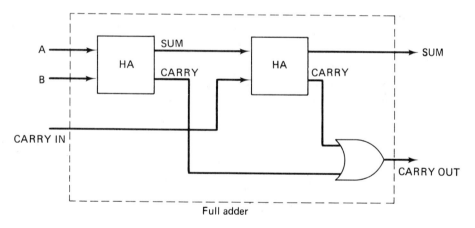

Figure 6.17

6.6　In the adder and subtractor circuits discussed in this chapter we gave no consideration to the possibility of *overflow*. Overflow occurs when the two numbers being added or subtracted produce a result that contains more bits than the capacity of the accumulator. For example, using 4-bit registers, including a sign bit, numbers ranging from $+7$ to -7 (in 2's complement) can be stored. Therefore, if the result of an addition or subtraction exceeds $+7$ or -7, we would say that an overflow had occurred. When an overflow occurs, the results are useless since they cannot be stored correctly in the accumulator register. To illustrate, add $+5$ (0101) and $+4$ (0100), which results in 1001. This 1001 would be interpreted incorrectly as a negative number since there is a 1 in the sign-bit position.

In computers and calculators there are usually circuits that are used to detect an overflow condition. There are several ways to do this. One method that can be used for the adder that operates in the 2's-complement system works as follows:

1. Examine the sign bits of the two numbers being added.

2. Examine the sign bit of the result.

3. Overflow occurs whenever the numbers being added are *both positive* and the sign bit of the result is 1 *or* when the numbers are *both negative* and the sign bit of the result is 0.

This method can be verified by trying several examples.

The reader should try the following cases for his own clarification: (1) $5 + 4$; (2) $-4 + (-6)$; (3) $3 + 2$. Cases 1 and 2 will produce an overflow and case 3 will not. Thus, by examining the sign bits a logic circuit can be designed that will produce a 1 output whenever the overflow condition occurs. Design this overflow circuit for the adder of Figure 6.6.

6.7 Design a look-ahead carry circuit for the adder of Figure 6.6 which generates the carry C_3 to be fed to the FA of the MSB position based on the values of A_0, B_0, C_0, A_1, B_1, A_2, and B_2. In other words, derive an expression for C_3 in terms of A_0, B_0, C_0, A_1, B_1, A_2, and B_2. (*Hint:* Begin by writing the expression for C_1 in terms of A_0, B_0, and C_0. Then write the expression for C_2 in terms of A_1, B_1, and C_1. Substitute the expression for C_1 into the expression for C_2. Then write the expression for C_3 in terms of A_2, B_2, and C_2. Substitute the expression for C_2 into the expression for C_3. Simplify the final expression for C_3 and put it in sum-of-products form. Implement the circuit.) How many levels of gates must the input signals propagate through before producing an output C_3? How does this compare to the arrangement of Figure 6.6?

6.8 What is the maximum propagation delay for the adder of Figure 6.7(b)? Assume that the maximum propagation delay for any one gate is 20 ns.

6.9 In the subtractor of Figure 6.9, what is the reason for using the $\bar{B}$ outputs from the B register and setting $C_0 = 1$?

6.10 The adder/subtractor diagram of Figure 6.10 is not complete since it does not show the connections required for transferring data into the B register and for transferring the sum outputs into the accumulator. The data for the B register usually come from memory via the outputs of another register, called the *memory buffer register* (MBR). Show the connections needed for providing the parallel transfer from the MBR into the B register and the transfer of the sum outputs into the accumulator.

6.11 For the circuit of Figure 6.10, determine the sum outputs for the following cases:
(a) A register $= 0101$ $(+5)$, B register $= 1110$ (-2); SUBTRACT $= 1$, ADD $= 0$.
(b) A register $= 1100$ (-4), B register $= 1110$ (-2); SUBTRACT $= 0$, ADD $= 1$.

6.12 For the serial adder of Figure 6.11, assume that the propagation delay of the FA is 50 ns. Also assume that the FFs each have a propagation delay of 20 ns. Determine the maximum frequency of clock pulses that can be used. Using this frequency, what is the total time required to perform the complete addition once the two numbers are in the A and B registers?

6.13 Show the sequence of states for each FF and FA output for the serial adder of Figure 6.11 during the addition of 1010 and 0111. Refer to Figure 6.12 for an example of how it is done.

6.14 Convert the serial adder of Figure 6.11 to a 2's-complement subtractor. Place 0101 ($+5$) in the A register and 0011 ($+3$) in the B register and follow through the circuit operation by applying the four successive clock pulses. The final result of 0010 ($+2$) should appear in the A register at the completion of the fourth pulse.

6.15 Design a serial adder/subtractor that will perform an addition when ADD $= 1$ and SUBTRACT $= 0$ and will perform a subtraction when ADD $= 0$ and SUBTRACT $= 1$.

61.6 What is the function of the logic gates in the BCD adder circuit of Figure 6.13?

6.17 Apply the BCD code for 376 into the A register of Figure 6.14 and apply the BCD code for 469 to the B register. Determine the Σ outputs. Each BCD adder is like the one in Figure 6.13.

6.18 Show the contents of registers A, B, and X and the adder outputs S_0–S_7 after each clock pulse during the process of multiplying 0111 (multiplicand) and 1001 (multiplier) using the circuit of Figure 6.15.

7

COUNTERS AND COUNTER
APPLICATIONS

In Chapter 4 we saw how flip-flops could be used in binary counters such that the output states of the FFs represent a binary number equal to the number of incoming clock pulses that have occurred. At that time we examined only the basic counter circuit in order to demonstrate the principle of binary counting. In practice there are many types of digital counters used in a wide variety of applications, including pulse counting, frequency division, and the sequencing and timing of operations. We will examine several of the most widely used types of counter circuits in detail. Even though many of them are available as off-the-shelf ICs, it is still worthwhile to study their internal operation because the techniques and principles employed are important and may be extended to other applications.

7.1 ASYNCHRONOUS (RIPPLE) COUNTERS

Figure 7.1 shows a 4-bit binary counter circuit such as the one discussed in Chapter 4. Recall the following points concerning its operation:

1. The clock pulses are applied only to the *CLK* input of FF *A*. Thus, FF *A* will toggle (change to its opposite state) each time the clock pulses make a negative (HIGH to LOW) transition. Note that $J = K = 1$ for all FFs.

2. The normal output of FF *A* acts as the *CLK* input for FF *B*, so FF *B* will toggle each time the *A* output goes from 1 to 0. Similarly, FF *C* will toggle when *B* goes from 1 to 0 and FF *D* will toggle when *C* goes from 1 to 0.

3. The table in Figure 7.1 shows the sequence of binary states that the FFs will follow as clock pulses are continuously applied. If we let the FF

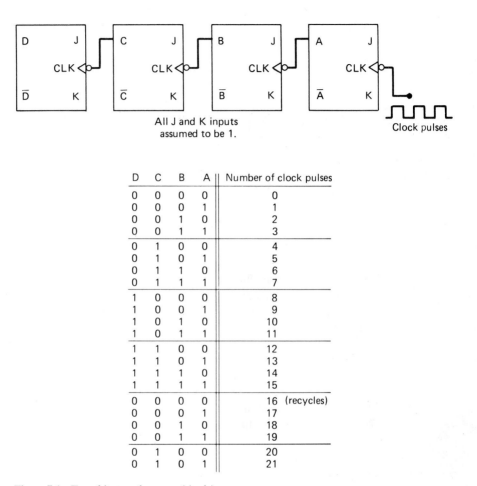

D	C	B	A	Number of clock pulses
0	0	0	0	0
0	0	0	1	1
0	0	1	0	2
0	0	1	1	3
0	1	0	0	4
0	1	0	1	5
0	1	1	0	6
0	1	1	1	7
1	0	0	0	8
1	0	0	1	9
1	0	1	0	10
1	0	1	1	11
1	1	0	0	12
1	1	0	1	13
1	1	1	0	14
1	1	1	1	15
0	0	0	0	16 (recycles)
0	0	0	1	17
0	0	1	0	18
0	0	1	1	19
0	1	0	0	20
0	1	0	1	21

Figure 7.1   Four-bit asynchronous (ripple) counter.

outputs D, C, B, and A represent a binary number, with D being the MSB and A the LSB, then a binary counting sequence from 0000 to 1111 is produced.

4. After the 15th clock pulse has occurred, the counter FFs are in the 1111 condition. On the 16th clock pulse FF A goes from 1 to 0, which causes FF B to go from 1 to 0, and so on until the counter is in the 0000 state. In other words, the counter has gone through one complete cycle (0000 through 1111) and has *recycled* back to 0000, from where it will begin a new counting cycle as subsequent clock pulses are applied.

This type of counter, where each FF output serves as the *CLK* input signal for the next FF, is referred to as an *asynchronous counter*. This is because all the FFs do *not* change states in exact synchronism with the clock pulses; only FF *A* responds to the clock pulses. FF *B* has to wait for FF *A* to change states before it is triggered; FF *C* has to wait for FF *B*; and so on. Thus, there is a delay between the responses of each FF. In modern FFs this delay may be very small (typically $10 - 40$ ns), but in some cases, as we shall see, it can be troublesome. Because of the manner in which this type of counter operates, it is also commonly referred to as a *ripple counter*. In the following discussions we will use the terms "asynchronous counter" and "ripple counter" interchangeably.

EXAMPLE 7.1   The counter in Figure 7.1 starts off in the 0000 state and then clock pulses are applied. Some time later the clock pulses are removed and the counter FFs read 0011. How many clock pulses occurred?

Solution: The apparent answer seems to be 3, since 0011 is the binary equivalent of 3. However, with the information given there is no way to tell whether the counter has recycled or not. This means that there could have been 19 clock pulses; the first 16 pulses bring the counter back to 0000 and the last 3 bring it to 0011. There could have been 35 pulses (two complete cycles and then 3 more), or 51 pulses, and so on.

MOD Number

The counter in Figure 7.1 has 16 distinctly different states (0000 through 1111). Thus, it is a *MOD-16 ripple counter*. Recall that the MOD number is always equal to the number of states which the counter goes through in each complete cycle before it recycles back to its starting state. The MOD number can be increased simply by adding more FFs to the counter. That is,

$$\text{MOD number} = 2^N \qquad (7.1)$$

where N is the number of FFs connected in the arrangement of Figure 7.1. For example, if five FFs were used, we would have a MOD-32 counter ($2^5 = 32$), which means that it would have 32 distinct states (00000 through 11111). It should be noted that the maximum binary number which the counter can reach is always $2^N - 1$. Thus, a 4-FF counter can count as high as $1111_2 = 2^4 - 1 = 15_{10}$; a 5-FF counter can go up to $11111_2 = 2^5 - 1 = 31_{10}$; and so on.

EXAMPLE 7.2   A counter is needed that will count the number of items passing on a conveyor belt. A photocell and light source combination is used to generate a single pulse each time an item crosses its path. The counter has to be able to count as many as one thousand items. How many FFs are required?

Solution: It is a simple matter to determine what value of N is needed so that $2^N \geq 1000$. Since $2^9 = 512$, 9 FFs will not be enough. $2^{10} = 1024$, so 10 FFs would produce a counter that could count as high as $1111111111_2 = 1023_{10}$. Therefore, we should use 10 FFs. We could use more than 10, but it would be a waste of FFs, since any FF past the tenth one will never be triggered.

Frequency Division

In Chapter 4 we saw that in the basic counter each FF provides an output waveform that is exactly *half* the frequency of the waveform at its *CLK* input. To illustrate, suppose that the clock-pulse-repetition rate in Figure 7.1 were 16 kHz. This is applied to the *CLK* input of FFA; thus, the waveform at output *A* is an 8-kHz squarewave, at output *B* it is 4 kHz, at output *C* it is 2 kHz, and at output *D* it is 1 kHz. Notice that the output of FF *D* has a frequency equal to the original clock frequency divided by 16. In general, for any counter the output from the last FF divides the input clock frequency by the MOD number of the counter. For example, a MOD-16 counter could also be called a *divide-by-16 counter*.

EXAMPLE 7.3   The first step involved in building a digital clock* is to take the 60-Hz power-line waveform and feed it into a shaping circuit to produce a squarewave as illustrated in Figure 7.2. The 60-Hz squarewave is then put into a MOD-60 counter, which is used to divide the 60-Hz frequency by exactly 60 to produce a 1-Hz waveform. This 1-Hz waveform is fed to a series of counters, which then count seconds, minutes, hours, etc. How many FFs are required for the MOD-60 counter?

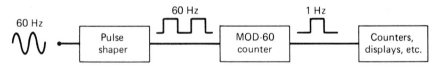

Figure 7.2

Solution: There is no integer power of 2 that will equal 60. The closest is $2^6 = 64$. Thus, a counter using six FFs would act as a MOD-64. Obviously, this will not satisfy the requirement. It seems that there is no solution using a counter of the type shown in Figure 7.1. This is partly true; in a later section we will see how to modify this basic binary counter so that virtually *any* MOD number can be obtained and we will not be limited to values of 2^N.

7.2 SELF-STOPPING RIPPLE COUNTER

It is possible to take a counter such as that of Figure 7.1 and modify it so that it will only count up to a certain binary value and then stop counting even if input pulses are continuously applied. For example, the 4-bit ripple counter would normally count up to 1111 (15), then recycle back to 0000, and so on. If we wanted it to count to only 1001 (decimal 9) and then stop, it could be accomplished as

---

*Here we are talking about a clock that indicates time in hours, minutes, etc.

shown in Figure 7.3. Here a NAND gate has been added to the basic ripple counter. The operation can be described as follows:

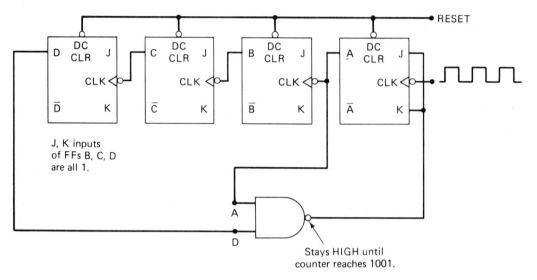

Figure 7.3 Self-stopping ripple counter that stops at 1001.

1. Assume that the counter FFs are initially all 0, so the counter is in the 0000 state.

2. Since the A and D FF outputs are fed to the NAND gate, the NAND output will initially be 1. This keeps the J and K inputs of FF A at the 1 level. Thus, initially all FFs are in the $J = K = 1$ toggle mode.

3. As pulses are applied the counter will count in its normal manner. The ninth input pulse brings the counter to the 1001 state. In this state both A and D are HIGH, thereby producing a 0 out of the NAND gate to the J and K inputs of FF A.

4. With its J and K inputs now both 0, FF A cannot toggle on any of the subsequent input pulses. Since A cannot change states, it is obvious that B, C, and D cannot change, so the counter simply remains in the 1001 count. This condition will persist until FFs A and D are cleared to the 0 state; this can be done by applying an appropriate momentary signal to their DC CLEAR inputs. The counter is then ready to count from 0000 to 1001, where it will again stop.

It is a relatively simple matter to stop the counter at any desired number by changing the inputs to the NAND gate. For example, if we used a three-input gate with inputs from FFs A, B, and D, the counter would stop at 1011. In general, to stop the counter at a desired count we must connect to the NAND inputs those FF outputs which will be HIGH at that count. Self-stopping counters are used in applications where it is necessary to produce a change in a logic level when the number of incoming pulses reaches a certain number. The NAND-gate output exhibits the required level change, going from 1 to 0 when the desired count is reached.

7.3   COUNTERS WITH MOD NUMBERS < 2^N

The basic ripple counter of Figure 7.1 is limited to MOD numbers that are equal to 2^N, where N is the number of FFs. This value is actually the maximum MOD number that can be obtained using N FFs. The basic counter can be modified to produce MOD numbers less than 2^N by allowing the counter to *skip states* that are normally part of the counting sequence. One of the most common methods for doing this is illustrated in Figure 7.4 where a 3-bit ripple counter is shown. Disregarding the NAND gate for a moment we can see that the counter is a MOD-8 binary counter which will count in sequence from 000 to 111. However, the presence of the NAND gate will alter this sequence as follows:

1. The NAND output is connected to the DC CLEAR inputs of each FF. As long as the NAND output is HIGH, it will have no affect on the counter. When it goes LOW, however, it will clear all the FFs so that the counter immediately goes to the 000 state.

2. The inputs to the NAND gate are the outputs of the B and C FFs, so the NAND output will go LOW whenever $B = C = 1$. This condition will occur when the counter goes from the 101 state to the 110 state (input pulse 6 on waveforms). The LOW at the NAND output will immediately (generally within a few nanoseconds) clear the counter to the 000 state. Once the FFs have been cleared, the NAND output goes back HIGH, since the $B = C = 1$ condition no longer exists.

3. The counting sequence is, therefore,

$$
\begin{array}{l}
000 \\
001 \\
010 \\
011 \\
100 \\
101 \\
110
\end{array}
$$

110   (temporary state needed to clear counter)

Although the counter does go to the 110 state, it remains there for only a few nanoseconds before it recycles to 000. Thus, we can essentially say that this counter counts from 000 (zero) to 101 (five) and then recycles

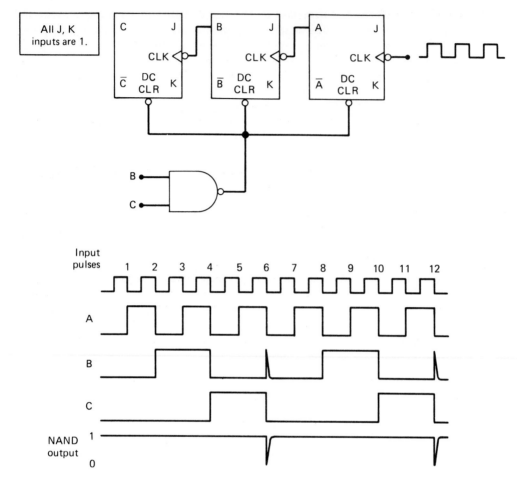

Figure 7.4   MOD-6 counter produced by clearing a MOD-8 counter when count of six (110) occurs.

to 000. It essentially skips 110 and 111 so that it only goes through six different states; thus, it is a MOD-6 counter.

It should be noted that the waveform at the *B* output contains a *spike* or *glitch* caused by the momentary occurrence of the 110 state before clearing. This glitch is very narrow and so would not produce any visible indication on indicator lights or numerical displays. It could, however, cause a problem if the *B* output is being used to drive other circuitry outside the counter. It should also be noted that the *C* output has a frequency equal to $\frac{1}{6}$ of the input frequency; in other words, this MOD-6 counter has divided the input frequency by *six*.

Changing the MOD Number

The counter of Figure 7.4 is a MOD-6 because of the choice of inputs to the NAND gate. Any desired MOD number can be obtained by changing these inputs. For example, using a three-input NAND gate with inputs A, B, and C, the counter would function normally until the 111 condition was reached, at which point it would immediately reset to the 000 state. Ignoring the temporary excursion into the 111 state, the counter would go from 000 through 110 and then recycle back to 000, resulting in a MOD-7 counter (seven states).

EXAMPLE 7.4   Determine the MOD number of the counter in Figure 7.5(a). Also determine the frequency at the D output.

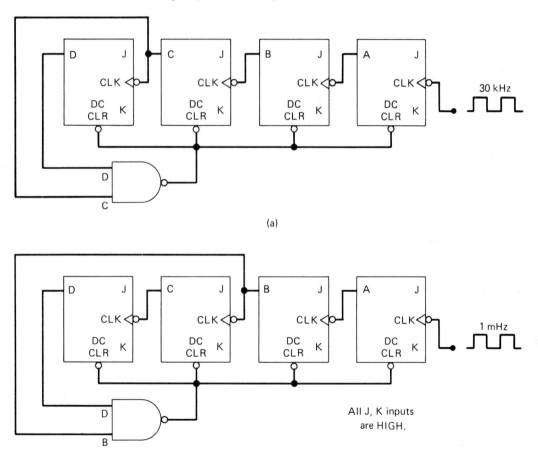

(a)

(b)

Figure 7.5   (a) MOD-12 ripple counter; (b) MOD-10 (decade) ripple counter.

Solution: This is a 4-bit counter, which would normally count from 0000 through 1111. The NAND inputs are D and C, which means that the counter will immediately recycle to 0000 when the 1100 (decimal 12) count is reached. Thus, the counter actually has *12* stable states 0000 through 1011 and is therefore a **MOD-12** counter. Since the input frequency is 30 kHz, the frequency at output D will be

$$\frac{30\ \text{kHz}}{12} = 2.5\ \text{kHz}$$

General Procedure

The general procedure for constructing a ripple counter of any MOD number should be readily apparent. It can be summarized as follows for obtaining a MOD-X counter, where X can be any integer:

   1.   Find the smallest number of FFs such that $2^N \geq X$ and connect them as a normal ripple counter.

   2.   Connect a NAND gate to the DC CLEAR inputs of *all* the FFs.*

   3.   The inputs to the NAND gate are taken from the normal outputs of those FFs which will be in the HIGH state when it is desired to reset the counter back to zero.

EXAMPLE 7.5   Construct a MOD-10 counter that will count from 0000 (zero) through 1001 (decimal 9).

Solution: $2^3 = 8$ and $2^4 = 16$; thus, four FFs are required. Since the counter is to have stable operation up to the count of 1001, it must be reset to zero when the count of 1010 is reached. Therefore, FF outputs D and B must be connected as the NAND-gate inputs. Figure 7.5(b) shows the arrangement.

Decade Counters

The MOD-10 counter of Example 7.5 is also referred to as a *decade counter*. In fact, a decade counter is any counter that has 10 distinct states, no matter what the sequence. A decade counter such as the one in Figure 7.5(b), which counts in sequence from 0000 (zero) through 1001 (decimal 9), is also commonly called a *BCD counter* because it uses only the 10 BCD code groups 0000, 0001, . . . , 1000, and 1001. To reiterate, any MOD-10 counter is a decade counter; and any decade counter that counts in binary from 0000 to 1001 is a BCD counter.

Decade counters, especially the BCD type, find widespread use in applications

---

*The reason for clearing *all* the FFs is to prevent any FF from toggling as a result of the preceding FF going from 1 to 0.

where pulses or events are to be counted and the results displayed on some type of decimal numerical readout. We shall examine this later in more detail. A decade counter is also often used for dividing a pulse frequency *exactly* by 10. The input pulses are applied to FF A and the output pulses are taken from the output of FF D, which has $\frac{1}{10}$ the frequency of the input.

EXAMPLE 7.6   In Example 7.3 a MOD-60 counter was needed to divide the 60-Hz line frequency down to 1 Hz. Construct an appropriate MOD-60 counter.

Solution: We can use the procedure outlined above using six FFs ($2^6 = 64$). The NAND inputs are taken from those FFs which will be HIGH at the count of sixty (111100). The result is shown in Figure 7.6(a). The output of FF F will be a waveform of 1 Hz.

An alternative approach is often used because of the availability of lower MOD counters in IC packages. IC counters up to MOD-16 are readily available as off-the-shelf items. As such, it is sometimes easier to combine two or more counters to obtain the desired overall MOD number. To illustrate, Figure 7.6(b) shows how a MOD-60 counter can be constructed from a MOD-10 and a MOD-6 counter. The 60-Hz input is applied to the input of the MOD-10 counter (to the *CLK* of FF A); the last FF in this decade counter (FF D) will have an output frequency of $\frac{60}{10} = 6$ Hz. This 6-Hz signal is then fed to the input of a MOD-6 counter, which will divide this frequency by 6 to obtain the desired 1 Hz. Although this approach requires seven FFs instead of the six used in the first solution, it is usually the one that is used because of the fewer interconnections that are required.

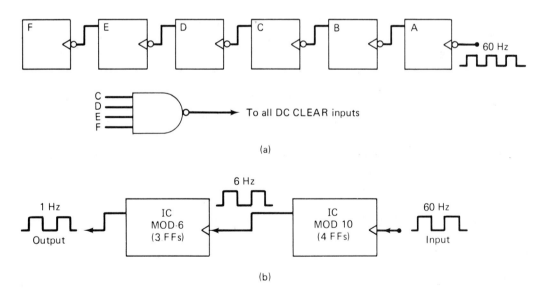

Figure 7.6   Two ways of obtaining MOD-60 counter.

7.4 ASYNCHRONOUS DOWN COUNTER

All the counters we have looked at this far have counted *upward* from zero; that is, they were *up counters*. It is a relatively simple matter to construct asynchronous (ripple) *down counters*, which will count downward from a maximum count to zero. Before looking at a ripple down counter, let us examine the count-down sequence for a 3-bit down counter:

	CBA	CBA	CBA
(7)	1 1 1	1 1 1	1 1 1
(6)	1 1 0	1 1 0	1 1 0
(5)	1 0 1	1 0 1	· · ·
(4)	1 0 0	1 0 0	· · ·
(3)	0 1 1	0 1 1	· · ·
(2)	0 1 0	0 1 0	etc.
(1)	0 0 1	0 0 1	
(0)	0 0 0	0 0 0	etc.

A, *B*, and *C* represent the FF output states as the counter goes through its sequence. It can be seen that the *A* FF (LSB) changes states (toggles) at each step

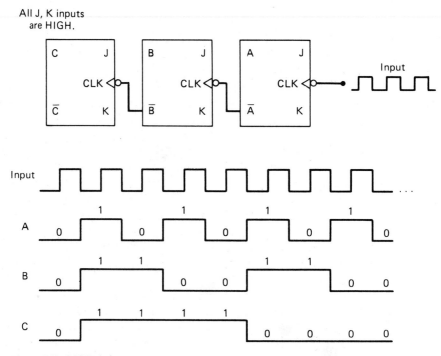

Figure 7.7 MOD-8 down counter.

in the sequence just as it does in the up counter. The B FF changes states each time A goes from LOW to HIGH; C changes states each time B goes from LOW to HIGH. Thus, in a down counter each FF, except the first, must toggle when the preceding FF goes from LOW to HIGH. If the FFs have CLK inputs that respond to negative transitions (HIGH to LOW), then an inverter can be placed in front of each CLK input; however, the same affect can be accomplished by driving each FF CLK input from the *inverted* output of the preceding FF. This is illustrated in Figure 7.7 for a MOD-8 down counter.

The input pulses are applied to the A FF; the $\bar{A}$ output serves as the CLK input for the B FF; the $\bar{B}$ output serves as the CLK input for the C FF. The waveforms at A, B, and C shows that B toggles whenever A goes LOW to HIGH (so $\bar{A}$ goes HIGH to LOW) and C toggles whenever B goes LOW to HIGH. This results in the desired down-counting sequence.

Down counters are not as widely used as up counters. Their major application is in situations where it must be known when a desired number of input pulses has occurred. In these situations the down counter is *preset* to the desired number and then allowed to count down as the pulses are applied. When the counter reaches the *zero* state it is detected by a logic gate whose output then indicates that the preset number of pulses has occurred. We shall discuss presettable counters in a later section.

7.5 ASYNCHRONOUS UP/DOWN COUNTER

In an up counter each FF is triggered by the *normal* output of the preceding FF; in a down counter each FF is triggered by the *inverted* output of the preceding FF. In both counters the first FF is triggered by the input pulses. It is possible to construct an up/down counter that will either count up or count down, depending on the status of control inputs. Figure 7.8 shows a 3-bit up/down counter whose operation is controlled by the COUNT-UP and COUNT-DOWN control inputs.

The logic gates are used to allow either the noninverted output or inverted output of one FF into the CLK input of the following FF, depending on the status of the control inputs. When the COUNT-UP line is held at 1 while the COUNT-DOWN line is at 0, the output of AND gate 2 will be disabled (output = 0), so it will have no affect on the OR gate output. AND 1 will be enabled; that is, it will allow A to pass through to the OR gate and into the CLK input of the B FF. Similarly, the B output will be gated into the CLK input of FF C. Thus, as input pulses are applied, the counter will count up.

When the opposite control condition is used with COUNT-UP = 0, COUNT-DOWN = 1 the upper AND gates (1 and 3) are disabled and the lower AND gates (2 and 4) are enabled, allowing $\bar{A}$ and $\bar{B}$ to pass into the CLK inputs of the following FFs. Thus, for this condition, the counter will count down as input pulses are applied.

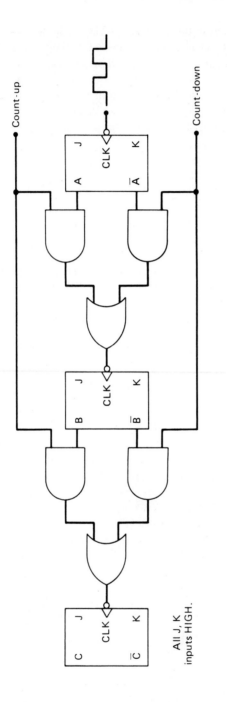

All J, K
inputs HIGH.

Count-up	Count-down	Operation
0	0	No counting
0	1	Counts down
1	0	Counts up
1	1	No counting

Figure 7.8 Asynchronous up/down counter.

When the control inputs are both 0 or both 1, the counter will not count up or down because the *CLK* inputs of *B* and *C* will be held constant at either 0 or 1. The *A* FF will keep toggling because it is always being clocked. These conditions are not normally used.

The up/down counter of Figure 7.8 is usually implemented using all NAND gates in place of the AND and OR gates. Using the techniques learned in Chapter 2 it is easily shown that each AND and OR gate can be replaced by a NAND gate without changing the overall logic operation.

7.6 PROPAGATION DELAY IN RIPPLE COUNTERS

Ripple counters are the simplest type of binary counters since they require the fewest amount of components to produce a given counting operation. They do, however, have one major drawback, which is caused by their basic principle of operation. Each FF is triggered by the transition at the output of the preceding FF. Because of the inherent propagation delay time (t_{pd}) of each FF, this means that the second FF will not respond until a time t_{pd} after the first FF receives an input pulse; the third FF will not respond until a time equal to $2 \times t_{pd}$ after the clock pulse occurs; and so on. In other words, the propagation delays of the FFs accumulate so that the *N*th FF cannot change states until a time equal to $N \times t_{pd}$ after the clock pulse occurs. This is illustrated in Figure 7.9, where the waveforms for a 3-bit ripple counter are shown.

The first set of waveforms in Figure 7.9(a) shows a situation where an input pulse occurs every 1000 ns (the clock period $T = 1000$ ns) and it is assumed that each FF has a propagation delay of 50 ns ($t_{pd} = 50$ ns). Notice that the *A* FF output toggles 50 ns after the falling edge of each input pulse. Similarly, *B* toggles 50 ns after *A* goes from 1 to 0 and *C* toggles 50 ns after *B* goes from 1 to 0. As a result, when the fourth input pulse occurs, the *C* output goes HIGH after a delay of 150 ns. In this situation the counter does operate properly in the sense that the FFs do eventually get to their correct states, representing the binary count. However, the situation worsens if the input pulses are applied at a much higher frequency.

The waveforms in Figure 7.9(b) show what happens if the input pulses occur once every 100 ns. Again, each FF output responds 50 ns after the 1-to-0 transition at its *CLK* inputs (note the change in the relative time scale). Of particular interest is the situation after the falling edge of the fourth input pulse where the *C* output does not go HIGH until 150 ns later, which is the same time that the *A* output goes HIGH in response to the fifth input pulse. In other words, the condition $C = 1$, $B = A = 0$ (count of 100), never appears because the input frequency is too high. This could cause a serious problem if this condition were supposed to be used to control some other operation in a digital system. Problems such as

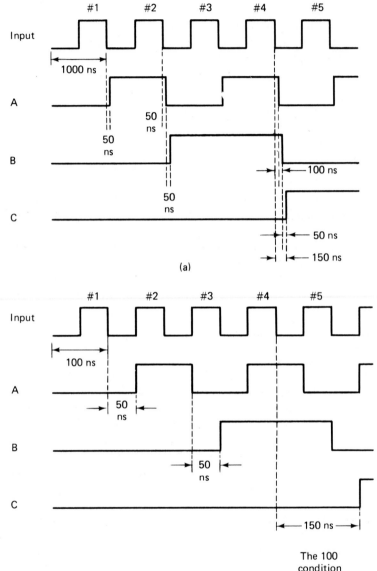

Figure 7.9 Waveforms of 3-bit ripple counter illustrating effects of FF propagation delays for different input pulse frequencies.

this can be avoided if the period between input pulses is made longer than the total propagation delay of the counter. That is,

$$T_{\text{clock}} \geq N \times t_{\text{pd}} \tag{7.2}$$

where N = number of FFs. Stated in terms of input-clock frequency, the maximum frequency that can be used is given by

$$f_{\text{max}} = \frac{1}{N \times t_{\text{pd}}} \tag{7.3}$$

For example, the 3-bit counter that has FFs with $t_{\text{pd}} = 50$ ns will have a maximum input frequency limit of

$$f_{\text{max}} = \frac{1}{3 \times 50 \text{ ns}} = 6.67 \text{ MHz}$$

Obviously, *as the number of FFs in the ripple counter is increased, the total delay will increase and f_{max} will be lower.* A 5-bit counter using 50-ns FFs will have $f_{\text{max}} = 4$ MHz.

Because of the problem described above, ripple counters cannot be used at very high frequencies. In addition, another problem related to propogation delays occurs when the ripple counter is decoded to produce control waveforms. This problem will be investigated later, when we discuss the decoding of counters.

7.7 SYNCHRONOUS (PARALLEL) COUNTERS

The problems encountered with ripple counters are caused by the accumulated FF propagation delays; stated another way, the FFs do not all change states simultaneously in synchronism with the input pulses. These limitations can be overcome with the use of *synchronous* or *parallel* counters in which all the FFs are triggered simultaneously (in parallel) by the clock input pulses. Since the input pulses are applied to all the FFs, some means must be used to control when each FF is to toggle or remain unaffected by a clock pulse. This is accomplished by using the J and K inputs and is illustrated in Figure 7.10(a) for a 4-bit, MOD-16 parallel counter.

Each of the FFs has the input pulses applied to its *CLK* input. The A FF has $J = K = 1$, so it will toggle on the falling edge of *each* input pulse. The B FF has its J and K inputs connected to the A output, so the value of A determines whether B will toggle on a particular input pulse. If $A = 0$ *before* the input pulse occurs, then $J_B = K_B = 0$ for the B FF, and it *will not* toggle when the pulse does occur (e.g., going from count of 2 to 3). If $A = 1$ *before* the input pulse occurs, then $J_B = K_B = 1$ and the B FF *will* toggle (e.g., going from count of 3 to 4).

Similarly, the C FF will toggle only if $A = B = 1$ before the input pulse occurs because of AND gate 1 (e.g., going from count of 3 to 4). The D FF will

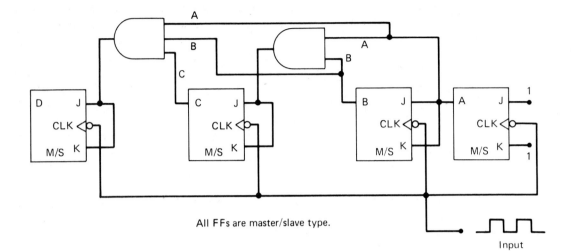

All FFs are master/slave type.

Input

Count	D	C	B	A
0	0	0	0	0
1	0	0	0	1
2	0	0	1	0
3	0	0	1	1
4	0	1	0	0
5	0	1	0	1
6	0	1	1	0
7	0	1	1	1
8	1	0	0	0
9	1	0	0	1
10	1	0	1	0
11	1	0	1	1
12	1	1	0	0
13	1	1	0	1
14	1	1	1	0
15	1	1	1	1

(a)

To
DC CLEAR
inputs of
each FF.

C

D

(b)

Figure 7.10 (a) Parallel MOD-16 counter; (b) NAND gate added to convert it to MOD-12 counter.

toggle only if $A = B = C = 1$ before the pulse occurs due to AND gate 2. Thus, the counter will progress through the binary counting sequence as shown in the table in Figure 7.10(a).

The most important characteristic of this parallel counter is the fact that all FFs will change states simultaneously in synchronism with the input pulses. The total propagation delay that must be allowed for is simply the t_{pd} of a *single* FF plus the delay needed for the levels to propagate through a *single* AND gate. This means that a much higher input-pulse frequency can be used than would be possible for a corresponding ripple counter. Of course, the parallel counter requires more logic circuitry and a greater number of connections.

EXAMPLE 7.7

(a) Determine f_{max} for the counter of Figure 7.10(a) if t_{pd} for each FF is 50 ns and t_{pd} for each AND gate is 20 ns. Compare this to f_{max} for a MOD-16 ripple counter.

(b) What has to be done to convert this counter to MOD-32?

(c) Determine f_{max} for the MOD-32 parallel counter.

Solution:

(a) The total delay that must be allowed between input clock pulses is equal to FF t_{pd} + AND gate t_{pd}. Thus, $T_{clock} \geq 50 + 20 = 70$ ns, so the parallel counter has

$$f_{max} = \frac{1}{70 \text{ ns}} = 14.3 \text{ MHz} \qquad \text{(parallel counter)}$$

A MOD-16 ripple counter uses four FFs with $t_{pd} = 50$ ns. Thus, f_{max} for the ripple counter is

$$f_{max} = \frac{1}{4 \times 50 \text{ ns}} = 5 \text{ MHz} \qquad \text{(ripple counter)}$$

(b) A fifth FF must be added since $2^5 = 32$. The *CLK* input of this FF is also fed by the input pulses. Its J and K inputs are fed by the output of a 4-input AND gate whose inputs are A, B, C, and D.

(c) f_{max} is still determined as in (a) regardless of the number of FFs in the parallel counter. Thus, f_{max} is still 14.3 MHz.

It should be pointed out that the parallel counter utilizes master/slave-type FFs. This is because the J and K inputs of a FF will sometimes be changing at approximately the same time as its *CLK* input is receiving a negative-going transition. Using M/S-type FFs eliminates the possibility of race problems causing unreliable triggering in these situations.

Varying the MOD Number

Parallel counters can be modified to have MOD numbers other than 2^N using the same technique used for ripple counters (Section 7.3). For example, the MOD-16 parallel counter in Figure 7.10(a) can be converted to a MOD-12 using the NAND gate of Figure 7.10(b). The output of the NAND gate is connected to the DC CLEARS of all the FFs in the counter. The MOD number of a parallel counter can also be varied by modifying the J-K gating circuits (see Problem 7.20).

7.8 PARALLEL DOWN AND UP/DOWN COUNTERS

In Section 7.4 we saw that a ripple counter could be made to count down by using the inverted outputs of each FF to drive the next FF in the counter. A parallel down counter can be constructed in a similar manner, that is, by using the inverted FF outputs to feed the various logic gates. For example, the parallel up counter of Figure 7.10 can be converted to a down counter by connecting the $\bar{A}$, $\bar{B}$, and $\bar{C}$ outputs to the AND gates in place of A, B, and C, respectively. The counter will then proceed through the following sequence as input pulses are applied:

```
(15)  1 1 1 1
(14)  1 1 1 0
(13)  1 1 0 1
(12)  1 1 0 0
       . . . .
       . . . .    recycle
       . . . .
 (3)  0 0 1 1
 (2)  0 0 1 0
 (1)  0 0 0 1
 (0)  0 0 0 0
```

To form a parallel up/down counter (see Figure 7.11) the control inputs (COUNT-UP and COUNT-DOWN) are used to control whether the normal FF outputs or the inverted FF outputs are fed to the J and K inputs of the following FFs. The counter in Figure 7.11 is a MOD-8 up/down counter which will count from 000 up to 111 when the COUNT-UP control input is 1 and from 111 down to 000 when the COUNT-DOWN control input is 1.

A logical 1 on the COUNT-UP line while COUNT-DOWN = 0 enables AND gates 1 and 2 and disables AND gates 3 and 4. This allows the A and B outputs through to the J and K inputs of the following FFs so that the counter will count up as pulses are applied. The opposite action takes place when COUNT-UP = 0 and COUNT-DOWN = 1.

Higher MOD counters of this type can be constructed by a simple extension of the arrangement shown in Figure 7.11; this is left as an exercise at the end of the chapter. It should be mentioned that when parallel counters are fabricated as integrated circuits the logic gates which are used are normally all NAND gates.

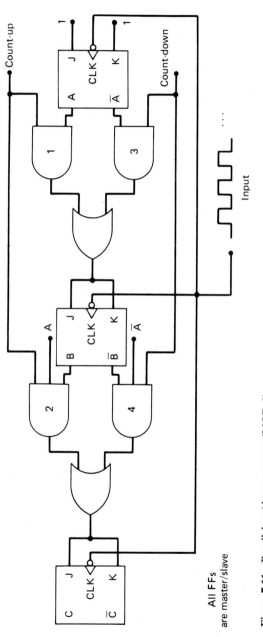

Figure 7.11 Parallel up/down counter (MOD-8).

All FFs
are master/slave

In other words, the gates in the counters of Figures 7.10 and 7.11 are replaced by their NAND-gate equivalents.

7.9 SYNCHRONOUS/ASYNCHRONOUS COUNTERS

Many counters can be constructed that are neither synchronous or asynchronous but rather a combination of the two. These *synchronous/asynchronous counters* represent a compromise between the speed of synchronous counters and the simplicity of asynchronous counters. An example is the BCD decade counter shown in Figure 7.12. Notice that the input pulses are applied only to the A FF, as in an asynchronous counter. Also notice that the *A* output drives the *CLK* inputs of both the *B* and *D* FFs so that the *B* and *D* FFs will trigger simultaneously as in a synchronous counter.

The operation of this counter is described as follows:

1. Assume that the counter is initally in the 0000 state. Thus, the *J* and *K* inputs of the *A*, *B*, and *C* FFs are all HIGH, so they are ready to toggle on a negative *CLK* transition. The *D* FF has $J = 0$, $K = 1$, so it will not be affected by any *CLK* transition.

2. The *A*, *B*, and *C* FFs will function as a normal ripple counter for the first *seven* input pulses. The *D* FF will remain LOW because its *J* input is at 0 whenever a HIGH-to-LOW transition occurs at its *CLK* input. Thus, the counter will advance from 0000 to 0111.

3. In the 0111 state the AND gate output is at 1, so the *D* FF has $J = K = 1$. When the *eighth* input pulse occurs, it will toggle *A* to the 0 state, which will in turn toggle *B* LOW and *D* HIGH. The *B* output transition will toggle *C* LOW. Thus, the counter is now in the 1000 state.

4. The *ninth* input pulse simply toggles *A* HIGH, bringing the counter to the 1001 state.

5. In the 1001 state the AND output is again 0, so the *D* FF has $J = 0$, $K = 1$, which means it will go LOW on the next negative transition at its *CLK* input. The *B* FF also has $J = 0$, $K = 1$, so it will remain in the LOW state. Thus, when the *tenth* input pulse occurs, the *A* FF will toggle LOW, which in turn will toggle *D* LOW, bringing the counter back to the 0000 state. The operation then returns to step 1 and repeats the sequence.

This BCD counter is available in IC form (Texas Instruments SN7490) and is widely used in medium-speed applications. Several other synchronous/asynchronous-type counters will be investigated in the problems at the end of the chapter.

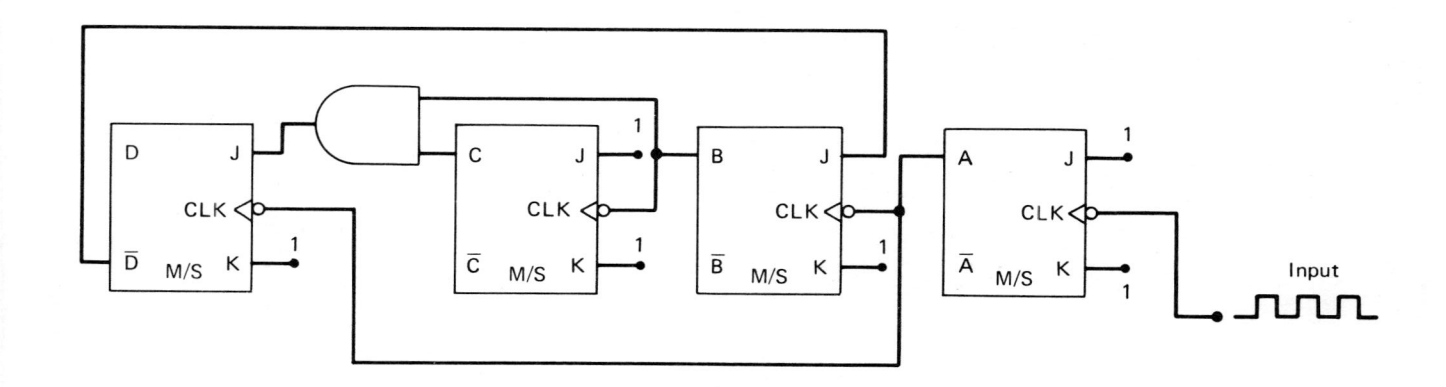

Count	D	C	B	A
0	0	0	0	0
1	0	0	0	1
2	0	0	1	0
3	0	0	1	1
4	0	1	0	0
5	0	1	0	1
6	0	1	1	0
7	0	1	1	1
8	1	0	0	0
9	1	0	0	1

Recycles

Figure 7.12 BCD decade counter.

7.10 DECODING A COUNTER

Digital counters are often used in applications where the count which the states of the FFs represent must somehow be determined or displayed. One of the simplest means for displaying the contents of a counter involves just connecting the output of each FF to a small indicator lamp. In this way the states of the FFs are visibly represented by the lamps (bright = 1, dark = 0) and the count can be mentally determined by decoding the binary states of the lamps. For instance, suppose that this method is used for the BCD counter of Figure 7.12 and the states of the lamps are respectively dark–bright–bright–dark. This would represent 0110, which we would mentally decode as decimal 6. Other combinations of lamp states would represent the other possible counts.

The indicator lamp method becomes inconvenient as the size (number of bits) of the counter increases, because it is much harder to mentally decode the displayed results. For this reason it would be preferable to develop a means for *electronically* decoding the contents of a counter and displaying the results in a form that would be immediately recognizable and would require no mental operations.

An even more important reason for electronic decoding of a counter occurs because of the many applications in which counters are used to control the timing or sequencing of operations *automatically* without human intervention. For example, a certain system operation might have to be initiated when a counter reaches the 101100 state (count of 44_{10}). A logic circuit can be used to decode for or detect when this particular count is present and then initiate the operation. Many operations may have to be controlled in this manner in a digital system. Clearly, human intervention in this process would be undesirable except in extremely slow systems.

Active-HIGH Decoding

A MOD-X counter has X different states; each state is a particular pattern of 0s and 1s stored in the counter FFs. A decoding network is a logic circuit that generates X different outputs, each of which detects (decodes) the presence of one particular state of the counter. The decoder outputs can be designed to produce either a HIGH or a LOW level when the detection occurs. An active-HIGH decoder produces HIGH outputs to indicate detection. Figure 7.13 shows the complete active-HIGH decoding logic for a MOD-8 counter. The decoder consists of 8 three-input AND gates. Each AND gate produces a HIGH output for one particular state of the counter.

For example, AND gate 0 has as its inputs the FF outputs $\bar{C}$, $\bar{B}$, and $\bar{A}$. Thus, its output will be LOW at all times *except* when $A = B = C = 0$, that is, on the count of 000 (zero). Similarly, AND gate 5 has as its inputs the FF outputs C, $\bar{B}$, and A, so its output will go HIGH only when $C = 1$, $B = 0$, and $A = 1$, that is, on the count of 101 (decimal 5). The rest of the AND gates perform in the same manner for the other possible counts. At any one time only one AND gate output

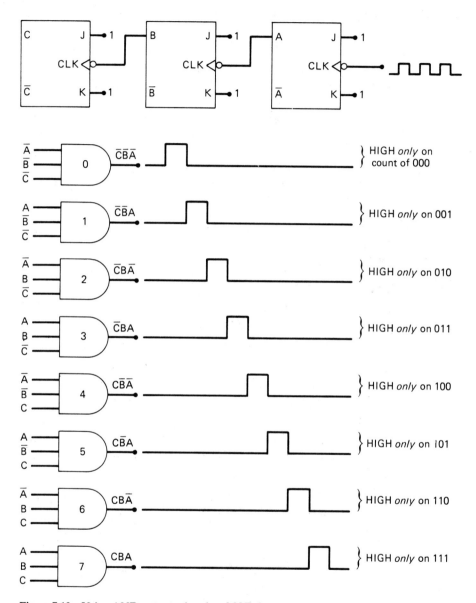

Figure 7.13   Using AND gates to decode a MOD-8 counter.

is HIGH, the one which is decoding for the particular count that is present in the counter. The waveforms in Figure 7.13 show this clearly.

The eight AND outputs can be used to control eight separate indicator lamps, which represent the decimal numbers 0–7. Only one lamp will be on at a given time, indicating the proper count.

The AND-gate decoder can be extended to counters with any number of states. The following example illustrates.

EXAMPLE 7.8 How many AND gates are required to completely decode all the states of a MOD-32 binary counter? What are the inputs to the gate that decodes for the count of 21?

Solution: A MOD-32 counter has 32 possible states. One AND gate is needed to decode for each state; therefore, the decoder requires 32 AND gates. Since $32 = 2^5$, the counter contains five FFs. Thus, each gate will have five inputs, one from each FF. To decode for the count of 21 that is 10101_2 requires AND-gate inputs of E, $\bar{D}$, C, $\bar{B}$, and A, where E is the MSB flip-flop.

Active-LOW Decoding

If NAND gates are used in place of AND gates the decoder outputs will produce a normally HIGH signal, which goes LOW only when the number being decoded occurs. Both types of decoders are used, depending on the type of circuits being driven by the decoder outputs.

EXAMPLE 7.9 Figure 7.14 shows a common situation in which a counter is used to help generate a control waveform which could be applied to devices such as a motor, solenoid valve, or heater. The MOD-16 counter cycles and recycles through its

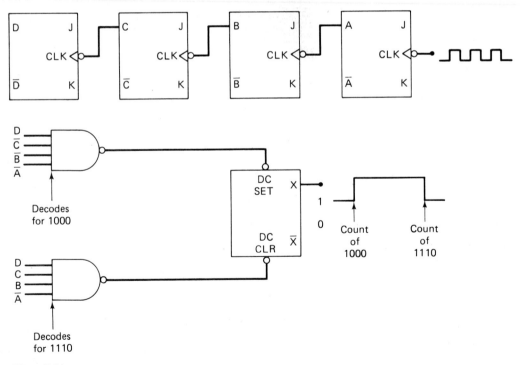

Figure 7.14

counting sequence. Each time it goes to the count of 8 (1000), the upper NAND gate will produce a LOW output, which sets FF X to the 1 state. FF X stays HIGH until the counter reaches the count of 14 (1110), at which time the lower NAND gate decodes it and produces a LOW output to clear X to the 0 state. Thus, the X output is HIGH between the counts of 8 and 14 for each cycle of the counter.

BCD Counter Decoding

A BCD counter has 10 states, which can be decoded using the techniques previously described. BCD decoders provide 10 outputs corresponding to the decimal digits 0 through 9 represented by the states of the counter FFs. These 10 outputs can be used to control 10 indicator lamps for a visual display. More often, instead of using 10 separate lamps, a single display device is used to display the decimal numbers 0 through 9. One such device, called a *nixie tube*, contains 10 very thin numerically shaped filaments stacked on top of each other. The BCD decoder outputs control which filament is illuminated. Another class of decimal displays contains seven small segments made of a material (usually incandescant filaments or light-emitting diodes) which emits light when an electric current is passed through it. The BCD decoder outputs control which segments are illuminated in order to produce a pattern representing one of the decimal digits.

We will go into more detail concerning these types of decoders and displays in Chapter 9. However, since BCD counters and their associated decoders and displays are very commonplace, we will use the decoder/display unit (see Figure 7.15) to represent the complete circuitry used to visually display the contents of a BCD counter as a decimal.

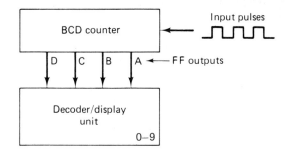

Figure 7.15 BCD counters usually have their count displayed on a single display device.

7.11   DECODING GLITCHES

In Section 7.6 the effects of FF propagation delays in ripple counters was discussed. As we saw then, the accumulated propagation delays serve to essentially limit the frequency response of ripple counters. The delays between FF transitions can also cause problems when decoding a ripple counter. The problem occurs in the form of glitches or spikes at the outputs of some of the decoding gates. This is illustrated in Figure 7.16 for a MOD-4 ripple counter.

The waveforms at the outputs of each FF and decoding gate are shown in the figure. Notice the propagation delay between the clock waveform and the A output waveform and between the A waveform and the B waveform. The glitches in the X_0 and X_2 decoding waveforms are caused by the delay between the A and B waveforms. X_0 is the output of the AND gate decoding for the normal 00 count. The 00 condition also occurs momentarily as the counter goes from the 01 to the 10 count, as shown by the waveforms. This is because B cannot change states until A goes LOW. This momentary 00 condition only lasts for several nanoseconds (depending on t_{pd} of FF B) but can be detected by the decoding gate if the gate's response is fast enough. Hence, the spike at the X_0 output.

A similar situation produces a glitch at the X_2 output. X_2 is decoding for the 10 condition, and this condition occurs momentarily as the counter goes from 11 to 00 in response to the fourth clock pulse, as shown in the waveforms. Again, this is due to the delay of FF B's response after A has gone LOW.

Although the situation is illustrated for a MOD-4 counter, the same type of situation can occur for *any* ripple counter. This is because ripple counters work on the "chain-reaction" principle, whereby each FF triggers the next one and so on. The spikes at the decoder outputs may or may not present a problem, depending on how the counter is being used. When the counter is being used only to count pulses and display the results, the decoding spikes are of no consequence because they are very short in duration and will not even show up on the display. However, when the counter is used to control other logic circuits, such as was done in Figure 7.14, the spikes can cause improper operation. For example, in Figure 7.14, a spike at the output of either decoding NAND gate would cause FF X to be set or cleared at the wrong time.

In situations where the decoding spikes cannot be tolerated, there are two basic solutions to the problem. The first possibility is to use a parallel counter instead of a ripple counter. Recall that in a parallel counter the FFs are all triggered at the same time by the clock pulses so that it appears that the conditions which produced the decoder spikes cannot occur. However, even in a parallel counter the spikes can occur because the FFs will not all necessarily have the same t_{pd}, especially when some FFs may be loaded more heavily than others.

Strobing

A more reliable method for eliminating the decoder spikes is to use a technique called *strobing*. This technique uses a signal called a *strobe-signal* to keep the decoding AND gates disabled (outputs at 0) until all the FFs have reached a stable state in response to the negative clock transition. This is illustrated in Figure 7.17(a), where the strobe signal is connected as an input to each decoding gate. The accompanying waveforms show that the strobe signal goes LOW when the clock

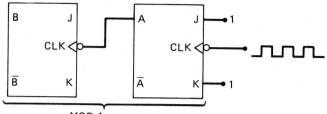

MOD-4 counter

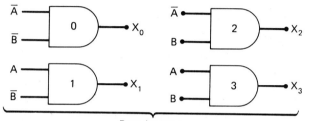

Decoder gates

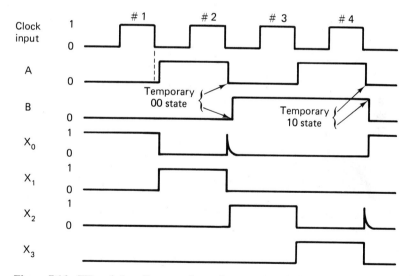

Figure 7.16 FF and decoding waveforms for a MOD-4 ripple counter showing glitches at X_0 and X_2 outputs.

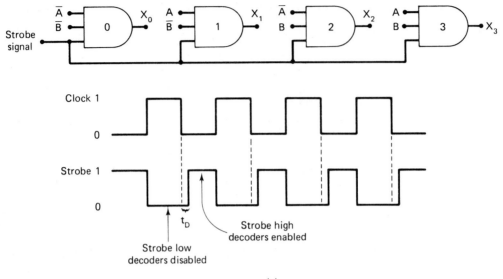

(a)

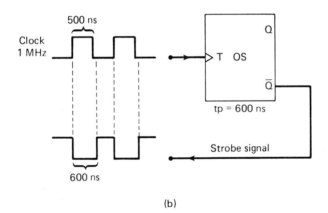

(b)

Figure 7.17   (a) Use of strobe signal to eliminate decoding spikes; (b) one possible way to generate strobe waveform.

pulse goes HIGH. During the time that the strobe is LOW, the decoding gates are kept LOW. The strobe signal goes HIGH some time t_D *after* the clock pulse goes LOW to enable the decoding gates. t_D is chosen to be greater than the total time it takes the counter to reach a stable count and of course depends on the FF delays and the number of FFs in the counter. In this way the decoding gate outputs will not contain any spikes because they are disabled during the time the FFs are in transition.

Figure 7.17(b) shows a simple circuit used to generate the strobe signal for a typical situation. The OS is triggered by the *positive* transition of the 1-MHz clock and its $\bar{Q}$ output goes LOW for $t_p = 600$ ns. Thus, the strobe output will stay LOW until 100 ns after the clock goes LOW. While the strobe signal is HIGH, the decoder outputs will be activated.

The strobe method is not used if a counter is only used for display purposes since the decoding spikes are too narrow to affect the display. The strobe signal is used when the counter is used in control applications like that of Figure 7.14, where the spikes could cause erroneous operation.

7.12 CASCADING BCD COUNTERS

BCD counters are used whenever pulses are to be counted and the results displayed in decimal. A single BCD counter can count from 0 through 9 and then recycles to 0. To count to larger decimal values, we can cascade BCD counters as illustrated in Figure 7.18. This arrangement operates as follows:

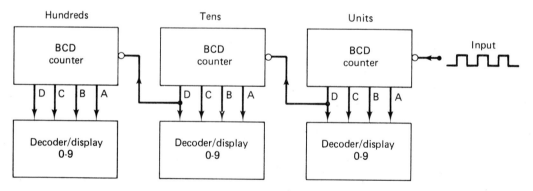

Figure 7.18 Cascading BCD counters to count and display numbers from 000 to 999.

1. Initially all counters are cleared to the zero state. Thus, the decimal display is 000.

2. As input pulses arrive, the units BCD counter advances one count per pulse. After nine pulses have occurred, the hundreds and tens BCD counters are still at zero and the units counter is at 9 (binary 1001). Thus, the decimal display reads 009.

3. On the tenth input pulse the units counter recycles to zero, causing its D FF output to go from 1 to 0. This 1-to-0 transition acts as the clock input for the tens counter and causes it to advance one count. Thus, after 10 input pulses, the decimal readout is 010.

4. As additional pulses occur, the units counter advances one count per pulse, and each time the units counter recycles to zero, it advances the tens counter one count. Thus, after 99 input pulses have occurred, the tens counter is at 9, as is the units counter. The decimal readout is thus 099.

5. On the hundredth input pulse, the units counter recycles to zero, which in turn causes the tens counter to recycle to zero. The D FF output of the tens counter thus makes a 1-to-0 transition which acts as the clock input for the hundreds counter and causes it to advance one count. Thus, after 100 pulses the decimal readout is 100.

6. This process continues up until 999 pulses. On the 1000th pulse, all the counters recycle back to zero.

It should be obvious that this arrangement can be expanded to any desired number of decimal digits simply by adding on more stages. For example, to count up to 999,999 will require six BCD counters and associated decoders and displays. In general, then, we need one BCD counter per decimal digit.

7.13 PRESETTABLE (PROGRAMMABLE) COUNTERS

All the up counters we have looked at thus far were assumed to start in the 000 . . . 0 state and all the down counters in the 111 . . . 1 state. This is accomplished by applying a momentary pulse to all the FF`DC CLEAR (or DC SET) inputs before the counting operation begins. Actually, a counter can be made to start in any desired state through the use of appropriate logic circuitry. Counters that have this capability are called *presettable* or *programmable counters.*

The most common method for presetting a counter is illustrated in Figure 7.19 for a MOD-8 ripple up counter. This method is called *jam entry* because the desired starting state is *jammed* into the DC SET and DC CLEAR inputs independently of what is happening at the *J, K* or *CLK* inputs. The desired preset count is determined by the preset inputs P_A, P_B, and P_C whose values are transferred into the counter FFs when the PRESET LOAD input is momentarily pulsed to the LOW level. When the PRESET LOAD input returns HIGH, the NAND gates are disabled (outputs all HIGH) and the counter is free to count input clock pulses starting from the count that has been preset into the FFs. A typical application of a presettable counter is examined in Problem 7.36.

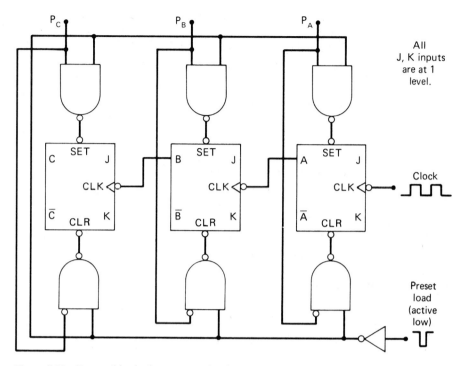

Figure 7.19   Presettable ripple counter using jam entry.

7.14   SHIFT-REGISTER COUNTERS

Shift registers can be arranged to form several types of counters. All *shift-register counters* use *feedback*, whereby the output of the last FF in the shift register is in some way connected to the first FF. The most widely used shift-register counters are the ring counter and the Johnson counter.

Ring Counter

The simplest shift-register counter is essentially a *circulating* shift register connected so that the last FF shifts its value into the first FF. This arrangement is shown in Figure 7.20 using master/slave D-type FFs (J-K FFs can also be used). The FFs are connected so that information shifts from left to right and back around from FF *D* to FF *A*. In most instances only a single 1 is in the register and it is made to circulate around the register as long as clock pulses are applied. For this reason it is called a *ring counter*.

The waveforms and sequence table in Figure 7.20 show the various states of

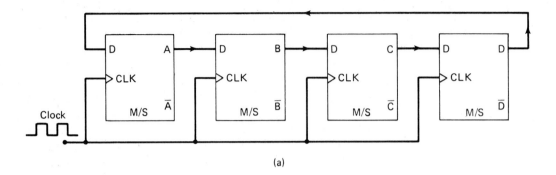

(a)

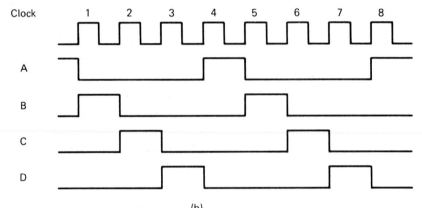

(b)

A	B	C	D		Count
1	0	0	0		0
0	1	0	0		1
0	0	1	0		2
0	0	0	1		3
1	0	0	0		4
0	1	0	0		5
0	0	1	0		6
0	0	0	1		7
:	:	:	:		:

(c)

Figure 7.20 (a) A 4-bit ring counter; (b) waveforms; (c) sequence table.

the FFs as pulses are applied, assuming a starting state of $A = 1$ and $B = C = D = 0$. After the first pulse, the 1 has shifted from A to B so that the counter is in the 0100 state. The second pulse produces the 0010 state, and the third pulse produces the 0001 state. On the *fourth* clock pulse the 1 from FF D is transferred to A, resulting in the 1000 state, which is, of course, the initial state. Subsequent pulses cause the sequence to repeat.

This counter functions as a MOD-4 counter since it has *four* distinct states before the sequence repeats. Although this circuit does not progress through the normal binary counting sequence, it is still a counter because each count corresponds to a particular state of the FFs. Note that each FF output waveform has a frequency equal to $\frac{1}{4}$ of the clock frequency, since this is a MOD-4 ring counter.

Ring counters can be constructed for any desired MOD number; a MOD-N ring counter uses N FFs connected in the arrangement of Figure 7.20. In general, a ring counter will require more FFs than a binary counter for the same MOD number; for example, a MOD-8 ring counter requires eight FFs while a MOD-8 binary counter requires only three FFs.

Despite the fact that it is less efficient in the use of FFs, a ring counter is still useful because it can be decoded without the use of decoding gates. The decoding signal for each state is obtained at the output of its corresponding FF. Compare the FF waveforms of the ring counter with the decoding waveforms in Figure 7.13. In some cases a ring counter might be a better choice than a binary counter with its associated decoding gates. This is especially true in applications where the counter is being used to control the sequencing of operations in a system.

Starting a Ring Counter

To operate properly a ring counter must start off with only one FF in the 1 state and all the other FFs at 0. When power is first applied to the circuit, there is only a remote possibility that the FFs will come up in such a state. Thus, it is necessary to preset the counter to the required starting state before clock pulses are applied. A simple way to accomplish this is to apply a momentary pulse to the DC SET input of one of the FFs and to the DC CLEAR inputs of all the others. This will place a single 1 into the ring counter.

Johnson Counter

The basic ring counter can be modified slightly to produce another type of shift register counter, which will have somewhat different properties. It is called a *Johnson* or *twisted-ring counter* and it is constructed exactly like a normal ring counter except that the *inverted* output of the last FF is connected to the input of the first FF. A 3-bit Johnson counter is shown in Figure 7.21. Note that the $\bar{C}$ output is connected back to the D input of FF A. This means that the inverse of the level stored in FF C will be transferred to FF A on the clock pulse.

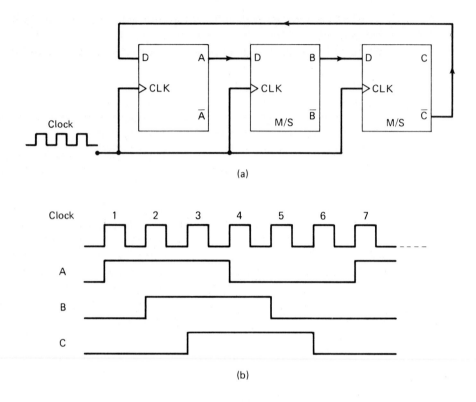

Figure 7.21   (a) MOD-6 Johnson counter; (b) waveforms; (c) sequence table.

The Johnson-counter operation is easy to analyze if we realize that on each positive clock-pulse transition the level at A shifts into B, the level at B shifts into C, and the *inverse* of the level at C shifts into A. Using these ideas and assuming that all FFs are initially 0, the waveforms and sequence table of Figure 7.21 can be generated.

Examination of the waveforms and sequence table reveals the following important points:

1. This counter has six distinct states: 000, 100, 110, 111, 011, and 001 before it repeats the sequence. Thus, it is a MOD-6 Johnson counter. Note that it does not count in a normal binary sequence.

2. The waveform of each FF is a squarewave (50 per cent duty cycle) at $\frac{1}{6}$ the frequency of the clock. In addition, the FF waveforms are shifted by one clock period with respect to each other.

The MOD number of a Johnson counter is always equal to *twice* the number of FFs. For example, five FFs connected in the same arrangement as Figure 7.21 would be a MOD-10 Johnson counter, where each FF output waveform is a squarewave equal to $\frac{1}{10}$ the clock frequency. Thus, it is possible to construct a counter with any *even* MOD number by using half that number of FFs in a Johnson-counter arrangement. *Odd* MOD numbers can be obtained by a modification of the basic Johnson-counter arrangement, which will be examined in one of the problems (7.41) at the end of the chapter.

Decoding a Johnson Counter

For a given MOD number, a Johnson counter requires only half the number of FFs that a ring counter requires. However, a Johnson counter requires decoding gates whereas a ring counter does not. As in the binary counter, the Johnson counter uses one logic gate to decode for each count, but each gate requires only two inputs, regardless of the number of FFs in the counter. Figure 7.22 shows the decoding gates for the six states of the Johnson counter of Figure 7.21.

Notice that each decoding gate only has two inputs, even though there are three FFs in the counter. This is because for each count, two of the three FFs are in a unique combination of states. For example, the combination $A = 0, C = 0$ occurs only once in the counting sequence, at the count of 0. Thus, AND gate 0 with inputs $\bar{A}$ and $\bar{C}$ can be used to decode for this count. This same characteristic is shared by all the other states in the sequence, as the reader can verify. In fact, for *any* Johnson counter, only two-input decoding gates are required.

Johnson counters represent a middle ground between ring counters and binary counters. A Johnson counter requires less FFs than a ring counter but generally more than a binary counter; it has more decoding circuitry than a ring counter but less than a binary counter. Thus, they sometimes represent a logical choice for certain applications.

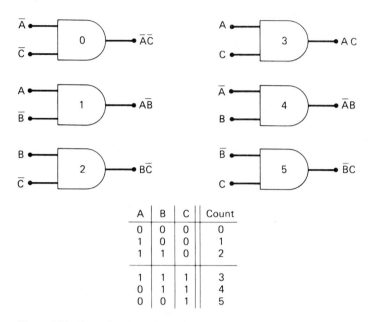

Figure 7.22   Decoding logic for MOD-6 Johnson counter.

7.15 COUNTER APPLICATIONS: FREQUENCY COUNTER

One of the most straightforward methods for measuring and displaying the frequency of a pulse signal uses the principle illustrated in Figure 7.23. Here a counter is being driven by the output of an AND gate. The inputs to the AND gate are the pulse signal whose frequency is to be measured and a SAMPLE pulse that is HIGH between time t_1 and t_2. It should be clear that the output of the AND gate will be held LOW except during the t_1–t_2 interval. During this interval (called the *sampling interval*) the pulses of unknown frequency will appear at the AND output and will be counted by the counter. After t_2 the AND output stays LOW, so the counter stops counting. Thus, the counter has counted the number of pulses that have occurred during the t_1–t_2 sampling interval, and the contents of the counter is a direct measure of the frequency of the pulse waveform.

EXAMPLE 7.10   The unknown frequency is 3792 Hz. The counter is cleared to the zero state prior to t_1. Determine the counter reading after a sampling interval of (a) 1 s, (b) 0.1 s, and (c) 10 ms.

Solution:
(a)   Within a sampling interval of 1 s there will be 3792 pulses entering the counter, so after t_2 the contents of the counter will read 3792.

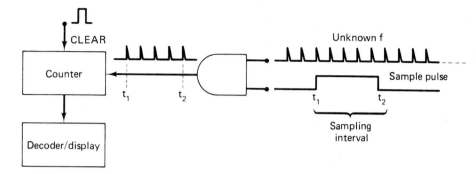

Figure 7.23   Basic frequency counter arrangement.

(b)   With a 0.1-s sampling interval the number of pulses passing through the AND gate into the counter will be 3792 pulses/second × 0.1 s = 379.2. This means that either 379 or 380 pulses will be counted, depending on what part of a pulse cycle that t_1 occurs.

(c)   With a 10 ms = 0.01 s sampling interval, the counter will read either 37 or 38.

The accuracy of this method is almost entirely dependent on the duration of the sampling interval, which must be very accurately controlled. A commonly used method for obtaining very accurate sample pulses is shown in Figure 7.24. A crystal-controlled oscillator is used to generate a very accurate 100-kHz wave-form, which is shaped into square pulses and fed to a series of decade counters which are being used to successively divide this 100-kHz frequency by 10. The frequencies at the outputs of each decade counter are as accurate (percentage-wise) as the crystal frequency.

The switch is used to select one of the decade counter outputs to be fed to a single FF to be divided by 2. For example, in switch position 1 the 1-Hz pulses are fed to FF Q, which is acting as a toggle FF so that its output will be a square-wave with a period of $T = 2$ s and a pulse duration of $t_p = T/2 = 1$ s. This pulse duration is the desired 1-s sampling interval. In position 2 the sampling interval would be 0.1 s, and so on for the other positions.

EXAMPLE 7.11   Assume that the counter in Figure 7.23 is made up of three cascaded BCD counters and their associated displays. If the unknown input frequency is in the range 1–10 kHz, what would be the best setting for the switch position in Figure 7.24?

Solution: With 3 BCD counters the total capacity of the counter is 999. A 10-kHz frequency would produce a count of 1000 if a 0.1-s sample interval were used. Thus, in order to use the full capacity of the counter, the switch should be set to position 2. If a 1-s sampling interval were used, the counter capacity would always be exceeded for frequencies in the range 1–10 kHz. If a lower interval were used, the counter would only count between 0 and 99; this would give a reading to only two significant figures and would be a waste of the counter's capacity.

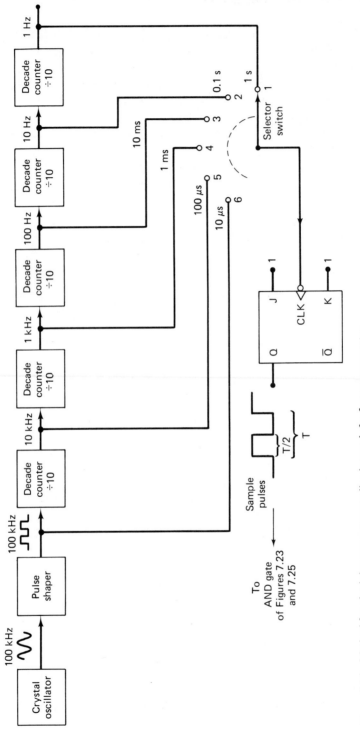

Figure 7.24 Method for obtaining accurate sampling intervals for frequency counter.

Figure 7.23 indicates that a clearing pulse is applied to the counter prior to the beginning of the sampling interval at t_1, so the counter begins each new measurement at zero. Figure 7.25 shows the frequency counter again, including the circuitry used for clearing the counter. The waveform timing diagrams are also shown. A step-by-step description of the complete operation follows:

1. Assume that FF X is in the 0 state (it has toggled to 0 on the falling edge of the previous sample pulse).

2. This LOW from X is fed to the AND gate, disabling its output, so no pulses are fed to the counter even when the first sample pulse occurs between t_1 and t_2.

3. At t_2 the falling edge of the first sample pulse toggles FF X to the 1 state (note that $J = K = 1$). This positive transition at X triggers the OS, which generates a 100-ns pulse to *clear* the counter. The counter now displays *zero*.

4. At t_3 the second sample pulse enables the AND gate (since X is now 1) and allows the unknown frequency into the counter to be counted until t_4.

5. At t_4 the sample pulse returns LOW and toggles X LOW, disabling the AND gate. The counter stops counting.

6. Between t_4 and t_6 the counter holds and displays the count that it had reached at t_4. Note that the third sample pulse does not enable the AND gate because FF X is LOW.

7. At t_6 the falling edge of the sample pulse toggles X HIGH and the operation follows the same sequence that began at t_2.

This frequency counter, then, goes through a repetitive sequence of counting, holding for display, clearing to zero, counting, and so on. A disadvantage of this method is that the display shows the resetting and counting action of the counter, thereby producing a blinking display, which can be irritating to the user. In Problem 7.45 we will investigate a means for eliminating the blinking by only changing the counter's contents *after* it is through counting.

Measurement of Period

The operating principle of the frequency counter can be modified and applied to the measurement of period rather than frequency. The basic idea is illustrated in Figure 7.26. An accurate 1-MHz reference frequency is gated into the counter/display for a time duration equal to T_x, the period of the signal being measured. The counter will count and display the value of T_x in units of 1 μs. For example, if T_x is 1.23 ms, the gate will allow 1230 pulses into the counter.

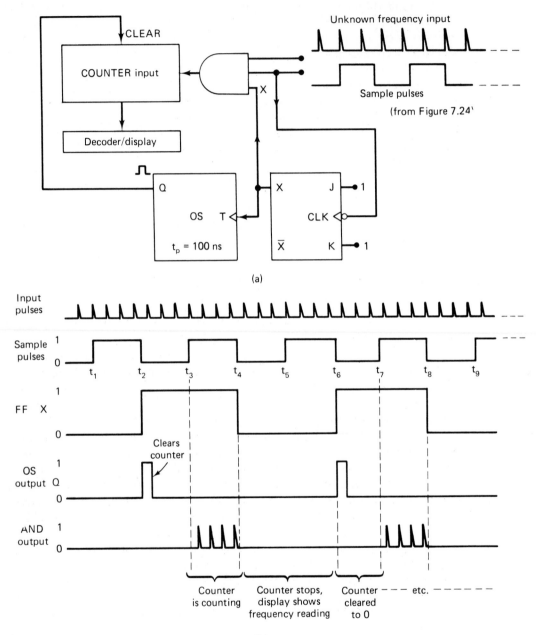

Figure 7.25 Frequency counter.

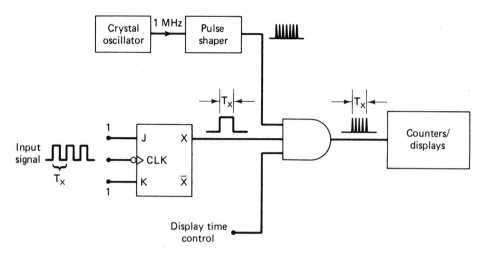

Figure 7.26 Measurement of period.

7.16 COUNTER APPLICATIONS: DIGITAL CLOCK

One of the most popular applications of counters is the digital clock, that is, a time clock which displays the time of day in hours, minutes, and sometimes seconds. In order to construct an accurate digital clock, a very closely controlled basic clock frequency is required. For battery-operated digital clocks (or watches) the basic frequency is normally obtained from a quartz-crystal oscillator. Digital clocks operated from the ac power line can use the 60-Hz power frequency as the basic clock frequency. In either case, the basic frequency has to be divided down to a frequency of 1 Hz or 1 pulse per second (pps). Figure 7.27 shows the basic block diagram for a digital clock operating from 60 Hz.

The 60-Hz signal is sent through a shaping circuit to produce square pulses at the rate of 60 pps. This 60-pps waveform is fed into a MOD-60 counter which is used to divide the 60 pps down to 1 pps. The 1-pps signal is fed into the SEC-ONDS section, which is used to count and display seconds from 0 through 59. The BCD counter advances one count per second. After 9 s the BCD counter recycles to 0, which triggers the MOD-6 counter and causes it to advance one count. This continues for 59 s when the MOD-6 counter is in the 101 (5) count and the BCD counter is at 1001 (9), so the display reads 59 s. The next pulse recycles the BCD counter to 0, which in turn recycles the MOD-6 counter to 0 (remember: the MOD-6 counts from 0 through 5).

The output of the MOD-6 counter in the SECONDS section has a frequency of 1 pulse per minute (the MOD-6 recycles every 60 s). This signal is fed to the MINUTES section, which counts and displays minutes from 0 through 59. The

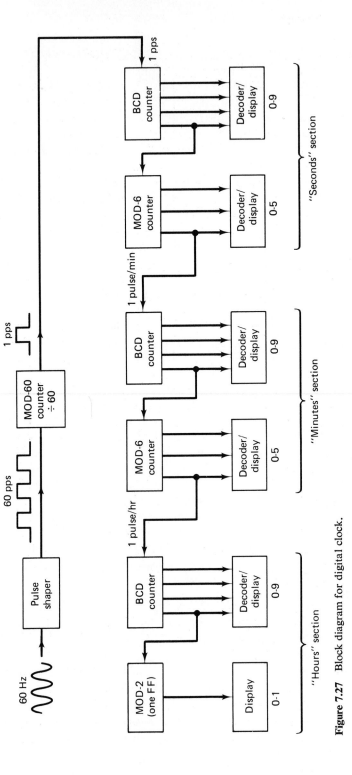

Figure 7.27 Block diagram for digital clock.

MINUTES section is identical to the SECONDS section and operates in exactly the same manner.

The output of the MOD-6 counter in the MINUTES section has a frequency of 1 pulse per hour (the MOD-6 recycles every 60 min). This signal is fed to the HOURS section, which counts and displays hours from 1 through 12. This section is different from the SECONDS and MINUTES sections in that it never goes to the zero state. The circuitry in this section is sufficiently unusual to warrant a closer investigation.

Figure 7.28 shows the circuitry contained in the HOURS section. FFs *A*, *B*, *C*, and *D* are connected as a BCD counter (same as Figure 7.12). The incoming pulses advance this BCD counter once per hour. For example, at 7 o'clock this counter will be in the 0111 state with the associated display indicating the numeral 7. When the BCD counter is in the 1001 state and the next input pulse occurs, the BCD counter recycles to 0000. The negative transition at *D* will toggle FF *X* to the 1 state, which will produce a numeral 1 on the FF *X* display while the BCD's display will be numeral 0. The combined displays thus will show "10," indicating 10 o'clock.

The next two input pulses will advance the BCD counter so that "11" and "12" are displayed at 11 o'clock and 12 o'clock, respectively. The next input pulse advances the BCD counter to the 0011 (decimal 3) state. In this state FFs *A* and *B* are both 1 as well as FF *X*. The NAND gate detects this condition and produces a LOW output, which immediately CLEARS all the FFs to 0 except for FFA, which remains HIGH. Thus, *X* is now 0 and the BCD counter is at 0001, resulting in a display of "01" for 1 o'clock.

Some of the following problems will provide more details of the clock circuitry and other counter applications.

QUESTIONS AND PROBLEMS

7.1 What does *asynchronous* mean in conjunction with counters?

7.2 How many FFs are needed to construct a MOD-128 binary counter?

7.3 If six FFs are connected as a ripple counter, what is the maximum count that can be represented?

7.4 How many possible states are there in a ripple counter composed of 10 FFs?

7.5 An 8-MHz squarewave drives a 5-bit binary counter. What is the frequency of the waveform at the output of the last FF of the counter? What is the duty cycle of this output?

7.6 Repeat Question 7.5 if the input is 8 MHz with a 20 per cent duty cycle.

7.7 A 5-bit ripple counter begins in the 00000 state. What will be the state of the counter after 144 input pulses?

7.8 Modify the counter of Figure 7.3 so that it stops at the count of 14.

7.9 To re-start the self-stopping counter of Figure 7.3, the counter must be returned to

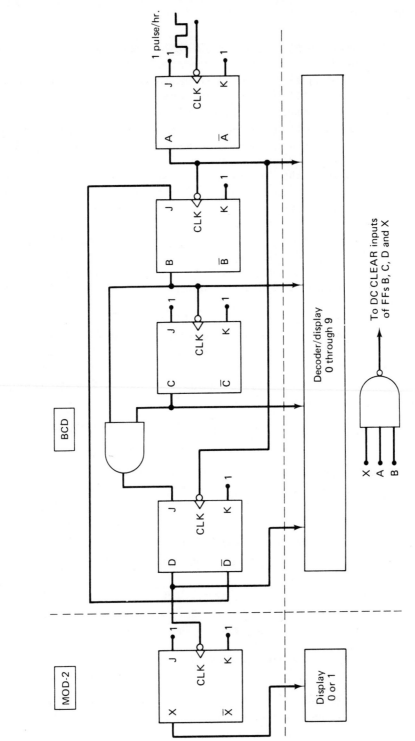

Figure 7.28 HOURS section of digital clock.

the *zero* state (0000). This is accomplished by applying a momentary pulse to the DC CLEAR inputs of *all* the FFs. Why is it necessary to apply this pulse to all the FFs rather than just those that are in the 1 state? (*Hint:* Each FF *CLK* input is triggered by the preceding FF's output.)

7.10 The self-stopping counter of Figure 7.3 can be modified so that it can be programmed to stop at any desired count from 0 to 15, as determined by the position of four toggle switches. Figure 7.29 shows the arrangement of the switches. The NAND output is connected to the *J* and *K* inputs of FF *A* as in Figure 7.3.

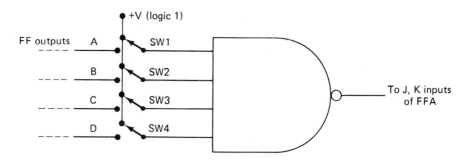

Figure 7.29

(a)   At what count will the counter stop if all the switches are in the up position except for SW3?
(b)   Repeat for SW1 and SW3 up and SW2 and SW4 down.
(c)   Repeat for all switches down.
(d)   Repeat for all switches up.

7.11 Design a programmable variable-MOD counter that can be programmed for any MOD number up to MOD-15 by the setting of four toggle switches.

7.12 Draw the waveforms for all the FFs in the decade counter of Figure 7.5(b) in response to a 1-kHz clock frequency. Show any glitches that might appear on any of the FF outputs. Determine the frequency at the *D* output.

7.13 When a NAND gate is used to clear a counter back to zero as in the counters of Figures 7.4–7.6, there may be a possible problem because the propagation delays from CLEAR input to FF output may vary from FF to FF. For example, suppose that one FF clears in 10 ns and another clears in 50 ns. When the faster FF clears to 0 it will cause the NAND gate output to go back HIGH, thereby terminating the CLEAR pulse and the slower FF may not be cleared. This problem is especially prevalent when the FF outputs are unevenly loaded, since, in general, the propagation delay of a FF increases as it is loaded.

A possible means for eliminating this problem is shown in Figure 7.30 for a MOD-6 counter. The SET-CLEAR FF is the cross-coupled NAND-gate type studied in Chapter 4. Analyze the operation of this circuit and explain how the clearing function is performed without the problem mentioned above.

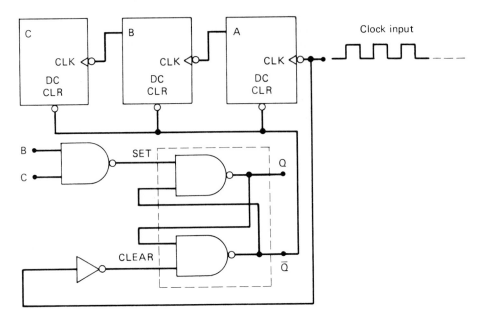

Figure 7.30

7.14 Design a MOD-6 counter that counts through the sequence 001, 010, 011, 100, 101, 110, and repeats. (*Hint:* Remember that a FF can have a DC SET as well as a DC CLEAR input.)

7.15 Design a MOD-6 down counter that counts from 101 (decimal 5) down to 000 (zero) using a NAND gate to return the counter to the 101 starting state at the appropriate time.

7.16 Modify the up/down counter of Figure 7.8 so that it counts up from 0 to 6 or counts down from 6 to 0, depending on the status of the COUNT-UP and COUNT-DOWN inputs.

7.17 A 4-bit ripple counter is driven by a 20-MHz clock signal. Draw the waveforms at the output of each FF if each FF has $t_{pd} = 20$ ns. Determine which counter states, if any, will not occur because of the propagation delays.

7.18 What is the maximum clock frequency that can be used with the counter of Problem 7.17? What would be f_{max} if the counter is expanded to 6 bits?

7.19 (a) Draw the circuit diagram for a MOD-64 parallel counter.
(b) Determine f_{max} for this counter if each FF has $t_{pd} = 20$ ns and each gate has $t_{pd} = 10$ ns.
(c) A more efficient gating arrangement can be used for this counter by deriving the inputs for each AND gate from the output of the preceding gate (see Figure 7.31). For example, the output of the first AND gate (equal to $A \cdot B$) can be fed to the input of the second AND gate, together with C, to produce $A \cdot B \cdot C$. Similarly, the output of the second AND gate (equal to $A \cdot B \cdot C$) is fed to the third AND gate along with D to produce $A \cdot B \cdot C \cdot D$, and so on. This arrangement requires only

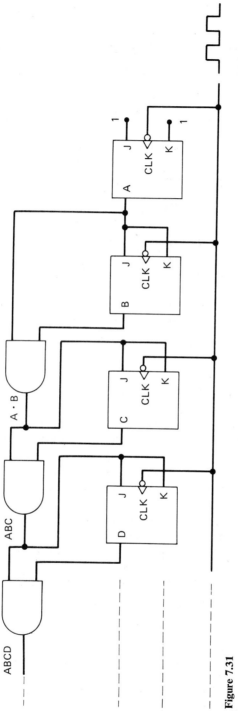

Figure 7.31

two-input AND gates. However, it increases the total propagation delay since the FF output signals must now propagate through more than one gate. Calculate the total delay for this arrangement and then calculate f_{max} and compare it to the value calculated in (b). Assume the same delays as given in (b).

7.20 Figure 7.32 shows a 4-bit parallel counter which is designed so that it does not sequence through the entire 16 binary states. Analyze its operation by drawing the waveforms at each FF output. Then determine the sequence which the counter goes through. Assume that all FFs are initially 0.

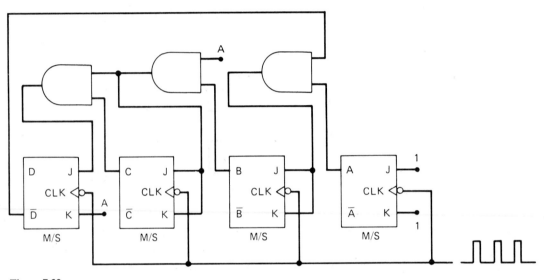

Figure 7.32

7.21 Why must master/slave FFs be used in synchronous (parallel) counters?

7.22 Draw the diagram for a MOD-16 *parallel* up/down counter.

7.23 In the parallel up/down counter, what will happen if both control inputs are at 0? What will happen if both control inputs are 1?

7.24 Figure 7.33 is another example of a combination synchronous/asynchronous counter. Determine the sequence that this counter goes through. Assume that all FFs are initially 0.

7.25 Draw the gates necessary to decode all the states of a MOD-16 counter using active-LOW outputs.

7.26 Draw the AND gates necessary to decode the 10 states of the BCD counter of Figure 7.12.

7.27 Figure 7.34 shows a counter being used to help generate control waveforms. Control waveforms 1 and 2 could be used for many purposes, including control of motors, solenoids, valves, and heaters. Determine the control waveforms assuming that all FFs are initially LOW.

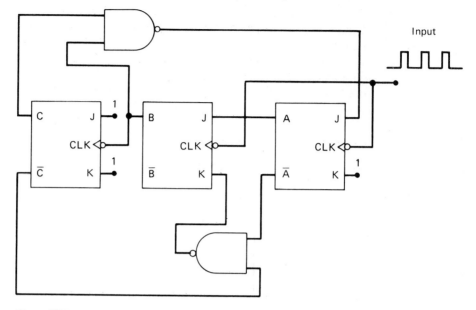

Figure 7.33

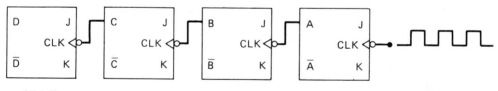

All J, K
inputs = 1

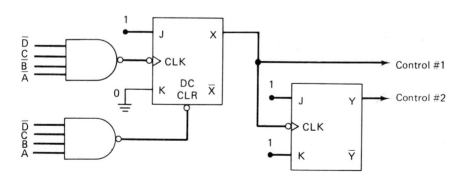

Figure 7.34

7.28   Draw the complete waveforms at the output of the decoding gates of a MOD-16 *ripple* counter, including any glitches or spikes that can occur due to the FF delays. Why are the gates which are decoding for *even* numbers the only ones that have glitches?

7.29   The circuit of Figure 7.14 might not work properly because of glitches at the output of the NAND gates. At what points in the counting sequence can a malfunction occur?

7.30   Repeat Problem 7.29 for the circuit of Figure 7.34.

7.31   How may the malfunctions determined in Problems 7.29 and 7.30 be eliminated?

7.32   Will decoding glitches usually occur for a parallel counter? Explain.

7.33   How many FFs are used in Figure 7.18? Indicate the states of each of these FFs after 795 pulses have occurred.

7.34   How many cascaded BCD counters are needed to be able to count up to 8000? How many FFs does this require? Compare this to the number of FFs required for a normal binary counter to count up to 8000. Since it uses more FFs, why is the cascaded BCD method used?

7.35   Draw the circuit diagram for a MOD-16 presettable ripple down counter.

7.36   Figure 7.35 shows how a presettable down counter can be used in a *programmable timer* circuit. The input clock frequency is an accurate 1 Hz derived from the 60-Hz line frequency after division by 60. Switches SW1–SW4 are used to preset the counter to a desired starting count when a momentary PRESET LOAD pulse is applied. The timer operation is initiated by depressing the START pushbutton switch. FF *Z* is used to eliminate effects of bounce in the START switch. The OS is used to provide a very narrow pulse to the PRESET LOAD input. The output of FF *X* will be a waveform that goes HIGH for a number of seconds equal to *one less* than the number set on the switches.

(a)   Assume that all FFs and the counter are in the 0 state and analyze and explain the circuit operation, showing waveforms when necessary, for the case where SW1 and SW4 are down and SW2 and SW3 are up. Be sure to explain the function of FF *X*.

(b)   Why can't the timer output be taken at the OR-gate output?

(c)   Why can't the START switch be used to trigger the OS directly?

(d)   What will happen if the START switch is held down too long? Add the necessary logic needed to ensure that holding the START switch down will not affect the timer operation.

7.37   Draw the diagram for a 5-bit ring counter using J-K flip-flops.

7.38   Combine the ring counter of Problem 7.37 with a *single* J-K FF to produce a MOD-10 counter. Determine the sequence of states for this counter. This is an example of a decade counter that is not a BCD counter.

7.39   In a certain industrial process five separate heaters are to be turned ON and OFF in sequence. A single 100-μs pulse starts the sequence by turning ON heater 1. Heater 2 must be turned ON and heater 1 OFF only after heater 1 has reached its proper temperature, which is indicated by the occurrence of a 100-μs pulse from a temperature-sensing circuit monitoring heater 1. Similarly, heater 3 must go ON

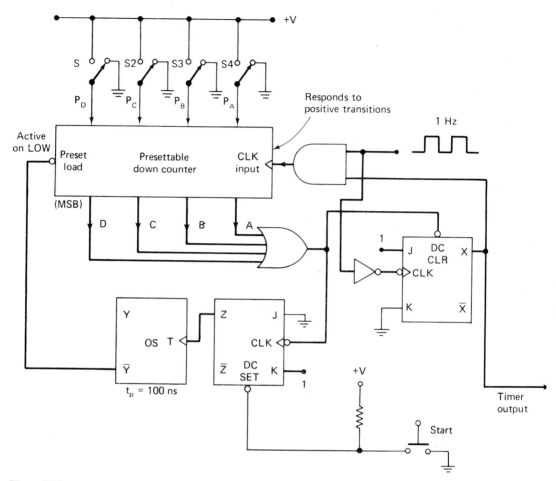

Figure 7.35

and heater 2 OFF when heater 2 reaches a certin temperature, and so on for heaters 4 and 5. Heater 5 stays ON indefinitely once it is turned ON. Each heater has a temperature-sensing circuit which generates a pulse when the heater reaches the proper temperature. Design a logic circuit that will operate as described above using a ring counter. Assume that the circuitry to generate the start pulse and temperature-sensing pulses is already available.

7.40 Draw the diagram for a MOD-10 Johnson counter using J-K flip-flops and determine its counting sequence. Draw the decoding circuit needed to decode each of the 10 states. This is another example of a decade counter that is not a BCD counter.

7.41 A Johnson counter always has an *even* MOD number. However, it can be modified, as illustrated in Figure 7.36, to have an *odd* MOD number. Assume an initial starting state of 000 and determine the counting sequence as pulses are applied.

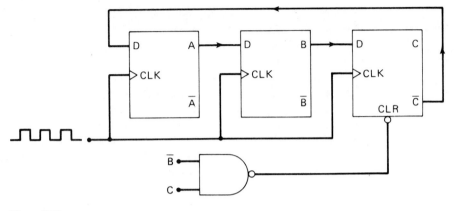

Figure 7.36

7.42 Design a MOD-9 Johnson counter. Draw its output waveforms.

7.43 Determine the frequency of the pulses at points w, x, y, and z in the circuit of Figure 7.37.

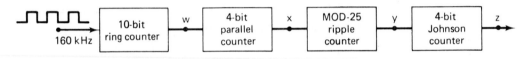

Figure 7.37

7.44 Indicate whether each of the following statements is *always true*, *sometimes true*, or *false*.

(a) A counter that can be programmed to count up to any number is called a presettable counter.

(b) A counter in which all FFs are clocked simultaneously is called a parallel or synchronous counter.

(c) A MOD-20 ring counter requires twice as many FFs as a MOD-20 Johnson counter.

(d) The easiest counter to decode is the Johnson counter.

(e) A MOD-7 counter can be formed by having a MOD-4 counter feeding a MOD-3 counter.

(f) A binary ripple counter (without gates) can only have a MOD number which is an integral power of 2.

(g) A 4-bit parallel down counter has less propagation delay than a 4-bit asynchronous up counter.

(h) In a normal Johnson counter the FF waveforms are always perfect square-waves.

(i) A straight binary counter requires the most decoding circuitry.

(j) Glitches may occur on the decoder outputs of a ripple counter only at very high clock frequencies.

(k) Cascading BCD counters usually require more FFs than using straight binary counters to count to the same number.

(l)   An up/down counter will count in a direction determined by the polarity of
the clock pulses.

(m)   A MOD-30 counter (any type) will always require more than four FFs but less
than 31.

7.45   The frequency counter of Figure 7.25 has the disadvantage of a blinking display.
This can be remedied through the use of D-type FFs, which are used to *store* the
contents of the counter at the end of the counting interval (t_3–t_4 in Figure 7.25)
and hold it for display until the end of the next counting interval (t_7–t_8). Figure 7.38
shows this modified frequency counter, which is the same as Figure 7.25 except that
D FFs have been added between the counter FFs and the decoder/displays. Each

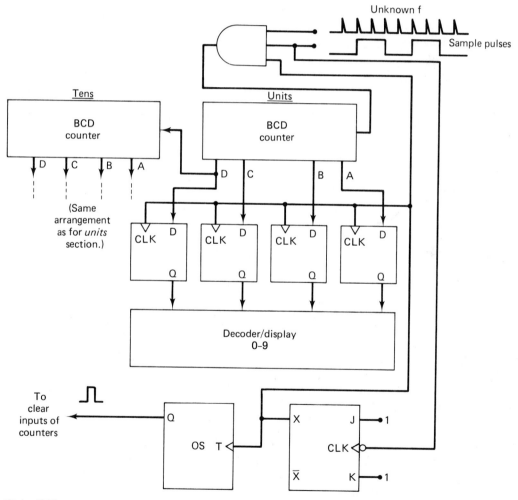

Figure 7.38

counter FF output is fed to a D FF input. The D FF outputs drive the decoder/display. Analyze this circuit and determine how its operation differs from Figure 7.25, especially concerning the readings of the displays.

7.46 Draw the complete circuit diagram for the SECONDS section of the digital clock of Figure 7.26, showing all FFs and necessary gates.

7.47 In the digital clock of Figure 7.27 there is no provision for setting the clock to any desired time. Devise a mechanical switching arrangement whereby the hours and minutes sections can be manually set using the 1-pps signal.

7.48 Modify the hours section of the digital clock so that it counts and displays "military time," that is, it counts and displays hours from 00 to 23.

INTEGRATED-CIRCUIT LOGIC FAMILIES

Almost all modern digital systems utilize digital integrated circuits (ICs) because they result in an increase in reliability and a reduction in weight and size. Digital IC technology has advanced rapidly from small-scale integration (SSI), with less than 13 equivalent logic gates per chip, through medium-scale integration (MSI), with between 13 and 99 equivalent logic gates per chip, to large-scale integration (LSI), which covers the range beyond 100 gates per chip. With the widespread use of ICs comes the necessity for becoming familiar with the characteristics of the most commonly used logic families. (A logic family refers to a specific class of logic circuits that are manufactured using the same manufacturing techniques.) In this chapter we will examine the TTL, ECL, N-MOS, P-MOS, CMOS, SOS, and I²L logic families in sufficient detail so that the relative advantages and disadvantages of each are understood.

The various logic families can be placed into two broad categories according to the IC fabrication process: *bipolar* and *metal-oxide-semiconductor* (*MOS*). The bipolar families utilize the bipolar transistor (NPN and PNP) as their principle circuit element. TTL, ECL, and I²L are bipolar families. The MOS families use MOS field-effect transistors (MOSFETs) as their principal circuit element. P-MOS, N-MOS, CMOS, and SOS are all MOS logic families.

In general, the MOS families tend to be well suited for MSI and LSI devices because the MOS circuits require less chip area and consume less power than their bipolar counterparts. On the other hand, MOS families tend to operate at slower speeds than bipolar and require special handling and storage precautions.

8.1 DIGITAL IC TERMINOLOGY

Although there are many digital IC manufacturers, much of the nomenclature and terminology is fairly standardized. The most useful terms are defined and discussed below.

Current and Voltage Parameters

$V_{IH}[V_{in(1)}]$ *high-level input voltage*: The voltage level required for a logical 1 at an *input*. Any voltage below this level will not be accepted as a HIGH by the logic circuit.

$V_{IL}[V_{in(0)}]$ *low-level input voltage:* The voltage level required for a logical 0 at an *input*. Any voltage above this level will not be accepted as a LOW by the logic circuit.

$V_{OH}[V_{out(1)}]$ *high-level output voltage:* The voltage level at a logic circuit *output* in the logical 1 state. The minimum value of V_{OH} is usually specified.

$V_{OL}[V_{out(0)}]$ *low-level output voltage:* The voltage level at a logic circuit *output* in the logical 0 state. The maximum value of V_{OL} is usually specified.

$I_{IH}[I_{in(1)}]$ *high-level input current:* The current that flows into an input when a specified high-level voltage is applied to that input.

$I_{IL}[I_{in(0)}]$ *low-level input current:* The current that flows into an input when a specified low-level voltage is applied to that input.

$I_{OH}[I_{out(1)}]$ *high-level output current:* The current that flows from an output in the logical 1 state under specified load conditions.

$I_{OL}[I_{out(0)}]$ *low-level output current:* The current that flows from an output in the logical 0 state under specified load conditions.

Fan-Out

In general, a logic-circuit output is required to drive several logic inputs. The *fan-out* (also called *loading factor*) is defined as the *maximum* number of standard logic inputs that an output can drive reliably. For example, a logic gate that is specified to have a fan-out of 10 can drive 10 standard logic inputs. If this number is exceeded, the output logic-level voltages cannot be guaranteed.

Transition Times

Some digital circuits respond to logic levels at their inputs, but others are activated by the rapid change in voltage. In the latter circuit type it is essential that the input signals have sufficiently fast level transitions or the circuit may not respond properly. For this reason the rise time t_R and fall time t_F of a logic output is often specified. The values of t_R and t_F are not necessarily equal, and both are dependent on the amount of loading placed on a logic output.

Propagation Delays

A logic signal always experiences a delay in going through a circuit. The two propagation delay times are defined as

t_{PLH}: *delay time in going from logical 0 to logical 1 state (LOW to HIGH).*

t_{PHL}: *delay time in going from logical 1 to logical 0 state (HIGH to LOW).*

Figure 8.1 illustrates these propagation delays. Note that t_{PHL} is the delay in the *output's* response as it goes to the 0 state, and vice versa for t_{PLH}.

In general, t_{PHL} and t_{PLH} are not the same value, and both will vary depending on loading conditions. The values of propagation times are used as a measure of the relative speed of logic circuits. For example, a logic circuit with values of 10 ns is a faster logic circuit than one with values of 20 ns.

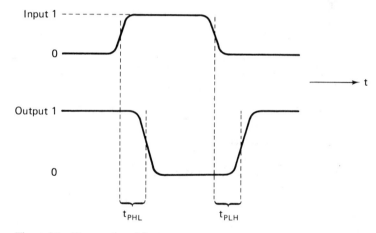

Figure 8.1   Propagation delays.

Power Requirements

The amount of power required by an IC is an important characteristic and is always specified on the manufacturer's data sheet. Sometimes it is given directly as average power dissipation P_D. More often it is indirectly specified in terms of the current drain from the IC power supply. This current is typically symbolized as I_{CC}. When the value for I_{CC} is known, the power drawn by the supply is obtained simply by multiplying I_{CC} by the power-supply voltage.

For some ICs the value of supply current I_{CC} will be different for the two logic states. In such cases two values for I_{CC} are specified. I_{CCH} [or $I_{CC(1)}$] is the supply current when *all outputs* on the IC chip are HIGH; I_{CCL} [or $I_{CC(0)}$] is the supply current when *all outputs* are LOW.

Noise Immunity

Stray electrical and magnetic fields can induce voltages on the connecting wires between logic circuits. These unwanted, spurious signals are called *noise* and can sometimes cause the voltage at the input to a logic circuit to drop below $V_{IH}(min)$ or rise above $V_{IL}(max)$, which could produce unreliable operation. The *noise immunity* of a logic circuit refers to the circuit's ability to tolerate noise voltages on its inputs. A quantitative measure of noise immunity is called *noise margin* and is illustrated in Figure 8.2.

Figure 8.2(a) is a diagram showing the range of voltages that can occur at a logic circuit output. Any voltages greater than $V_{OH}(min)$ are considered a logic 1, and any voltages lower than $V_{OL}(max)$ are considered a logic 0. Voltages in the indeterminate range should not appear at a logic circuit output under normal conditions. Figure 8.2(b) shows the voltage requirements at a logic circuit input. The logic circuit will respond to any input greater than $V_{IH}(min)$ as a logic 1, and will respond to voltages lower than $V_{IL}(max)$ as a logic 0. Voltages in the indeterminate range will produce an unpredictable response and should not be used.

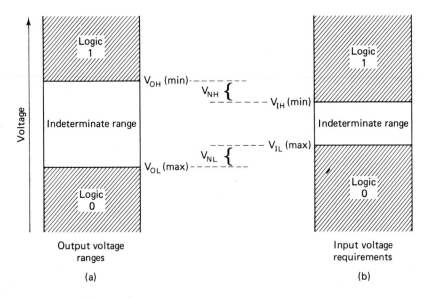

Figure 8.2 Noise margins.

The *high-state noise margin* V_{NH} is defined as

$$V_{NH} = V_{OH}(min) - V_{IH}(min) \tag{8.1}$$

as illustrated in Figure 8.2. V_{NH} is the difference between the lowest possible HIGH output and the minimum input voltage required for a HIGH. When a HIGH logic output is driving a logic circuit input, any negative noise spikes greater than

V_{NH} appearing on the signal line will cause the voltage to drop into the indeterminate range, where unpredictable operation can occur.

The *low-state noise margin* V_{NL} is defined as

$$V_{NL} = V_{IL}(\text{max}) - V_{OL}(\text{max}) \tag{8.2}$$

and it is the difference between the largest possible LOW output and the maximum input voltage required for a LOW. When a LOW logic output is driving a logic input, any positive noise spikes greater than V_{NL} will cause the voltage to rise into the indeterminate range.

EXAMPLE 8.1   The input/output voltage parameters for the standard TTL family are listed below. Calculate the noise margins.

Parameter	Min (V)	Typical (V)	Max (V)
V_{OH}	2.4	3.6	
V_{OL}		0.2	0.4
V_{IH}	2.0*		
V_{IL}			0.8*

*Normally only the minimum V_{IH} and maximum V_{IL} values are given.

Solution:

$$V_{NH} = V_{OH}\,(\text{min}) - V_{IH}\,(\text{min})$$
$$= 2.4\,\text{V} - 2.0\,\text{V} = 0.4\,\text{V}$$
$$V_{NL} = V_{IL}\,(\text{max}) - V_{OL}\,(\text{max})$$
$$= 0.8\,\text{V} - 0.4\,\text{V} = 0.4\,\text{V}$$

These values are actually *worst-case guaranteed* noise margins. In typical operation, the noise margins might be around 1 V each.

AC Noise Margin

Strictly speaking, the noise margins predicted by expressions (8.1) and (8.2) are termed dc noise margins. The term "dc noise margin" might seem somewhat inappropriate when dealing with noise, which is generally thought of as an ac signal of the transient variety. However, in today's high-speed integrated circuits, a pulse width of 1 μs is extremely long and may be treated as dc as far as the response of a logic circuit is concerned. As pulse widths decrease to the low-nanosecond region, a limit is reached where the *pulse duration* is too short for the circuit to respond. At this point, the pulse amplitude would have to be increased to produce a change in the circuit output. What this means is that a logic circuit can tolerate a large noise amplitude if the noise is of a very short duration. In other words, a logic circuit's *ac noise margins* are generally substantially greater than its dc noise margins given by (8.1) and (8.2). Manufacturers generally supply ac-noise-margin information in the form of a graph such as that in Figure 8.3. Note that the noise margins are constant for pulse widths greater than 10 ns but increase rapidly for narrower pulses.

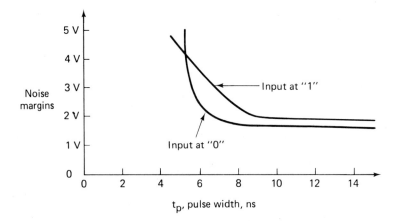

Figure 8.3   Typical ac noise-immunity graphs.

Current-Sourcing and Current-Sinking Logic*

Logic families can be categorized according to how current flows from the output of one logic circuit to the input of another. Figure 8.4(a) illustrates *current-sourcing* logic. When the output of gate 1 is in the HIGH state, it supplies a current I_{IH} to the input of gate 2, which acts essentially as a resistance to ground. Thus, the output of gate 1 is acting as a source of current for the gate 2 input.

Current-sinking logic is illustrated in Figure 8.4(b). Here the input circuitry of gate 2 is represented as a resistance tied to $+V_{CC}$, the positive terminal of a power supply. When the gate 1 output goes to its LOW state, current will flow in the direction shown from the input circuit of gate 2 back through the output resistance of gate 1 to ground. In other words, in the LOW state the circuit driving an input of gate 2 must be able to *sink* a current, I_{IL}, coming from that input.

The distinction between current-sourcing and current-sinking logic circuits is an important one which will become more apparent as we examine the various logic families.

---

*In this text we will assume the *conventional* direction for current, that is, the direction in which *positive* charges would move.

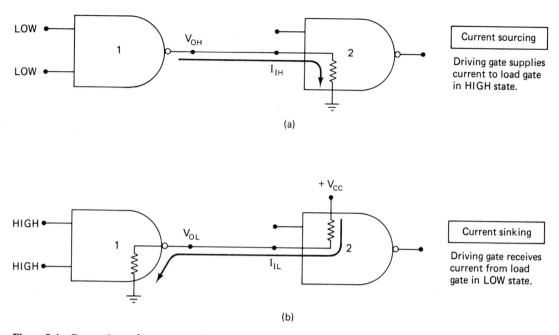

Figure 8.4   Comparison of current-sourcing and current-sinking actions.

8.2   THE TTL LOGIC FAMILY

At this writing, the most widely used family of logic circuits is the transistor–transistor logic (TTL)* family. Figure 8.5(a) shows the circuit diagram for the basic TTL logic gate. Notice the *multiple-emitter* input transistor Q_1 and the *totem-pole* arrangement of output transistors Q_3 and Q_4. The TTL family uses *bipolar* transistors and so falls into the category of bipolar logic families.

Circuit Operation

The analysis can be simplified somewhat by using the diode equivalent of the multiple-emitter transistor Q_1 [see Figure 8.5(b)]. Diodes D_2 and D_3 represent the two base–emitter diodes and D_4 is the collector–base diode. When inputs A and B are both HIGH, diodes D_2 and D_3 will not conduct. Thus, the current from the +5-V supply through R_1 will flow through D_4 into the base of Q_2, thereby turning Q_2 ON. Emitter current from Q_2 then turns Q_4 ON, while the low voltage at Q_2's collector keeps Q_3 OFF (this is ensured by diode D_1). With Q_4

*"TTL" is often referred to as "T²L."

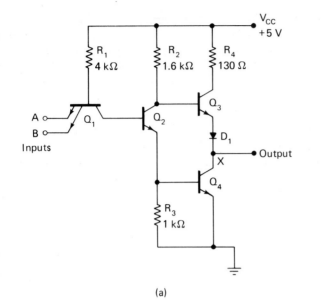

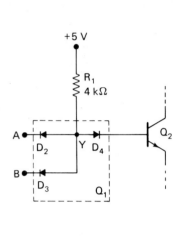

(a)   (b)

Input conditions	Output conditions
A and B both HIGH (> 2 V). Currents at inputs very small.	Q_4 ON (saturated) so that V_X is LOW (< 0.4 V) Q_3 OFF
Either or both inputs LOW (< 0.4 V). Current flows back through inputs (1.1 mA).	Q_4 OFF Q_3 — acting as emitter-follower so that V_X is HIGH (3.6 V)

Figure 8.5   (a) Basic TTL NAND gate; (b) diode equivalent circuit for Q_1.

ON, the voltage at output X will be very low. Thus, with A and B both HIGH, the output will be LOW.

If one (or both) of the inputs is at a LOW voltage, its corresponding input diode will be forward-biased. Current will flow from the $+5$-V supply through R_1 and the input diode to ground. This will clamp point Y [see Figure 8.5(b)] to a voltage that is too small to turn ON D_4 and Q_2. Thus, Q_2 will turn OFF, which in turn causes Q_4 to turn OFF. The high voltage at the collector of Q_2 now allows Q_3 to conduct. Q_3 will act as an emitter follower, producing a high voltage at output terminal X, which will be typically around 3.6 V (5 V minus the 0.7 V voltage drops at base–emitter of Q_3 and diode D_1). Thus, with *any* input LOW, the output will be HIGH.

Clearly, this circuit functions as a NAND gate since a LOW output occurs only when all inputs are HIGH. The table in Figure 8.5 summarizes the operation of the gate for both output states. This table also indicates that with a HIGH input voltage, only a very small input current flows since the emitter–base junctions are reverse-biased. With a LOW input voltage, current will flow back through the corresponding emitter lead to the input. These are characteristics of *current-sinking* logic circuits.

TTL logic circuits are current-sinking circuits in which outputs *receive* current from the inputs they are driving in the LOW state. Figure 8.6 shows the output of one TTL gate driving the input of another. When the driving gate output is LOW, transistor Q_4 is saturated and Q_3 is OFF. The LOW voltage at X forward-biases the emitter of Q_1 and current flows, as shown, back through Q_4. The saturation collector current of Q_4 is provided by the gate being driven. Q_4 is acting as a current sink.

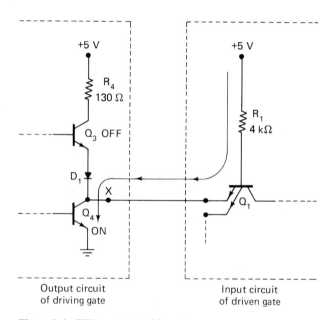

Output circuit   Input circuit
of driving gate   of driven gate

Figure 8.6   TTL current-sinking action.

Totem-Pole Output Circuit

Several points should be mentioned concerning the totem-pole arrangement of the TTL output circuit since it is not readily apparent why it is used. The same logic could be accomplished by eliminating Q_3 and D_1 and connecting the bottom of R_4 to the collector of Q_4. But this would mean that Q_4 would conduct a fairly heavy current in its saturation state (5 V/130 Ω ≈ 40 mA). With Q_3 present there will be no current through R_4 in the output LOW state. This is important because it keeps the circuit power dissipation down.

Another advantage of this arrangement occurs in the output HIGH state. Here Q_3 is acting as an emitter–follower with its associated low output impedance (typically 10 Ω). This low output impedance provides a short time constant for charging up any capacitive load on the output. This action (commonly called *active pull-up*) provides very fast rise-time waveforms at TTL outputs.

A disadvantage of the totem-pole output arrangement occurs during the transition from LOW to HIGH. Unfortunately, Q_4 turns OFF more slowly than Q_3 turns ON, so there is a period of a few nanoseconds during which both transistors are conducting and a relatively large current (30–40 mA) will be drawn from the 5-V supply. This can present a problem that will be examined later.

The TTL NAND gate has the basic structure of all TTL logic circuits. The other TTL gates and logic circuits are based on this same structure. All TTL circuits have inputs that are the emitters of a single- or multiple-emitter transistor, and so are current-sinking inputs. Most, but not all, TTL circuits have a totem-pole output configuration. The exceptions to this will be discussed shortly.

8.3 STANDARD TTL SERIES CHARACTERISTICS

In 1964 Texas Instruments introduced the first standard product line of TTL circuits. The 5400/7400 series, as it is called, has been one of the most widely used families of IC logic. We will simply refer to it as the 7400 series since the only difference between the 5400 and 7400 versions is that the 5400 series is meant for military use and can operate over a wider temperature and power-supply range. Many IC manufacturers now produce the 7400 line of ICs, although some use their own identification numbers. For example, Fairchild has a series of TTL ICs, which uses numbers such as 9N00, 9300, 9600, and so on. However, on the Fairchild specification sheets the equivalent 7400 series number is usually indicated.

The 7400 series operates reliably over the temperature range 0–70°C and with a supply voltage (V_{CC}) of from 4.75 to 5.25 V. The 5400 series is somewhat more flexible since it can tolerate a −55 to +125°C temperature range and a supply variation of 4.5–5.5 V. Both series typically have a fan-out of 10, indicating that they can reliably drive 10 other inputs.

7400 Voltage Levels

Table 8.1 lists the input and output voltage levels for the standard 7400 series. The minimum and maximum values shown are for worst-case conditions of power supply, temperature, and loading conditions. Inspection of the table reveals a guaranteed maximum logical 0 output $V_{OL} = 0.4$ V, which is 400 mV less than the logical 0 voltage needed at the input $V_{IL} = 0.8$ V. This means that the guar-

Table 8.1   Standard 7400 Series Voltage Levels

	Minimum	Typical	Maximum
V_{OL}	—	0.2	0.4
V_{OH}	2.4	3.6	—
V_{IL}	—	—	0.8
V_{IH}	2.0	—	—

anteed LOW-state dc noise margin is 400 mV. That is,

$$V_{NL} = V_{IL}(\text{max}) - V_{OL}(\text{max}) = 0.8 \text{ V} - 0.4 \text{ V} = 0.4 \text{ V} = 400 \text{ mV}$$

Similarly, the logical 1 output V_{OH} is a guaranteed minimum of 2.4 V, which is 400 mV greater than the logical 1 voltage needed at the input $V_{IH} = 2.0$ V. Thus, the HIGH-state dc noise margin is 400 mV.

$$V_{NH} = V_{OH}(\text{min}) - V_{IH}(\text{min}) = 2.4 \text{ V} - 2.0 \text{ V} = 0.4 \text{ V} = 400 \text{ mV}$$

Thus, the *guaranteed worst-case* dc noise margins for the 7400 series are both 400 mV. In actual operation the *typical* dc noise margins are somewhat higher ($V_{NL} = 1$ V and $V_{NH} = 1.6$ V).

Power Dissipation

The basic TTL logic circuit is the NAND gate of Figure 8.5. Typically, it draws an average supply current I_{CC} of 2 mA, resulting in a power dissipation of 2 mA $\times$ 5 V = 10 mW.

Propagation Delay

The basic TTL NAND gate has typical propagation delays of $t_{PLH} = 11$ ns and $t_{PHL} = 7$ ns, which is an *average* propagation delay of 9 ns.
   Table 8.2 summarizes these characteristics of the standard 7400 series.

Table 8.2   Standard 7400 Series Characteristics

Noise margins (worst-case)	$V_{NL} = V_{NH} = 400$ mV
Average power dissipation (basic gate)	$P_D = 10$ mW
Average propagation delay	9 ns
Typical fan-out	10

EXAMPLE 8.2   Refer to the data sheet for the 7400 quad two-input NAND-gate IC in Appendix III. Determine the typical average power dissipation and typical average propagation delay of a *single* NAND gate.

Solution: The first thing to note is that this IC contains *four* NAND gates. Under the "electrical characteristics" heading we can find the values of supply current I_{CC}. The typical values for I_{CC} are 4 mA and 12 mA in the HIGH state and LOW states, respectively. This gives an average supply current of 8 mA, which we can assume divides equally among the four gates. Thus, the average power dissipation of a single gate is 2 mA $\times$ 5 V $=$ **10 mW**.

The typical values for t_{PLH} and t_{PHL} are listed under "switching characteristics" as 11 ns and 7 ns, respectively. This gives an average delay time of **9 ns**.

8.4   OTHER TTL SERIES

The standard 7400 series ICs offer a combination of speed and power dissipation suited for many applications. ICs offered in this series include a wide variety of gates, flip-flops, and one-shots in the small-scale integration (SSI) line and shift registers, counters, decoders, memories, and arithmetic circuits in the medium-scale integration (MSI) line.

Besides the standard 7400 series, several other TTL series have been developed to provide a wider choice of speed and power-dissipation characteristics.

Low-Power TTL, 74L00 Series

Low-power TTL circuits designated as the 74L00 series have essentially the same basic circuit as the standard 7400 series except that all the resistor values are *increased*. The larger resistors reduce the power requirements but at the expense of longer propagation delays. A typical NAND gate in this series has an average power dissipation of 1 mW and an average propagation delay of 33 ns.

The 74L00 series is ideal for applications in which power dissipation is more critical than speed. Low-frequency, battery-operated circuits such as calculators are well suited for this TTL series.

High-Speed TTL, 74H00 Series

The 74H00 series is a high-speed TTL series. The basic circuitry for this series is essentially the same as the standard 7400 series except that *smaller* resistor values are used and the emitter–follower transistor Q_4 is replaced by a Darlington pair. These differences result in a much faster switching speed with an average propagation delay of 6 ns. However, the increased speed is accomplished at the expense of increased power dissipation. The basic NAND gate in this series has an average P_D of 23 mW.

Schottky TTL, 74S00 Series

The 74S00 series has the highest speed available in the TTL line. It achieves this performance by using a Schottky barrier diode (SBD) connected as a clamp from base to collector of each circuit transistor. The characteristics of the SBD that make it useful are its low forward voltage drop (typically 0.25 V) and its fast switching speed. Figure 8.7(a) shows a Schottky-clamped transistor. The presence of the Schottky diode prevents the transistor's collector–base junction from becoming forward-biased by more than 0.25 V when the onset of saturation occurs. As a result, the transistor will not go as deeply into saturation, and will therefore turn OFF much faster. This reduces the average propagation delay to 3 ns for a typical NAND gate. The 74S00 series also uses smaller resistor values to increase switching speed. This produces an increase in average power dissipation to 23 mW per gate. Since it has essentially the same P_D as the 74H00 series while performing at a higher speed, it is the most widely used TTL series in applications where high speed is important.

Figure 8.7(b) shows the common symbol used to represent a Schottky-clamped transistor.

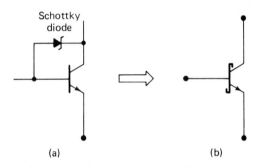

Figure 8.7 (a) Schottky diode connected to collector-base of transistor. (b) Symbol for Schottky-clamped transistor.

Low-Power Schottky TTL, 74LS00 Series

Another Schottky-clamped TTL series uses larger resistor values to decrease power dissipation. The 74LS00 series has a typical P_D of only 2 mW per gate, which is the lowest for TTL except for the 74L00 series. The larger resistances cause an increase in propagation delay to approximately 9.5 ns. Thus, this series has about the same speed as the standard 7400 series while requiring much less power. This has resulted in the 74LS00 series starting to take over many of the applications areas previously dominated by the 7400 series. As the cost of 74LS00 devices continues to come down, it is probable that it will become the major TTL series.

Table 8.3 gives a comparison of the five TTL series. The maximum clock-rate entry is the maximum frequency at which FFs, counters, and so on, will operate reliably.

Table 8.3

Series	Average Delay (ns)	Power (mW)	Maximum Clock Rate	Fan-Out*
7400	9	10	35 MHz	10
74L00	33	1	3 MHz	10
74H00	6	23	50 MHz	10
74S00	3	23	125 MHz	10
74LS00	9.5	2	45 MHz	10

*Assumes output is driving inputs of same series.

8.5 TTL LOADING RULES

In designing digital systems using TTL devices it is important to know how to determine and use the fan-out or drive capability of each circuit. Figure 8.8(a) shows a single TTL output in the LOW state connected to several TTL inputs. Transistor Q_4 is ON and is acting as a current sink for all the currents (I_{IL}) coming back from each input. Although Q_4 is saturated, its ON-state resistance is some value other than zero, so the current I_{OL} produces an output voltage drop V_{OL}. The value of V_{OL} must not exceed 0.4 V for TTL (Table 8.1), and this limits the value of I_{OL} and thus the number of loads that can be driven.

The HIGH-state situation is shown in Figure 8.8(b). Here Q_3 is acting as an emitter–follower and is sourcing (supplying) current to each TTL input. These currents (I_{IH}) are just reverse-bias leakage currents, since the TTL input emitter–base junctions are reverse-biased. If too many loads are driven, however, the total output current I_{OH} can become too large, causing larger drops across R_2, Q_3, and D_1, thereby lowering V_{OH} below the minimum allowable 2.4 V (Table 8.1).

Unit Loads

In order to simplify designing with TTL circuits, the manufacturers have established standardized input and output loading factors in terms of *current*. These currents are called unit loads (UL) and are defined as follows:

$$1 \text{ units load (UL)} = \begin{cases} 40 \ \mu\text{A in the HIGH state} \\ 1.6 \text{ mA in the LOW state} \end{cases}$$

These unit-load factors are used to express the output drive capabilities and input requirements for TTL circuits in any of the five TTL series. The following examples illustrate how they are used.

EXAMPLE 8.3   Refer to the data sheet for the 7400 quad NAND gate IC in Appendix III. Determine its input and output loading factors in terms of unit loads.

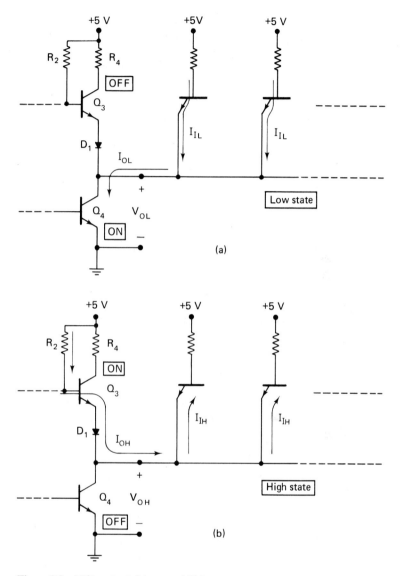

Figure 8.8   TTL output-drive capabilities.

Solution: The maximum input current parameters for the IC are listed as $I_{IH} = 40 \ \mu A$*
and $I_{IL} = -1.6$ mA (the negative sign simply indicates that the input current in the
LOW state actually flows out of the input terminal). Thus, a 7400 NAND gate

*The second entry for I_{IH} (1 mA) is measured at the onset of reverse breakdown ($V_{IH} = $
5.5 V). This value of I_{IH} is not used to determine the input current requirements.

has an input loading factor of 1 UL in both the HIGH and LOW states. In other words, any input to one of these gates acts as 1 UL.

The output drive capabilities of this IC are given under "recommended operating conditions," where a fan-out of 10 UL is indicated for each output. This means that each output can reliably drive a number of inputs whose total loading factor equals 10 UL. For example, one of the NAND-gate outputs (fan-out = 10 UL) can drive 10 other NAND inputs (each input = 1 UL).

EXAMPLE 8.4   Determine the maximum output currents for the 7400 NAND gate in both states.

Solution: We saw in Example 8.3 that the 7400 NAND output has a fan-out of 10 UL. In the HIGH state 1 UL is 40 μA, so this gate output can *supply* 10 $\times$ 40 μA = 400 μA = .4 mA while still maintaining $V_{OH} \geq 2.4$ V.

In the low state 1 UL = 1.6 mA, so this gate output can *sink* 10 $\times$ 1.6 mA = 16 mA while maintaining $V_{OL} \leq 0.4$ V. These values of I_{OH} and I_{OL} could also be found listed under the test conditions used for measuring V_{OH} and V_{OL} (refer to the 7400 data sheet).

It should be pointed out that these maximum current values are not determined by the limitations of the output transistors, but rather by the necessity for maintaining proper logic levels. The totem-pole output transistors can safely handle much more current than the maximum values above, if voltage levels are not important.

EXAMPLE 8.5

    (a)   Determine the fan-out of a 74S00 NAND gate.

    (b)   How many 74S00 inputs can a 74S00 output drive?

    (c)   How many 74S00 inputs can a 7400 output drive?

Solution:

    (a)   The data sheet for the 74S00 IC in Appendix III does not specify the fan-out explicitly. However, the output current values can be obtained under the test conditions for V_{OH} and V_{OL}. They are given as

$$I_{OH} = 1 \text{ mA}; I_{OL} = 20 \text{ mA}$$

Since $I_{OH} = 1$ mA, this means that the output can supply 1 mA of current in the HIGH state. One UL in the HIGH state is 40 μA. Thus, one 74S00 output can drive 1 mA/40 μA = 25 UL. That is,

$$\text{fan-out (HIGH state)} = \frac{I_{OH}}{40 \ \mu\text{A}}$$

$$= \frac{1 \text{ mA}}{40 \ \mu\text{A}} = 25 \text{ UL}$$

Since $I_{OL} = 20$ mA the 74S00 output can sink 20 mA in the LOW state. One UL in the LOW state is 1.6 mA. Thus,

$$\text{fan-out (LOW state)} = \frac{I_{OL}}{1.6 \text{ mA}}$$

$$= \frac{20 \text{ mA}}{1.6 \text{ mA}} = 12.5 \text{ UL}$$

Notice that the fan-outs are different for the two output states.

(b) The 74S00 inputs each represent a load of 1.25 UL, as indicated on its data sheet. Since the 74S00 has a fan-out of 12.5 UL in the LOW state, then it can safely drive 12.5/1.25 = 10 other 74S00 inputs.

(c) The 7400 NAND output has a fan-out of 10 UL. Therefore, it can drive 10/1.25 = 8 different 74S00 inputs.

EXAMPLE 8.6 The output of a 7400 NAND gate is providing the clock signal to a parallel counter consisting of J-K FFs. The 7473 IC (Appendix III) is used for the J-K FFs. What is the maximum number of FFs that this counter can have?

Solution: The input loading factor for a 7473 clock input can be determined from the data sheet by noting that $I_{IL} = 3.2$ mA at the clock input. This represents 3.2/1.6 = 2 UL. Thus, one 7400 NAND gate (fan-out of 10 UL) can drive 10/2 = 5 clock inputs.

Table 8.4 gives the typical input and output loading factors for the five TTL series. These values are typical, so there may be some variations, depending on the particular device or manufacturer. The device data sheet should be consulted to determine exact values.

Table 8.4

TTL Series	Input Loading		Fan-Out (UL)	
	High	Low	High	Low
7400	1 UL	1 UL	10 UL	10 UL
74H00	1.25 UL	1.25 UL	12.5 UL	12.5 UL
74L00	0.5 UL	≈0.1 UL*	10 UL	2.5 UL
74S00	1.25 UL	1.25 UL	25 UL	12.5 UL
74LS00	0.5 UL	≈0.25 UL*	10 UL	5 UL

$$1 \text{ UL} = \begin{cases} 40 \text{ } \mu\text{A (HIGH)} \\ 1.6 \text{ mA (LOW)} \end{cases}$$

*Actual I_{IL} for 74L00 series is 0.18 mA, slightly more than 0.1 UL, and actual I_{IL} for 74LS00 series is 0.36 mA, slightly less than 0.25 UL.

Note that the higher-speed series (74H00, 74S00) have larger input loading factors and fan-outs than the standard 7400 series. On the other hand, the lower power series (74L00, 74LS00) have lower input loading requirements and lower fan-outs.

EXAMPLE 8.7   How many 7400 NAND-gate inputs can be driven by a 74LS00 NAND-gate output?

Solution: The 74LS00 has a fan-out of 5 UL in the LOW state while the 7400 has an input load factor of 1 UL in the LOW state. Thus, the 74LS00 can drive *five* 7400 inputs.

EXAMPLE 8.8   What is the maximum value of I_{IL} for a 74LS00 input?

Solution: Table 8.4 indicates that a 74LS00 has an input load factor of 0.25 UL in the LOW state. Thus, we should expect

$$I_{IL} = 0.25 \times 1.6 \, \text{mA} = 0.4 \, \text{mA}$$

If we look at the data sheet for the 74LS00 NAND-gate chip in Appendix III, we will find I_{IL} (max) $= -0.36$ mA, which is slightly less than predicted.

Tied-Together Inputs

If two or more multiple-emitter inputs to a TTL logic circuit are connected together to act as a single input, the amount of input loading will be different than for a single input. In the input LOW state, the tied-together inputs will have the *same* input loading factor as a single input. This is because I_{IL} is determined by the R_1 bias resistor in the TTL circuit, and it remains the same no matter how many inputs are driven LOW. In the input HIGH state, however, the tied-together inputs will have an input loading factor equal to the *sum* of the individual input loading factors. This is because the input signal has to supply the reverse-bias leakage currents for each emitter–base junction path.

For example, if two inputs of a 7400 NAND gate are tied together as a single input, this single input represents a loading factor of 1 UL (1.6 mA) in the LOW state and 2 UL (80 μA) in the HIGH state. In summary, for HIGH level loading count *every* input being driven as 1 UL; for LOW level loading count each separate gate input as 1 UL and any tied-together multiple-emitter inputs as 1 UL.*

EXAMPLE 8.9   Determine the number of ULs being driven by the gate 1 output in Figure 8.9.

Solution: The loading factors for the two states are calculated as shown in Figure 8.9.

8.6 OTHER TTL PROPERTIES

Several other characteristics of TTL logic must be understood if one is to intelligently use TTL in a digital-system application.

---

*Some TTL gates, notably ORs, NORs, and EX-ORs, have single-emitter inputs. These inputs, tied together or not, are always counted as separate ULs in either logic state.

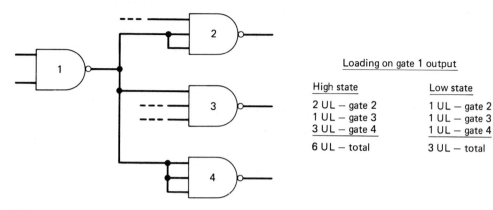

Figure 8.9

Unconnected Inputs

Any input to a TTL circuit that is left disconnected (open) acts exactly like a logical 1 applied to that input, because in either case the emitter–base junction at the input will not be forward-biased. This means that on *any* TTL IC, *all* the inputs are 1s if they are not connected to some logic signal or to ground.

Unused Inputs

Frequently, all the inputs on a TTL IC are not being used in a particular application. A common example of this is when all the inputs to a logic gate are not needed for the required logic function. For example, suppose that we needed the logic operation $\overline{AB}$ and we were using a chip that had a three-input NAND gate. The possible ways of accomplishing this are shown in Figure 8.10.

In Figure 8.10(a) the unused input is left disconnected, which means that it acts as a logical 1. The NAND-gate output is therefore $x = \overline{A \cdot B \cdot 1} = \overline{A \cdot B}$, which is the desired result. Although the logic is correct, it is usually undesirable to leave an input disconnected because it will act like an antenna which is liable to

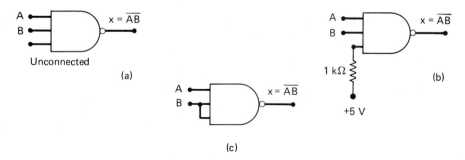

Figure 8.10   Three ways to handle unused logic inputs.

pick up stray radiated signals that could cause the gate to operate improperly. A better technique is shown in Figure 8.10(b). Here the unused input is connected to +5 V through a 1-kΩ resistor, so the logic level is a 1. The 1-kΩ resistor is simply for current protection of the emitter–base junctions of the gate inputs in case of spikes on the power-supply line. This same technique can be used for AND gates since a 1 on an unused input will not affect the output. As many as 30 unused inputs can share the same 1-kΩ resistor tied to V_{CC}.

A third possibility is shown in Figure 8.10(c), where the unused input is tied to a used input. This is satisfactory provided that the circuit driving input B is not going to have its fan-out exceeded. This technique can be used for *any* type of gate.

For OR gates and NOR gates the unused inputs cannot be left disconnected or tied to +5 V since this would produce a constant-output logic level (1 for OR, 0 for NOR) regardless of the other inputs. Instead, for these gates the unused inputs must either be connected to ground (0 V) for a logic 0 or they can be tied to a used input as in Figure 8.10(c).

Biasing TTL Inputs to 0

Occasionally, the situation arises where a TTL input must be held normally LOW and then caused to go HIGH by the actuation of a mechanical switch. This is illustrated in Figure 8.11 for the input to a one-shot. This OS triggers on a positive

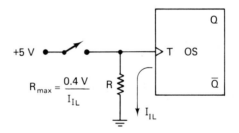

Figure 8.11

transition that occurs when the switch is momentarily closed. The resistor R serves to keep the T input LOW while the switch is open. Care must be taken to keep the value of R low enough so that the voltage developed across it by the current I_{IL} from the OS input will not exceed 0.4 V, the maximum allowable LOW voltage (V_{IL}). Thus, the largest value of R is given by

$$I_{IL} \times R_{max} = 0.4 \text{ V}$$

$$R_{max} = \frac{0.4 \text{ V}}{I_{IL}} \tag{8.3}$$

For example, if I_{IL} is 1.6 mA (1 UL) this gives $R_{max} = 250\ \Omega$. There is no limit on the smallest value of R except for the current drain on the 5-V supply when the switch is closed.

Transition Times of Inputs

The input signals that drive TTL circuits must have relatively fast transitions for reliable operation. If the input rise times or fall times are greater than 1 μs, there is a possibility that oscillations might occur on the output, as shown in Figure 8.12(a). These oscillations could cause serious problems if this output is being fed to a FF, OS, or counter.

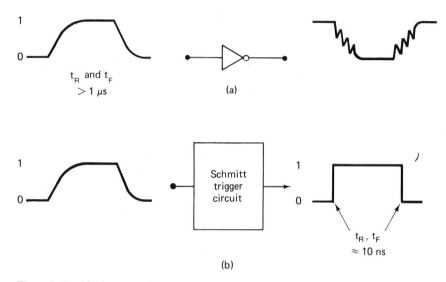

Figure 8.12 (a) Slow t_R and t_F cause TTL outputs to oscillate; (b) Schmitt trigger used to reduce t_R and t_F.

A slow signal can be sharpened up by passing it through a circuit called a *Schmitt trigger*, as in Figure 8.12(b). The Schmitt-trigger circuit produces fast transitions on the output (typically 10 ns) independent of the input rise and fall times. This output can then be fed to the TTL circuit.

Some TTL logic circuits are designed with Schmitt triggers already built in, so they can handle slow input signal transitions with no problems. The 7413 IC is an example of this type of TTL circuit. It contains two four-input NAND gates which will respond properly to slow-changing input signals.

Current Transients

TTL logic circuits suffer from internally generated current transients or spikes because of the totem-pole output structure. When the output is switching from the LOW state to the HIGH state (see Figure 8.13), the two output transistors

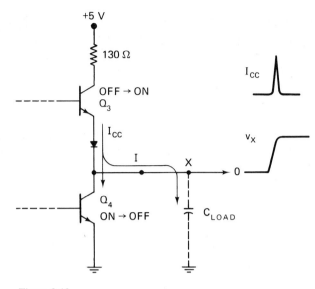

Figure 8.13

are changing states; Q_3 OFF to ON and Q_4 ON to OFF. Since Q_4 is changing from the saturated condition it will take longer than Q_3 to switch states. Thus, there is a short interval of time (about 10 ns) during the switching transition where both transistors are conducting and a relatively large surge of current (30–50 mA) is drawn from the +5-V supply. The duration of this current transient is extended by the effects of any load capacitance on the circuit output. This capacitance consists of stray wiring capacitance and the input capacitance of any load circuits and must be charged up to the HIGH-state output voltage. This over-all effect can be summarized as follows: *Whenever a totem-pole TTL output goes from LOW to HIGH a high-amplitude current spike is drawn from the V_{CC} supply.*

In a complex digital circuit or system there may be many TTL outputs switching states at the same time, each one drawing a narrow spike of current from the power supply. The accumulative effect of all these current spikes will be to produce a voltage spike on the common V_{CC} line, mostly due to the distributed inductance on the supply line [remember: $V = L(di/dt)$ for inductance and di/dt is very large for a 10-ns current spike]. This voltage spike can cause serious malfunctions during switching transitions unless some type of filtering is used. The most common technique uses small *RF* capacitors connected from V_{CC} to GROUND to essentially "short-out" these high-frequency spikes. This is called *power-supply decoupling*. A good rule of thumb requires that 2000 pF of capacitance should be used for each totem-pole output. The capacitors should be low inductance types such as ceramic disc and should be mounted with leads as short as possible to minimize lead inductance.

On printed circuit boards the total required capacitance should be uniformly distributed around the whole board. For example, with a printed circuit board that contains 20 TTL ICs with a total of 100 totem-pole outputs, a total capacitance of at least $2000\ pF \times 100 = 200{,}000\ pF$ would be required. This total capacitance should be uniformly distributed around the board connected from the V_{CC} line to GROUND such that no chip is more than *three* inches from a capacitor.

If TTL MSI chips are being used and it is difficult to determine the number of totem-pole outputs, be sure to use a 0.01- to 0.1-μF *RF* capacitor for each pair of MSI chips.

Manufacturers also recommend connecting one tantalum electrolytic capacitor of 2 to 20 μF from V_{CC} to GROUND, where power enters the printed circuit board to filter out low-frequency noise.

Protective Diodes

Many TTL series have protective diodes connected from each input to ground (see Figure 8.14). These diodes protect the input transistors from negative voltage swings that may be caused by ringing on the input signal. The diodes limit any negative excursions to about -0.7 V.

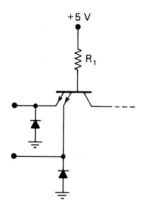

Figure 8.14 Protective diodes on TTL inputs.

8.7 TTL OPEN-COLLECTOR OUTPUTS

Consider the logic circuit of Figure 8.15(a). NAND gates 4 and 5 provide the AND function, which is ANDing the outputs of NAND gates 1, 2, and 3, so the final output X has the expression

$$X = \overline{AB} \cdot \overline{CD} \cdot \overline{EF}$$

The circuit of Figure 8.15(b) shows the same logic operation obtained by simply tying together the outputs of NAND gates 1, 2, and 3. In other words, the AND

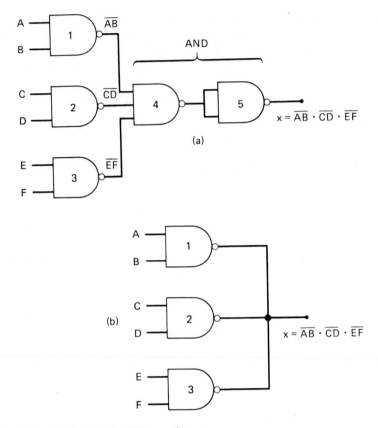

Figure 8.15 Wired-AND operation.

operation is performed by tying the outputs together. This can be reasoned as follows: With all outputs tied together, when any one of the gate outputs goes to the LOW state, the common output point must go LOW as a result of the "short-ing-to-ground" action of the Q_4 transistor in that gate. The common output point will be HIGH only when all gate outputs are in the HIGH state. Clearly, this is the AND operation.

The arrangement in Figure 8.15(b) has two advantages over the conventional arrangement of Figure 8.15(a). It requires fewer gates and it produces less propa-gation delay from input to output. This configuration is called the *wired-AND* operation because it produces the AND operation by connecting output wires together. It is sometimes misleadingly called the *wired-OR* operation.

Totem-Pole Outputs Cannot Be Wired-ANDed

In order to take advantage of the wired-AND configuration the outputs of two or more gates must be tied together with no harmful effects. Unfortunately, the totem-pole output circuitry of conventional TTL circuits prohibits tying outputs together. This is illustrated in Figure 8.16, where the totem-pole outputs of two separate gates are connected together at point X. Suppose that the gate A output is in the HIGH state (Q_{3A} ON, Q_{4A} OFF) and the gate B output is in the LOW state (Q_{3B} OFF, Q_{4B} ON). In this situation Q_{4B} is a very low resistance load on Q_{3A} and will draw a current which can go as high as 55 mA. This current can easily damage Q_{4B}, which is usually guaranteed to sink only 16 mA (I_{OL}). The situation is even worse when more than two TTL outputs are tied together.

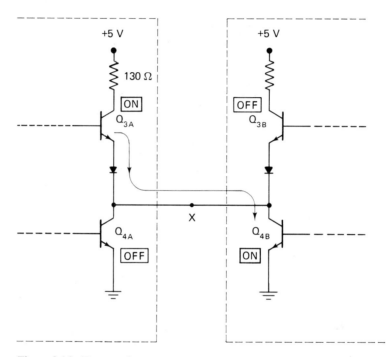

Figure 8.16 Totem-pole outputs tied together can produce damaging current through Q_4.

Open-Collector Outputs

It is for this reason that some TTL circuits are designed with *open-collector* outputs. As shown in Figure 8.17(a) the open-collector-type circuit eliminates Q_3, D_1, and R_4. The output is taken at Q_4's collector, which is unconnected. In the output LOW state, Q_4 is ON (has base current), and in the HIGH state it is OFF (essentially an open circuit). For proper operation an external *pull-up* collector resistor R_c should be connected as shown in Figure 8.17(b), so a high voltage level will appear at the output in the HIGH state.

With open-collector outputs, the wired-AND operation can be accomplished safely. Figure 8.18 shows three two-input *open-collector* NAND gates which are wired-ANDed together. Notice that the open-collector NAND gates have no

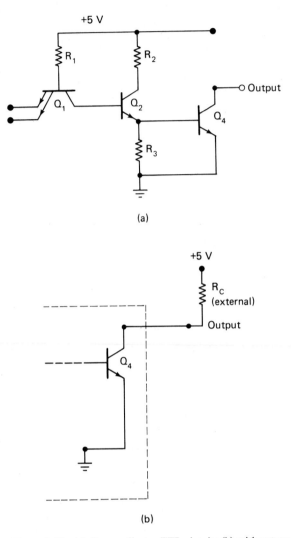

Figure 8.17 (a) Open-collector TTL circuit; (b) with external
pull-up resistor.

special symbol. Also notice the presence of the external pull-up resistor and the
sometimes-used wired-AND symbol.

Value of R_c

The value of R_c must be chosen so that when one gate output goes LOW while
the others are HIGH, the sink current through the LOW output does not exceed
its I_{OL} limit (usually 16 mA). A value of $R_c = 1$ kΩ will produce a sink current
of 5 mA through the LOW gate's output transistor. Of course, the output node

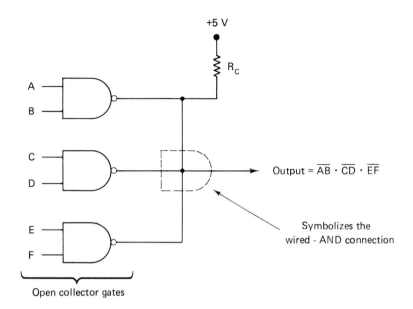

Figure 8.18   Wired-AND operation using open-collector gates.

is usually driving other TTL loads that will add to this sink current, whose total must not exceed I_{OL}. It might appear that a large value of R_c would be therefore advisable. However, it must be realized that any load capacitance will be charged up through R_c so that R_c should be made as small as possible to enhance switching time. Even with R_c minimized, this open-collector arrangement is much slower than totem-pole TTL outputs, which use Q_3 as a low impedance emitter–follower to charge up load capacitance. For this reason, open-collector circuits should not be used in applications where switching speed is a principal consideration.

EXAMPLE 8.10   The 7405 IC (see Appendix III) contains six inverters with open-collector outputs. These six inverters are connected in a wired-AND arrangement in Figure 8.19(a).
   (a)   Determine the logic expression for output X.
   (b)   Determine a value for R_c assuming that output X is to drive other circuits with a total loading factor of 4 UL.

Solution:
   (a)   Each inverter output is the inverse of its input. The wired-AND connection simply ANDs each inverter output. Thus,

$$X = \bar{A} \cdot \bar{B} \cdot \bar{C} \cdot \bar{D} \cdot \bar{E} \cdot \bar{F}$$

Using DeMorgan's theorem, this is equivalent to

$$A = \overline{A + B + C + D + E + F}$$

which is the NOR operation.

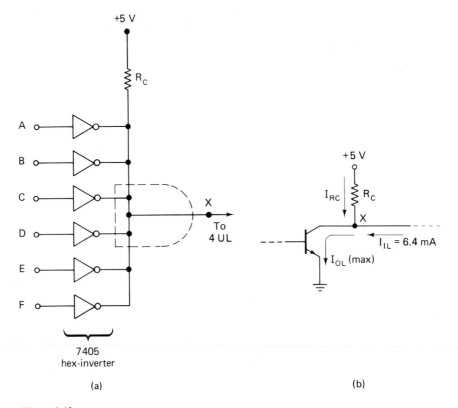

Figure 8.19

(b)   Since X is driving 4 UL, these loads will provide a LOW-state sink current of 4×1.6 mA $= 6.4$ mA [see Figure 8.19(b)]. Referring to the data sheet for the 7405, it is seen that each output has a fan-out of 10 UL. Thus, $I_{OL}(\max) = 10 \times 1.6 = 16$ mA, which means that the current through R_c in the LOW state must be limited to 16 mA $-$ 6.4 mA $= 9.6$ mA. The minimum value of R_c can now be found using Ohm's law.

$$R_c(\min) = \frac{V_{CC} - V_{OL}}{I_{RC}} = \frac{5 \text{ V} - 0.4 \text{ V}}{9.6 \text{ mA}} = 480 \ \Omega$$

To be on the safe side a value of 560 Ω (standard value) can be used.

8.8   TRI-STATE TTL

One of the newest developments in digital-integrated circuitry is TRI-STATE TTL (abbreviated TSL). This new form of TTL has a third stable output state, which is a high-impedance condition (Q_3 and Q_4 *both* in the OFF state) that allows large numbers of TTL circuits to communicate reliably with each other over common bus lines. We shall examine the applications of TSL in Chapter 9. For now we will briefly concern ourselves with the characteristics of TSL.

Figure 8.20a shows the logic symbol for a two-input TRI-STATE NAND gate. Notice that it has two conventional data inputs A and B. In addition, it has

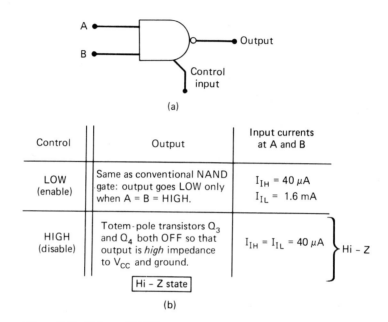

(a)

Control	Output	Input currents at A and B
LOW (enable)	Same as conventional NAND gate: output goes LOW only when A = B = HIGH.	$I_{IH} = 40\,\mu A$ $I_{IL} = 1.6\,mA$
HIGH (disable)	Totem-pole transistors Q_3 and Q_4 both OFF so that output is *high* impedance to V_{CC} and ground. $\boxed{\text{Hi – Z state}}$	$I_{IH} = I_{IL} = 40\,\mu A$ } Hi – Z

(b)

Figure 8.20 Tristate NAND gate.

a third input, called the CONTROL input, which is used to produce the third output state. The table in Figure 8.20b explains what the CONTROL input does. If the CONTROL is kept LOW, the gate operates in exactly the same manner as a conventional TTL NAND gate and will be at a LOW or HIGH voltage level, depending on A and B.

When the CONTROL input is HIGH, it causes the gate output to be *disabled* such that it no longer depends on A and B but rather acts as a high-impedance since Q_3 and Q_4 are both cut off. In essence, the output is almost an open circuit when the CONTROL input is HIGH. We call this the *Hi-Z* state.

The CONTROL input also affects the current requirements at the A and B inputs. When the CONTROL is LOW, the values of I_{IH} and I_{IL} are the normal TTL values 40 μA and 1.6 mA, respectively. In the disabled condition, when the CONTROL is HIGH, both I_{IH} and I_{IL} will be low leakage currents of 40 μA maximum. Thus, the inputs and output are all in a high impedance state when the TSL gate is disabled.

The CONTROL input is often called either the ENABLE input or DISABLE input, depending on the manufacturer's preference. We will normally use one or the other of these designations in our applications of tristate devices.

Figure 8.21 is a typical situation showing how TSL is used. Here we have three D-type FFs of the TSL variety whose outputs are tied together to a common line called a *bus*. Each FF is storing either a HIGH or LOW level at its Q output. It is desired to transmit the outputs of each FF one at a time over the bus line to a certain destination. This can be accomplished by enabling only one of the FF outputs at a time while disabling the other two via the TSL DISABLE inputs. For example, with DISABLE A held LOW and DISABLE B and C HIGH, only FF output A will be connected to the bus line since outputs B and C will be

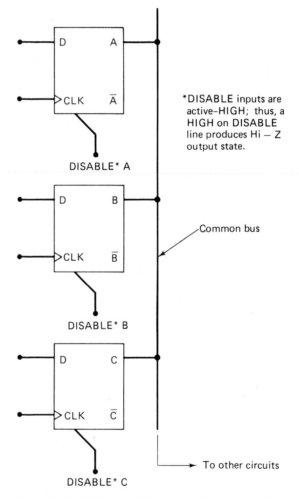

Figure 8.21 TSL D-type FFs connected to common bus

in their Hi-Z states. Note that even though these FF outputs are totem-pole circuits, no harmful currents will flow since only one FF output is enabled at one time.

The reader should understand and appreciate the difference between TSL and open-collector outputs. TSL employs the totem-pole arrangement so that low-impedance high-speed output switching can be realized in the enabled condition while the open-collector outputs, as we saw earlier, suffer from slower switching speed because a relatively large external collector pull-up resistor has to be used to limit output sink current to a safe value.

The tristate operation is also available in many MOS logic-family devices. In fact, the development of MOS tristate operation was a prime reason for the rapid development of microprocessor and microcomputer systems. We will not discuss tristate operation when we discuss MOS family characteristics because it is basically the same as we described here.

EXAMPLE 8.11   Figure 8.22 shows a TTL device called a *tristate buffer*. It is used to convert a totem-pole output signal to a tristate output. The 74125 does not change the logic level of input A; however, it does control whether the logic level at A reaches output X or not. When the ENABLE input is LOW, the tristate buffer is enabled and allows A to pass through to the output so that $X = A$. When ENABLE is HIGH, the buffer is disabled and output X is in its Hi-Z state.

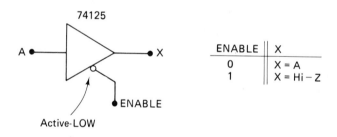

ENABLE	X
0	X = A
1	X = Hi − Z

Figure 8.22   Tristate buffer.

8.9   THE ECL DIGITAL IC FAMILY

The TTL logic family (with the exception of Schottky TTL) uses transistors operating in the saturated mode. As a result, their switching speed is limited by the storage delay time associated with a transistor that is driven into saturation. Another *bipolar* logic family has been developed that prevents transistor saturation, thereby increasing overall switching speed. This logic family is called *emitter-coupled logic* (ECL), and it operates on the principle of current switching whereby a fixed bias current less than $I_{C(sat)}$ is switched from one transistor's collector to another. Because of this current-mode operation, this logic form is also referred to as *current-mode logic* (CML).

Basic ECL Circuit

The basic circuit for emitter-coupled logic is essentially the differential amplifier configuration of Figure 8.23(a). The V_{EE} supply produces an essentially fixed current I_E which remains around 3 mA during normal operation. This current is allowed to flow through either Q_1 or Q_2, depending on the voltage level at V_{IN}. In other words, this current will switch between Q_1's collector and Q_2's collector as V_{IN} switches between its two logic levels of -1.7 V (logical 0 for ECL) and -0.8 V (logical 1 for ECL). The table in Figure 8.23(a) shows the resulting output voltages for these two conditions at V_{IN}. Two important points should be noted: (1) V_{C1} and V_{C2} are the *complements* of each other, and (2) the output voltage levels are not the same as the input logic levels.

The second point noted above is easily taken care of by connecting V_{C1} and V_{C2} to emitter–follower stages (Q_3 and Q_4), as shown in Figure 8.23(b). The emitter–followers perform two functions: (1) they subtract approximately 0.8 V from V_{C1} and V_{C2} to shift the output levels to the correct ECL logic levels, and (2) they provide a very low output impedance (typically 7 Ω), which provides for large fan-out and fast charging of load capacitance. This circuit produces two complementary outputs: V_{OUT1}, which equals $\overline{V_{IN}}$, and V_{OUT2}, which is equal to V_{IN}.

ECL OR/NOR Gate

The basic ECL circuit of Figure 8.23(b) can be used as an inverter if the output is taken at V_{OUT1}. This basic circuit can be expanded to more than one input by paralleling transistor Q_1 with other transistors for the other inputs, as in Figure 8.24(a). Here either Q_1 or Q_3 can cause the current to be switched out of Q_2, resulting in the two outputs V_{OUT1} and V_{OUT2} being the logical NOR and OR operations, respectively. This OR/NOR gate is symbolized in Figure 8.24(b) and is the fundamental ECL gate.

ECL Characteristics

The following are the most important characteristics of the ECL family of logic circuits:

1. The transistors never saturate, so switching speed is very high. Typical propagation delay time is 2 ns, which makes ECL a little faster than Schottky TTL (74S00 series). Although the 74S00 series is almost as fast as ECL, it requires a somewhat more complex fabrication process, so it is somewhat higher in cost.

2. The logic levels are nominally -0.8 V and -1.70 V for the logical 1 and 0, respectively.

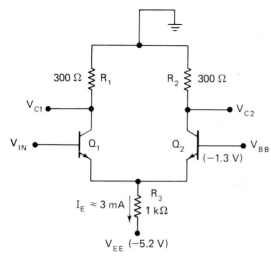

Operating States

V_{IN}	Outputs	
−1.7 V (logic 0)	$V_{C1} = 0$ V $V_{C2} = -0.9$ V	$\Big\}$ Q_2 conducts
−0.8 V (logic 1)	$V_{C1} = -0.9$ V $V_{C2} = 0$ V	$\Big\}$ Q_1 conducts

(a)

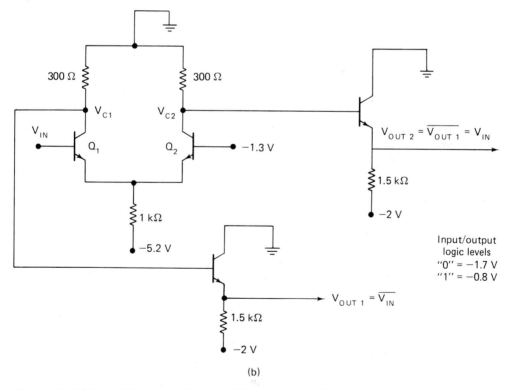

(b)

Figure 8.23 (a) Basic ECL circuit; (b) with addition of emitter followers.

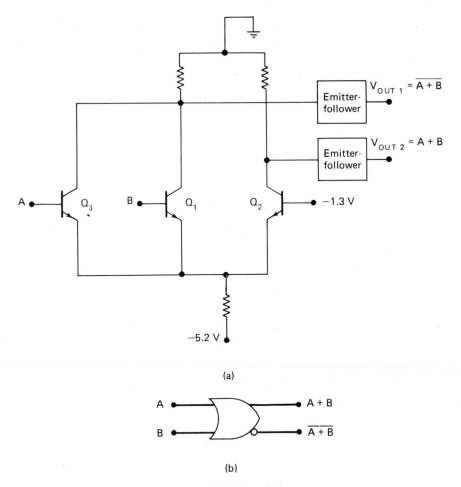

(a)

(b)

Figure 8.24   (a) ECL NOR/OR circuit; (b) logic symbol.

3. Worst-case ECL noise margins are approximately 250 mV. These low noise margins make ECL somewhat unreliable for use in heavy industrial environments.

4. An ECL logic block usually produces an output and its complement. This eliminates the need for inverters.

5. Fan-outs are typically around 25, owing to the low-impedance emitter–follower outputs.

6. Typical power dissipation for a basic ECL gate is 25 mW, just slightly higher than Schottky TTL.

7. The total current flow in an ECL circuit remains relatively constant regardless of its logic state. This helps to maintain an unvarying current drain on the circuit power supply even during switching transitions. Thus, no

noise spikes will be internally generated like those produced by TTL totem-pole circuits.

Table 8.5 shows how ECL compares to the TTL logic families.

Table 8.5

Logic Family	t_{pd} (ns)	P_D (mW)	Worst-Case Noise Margin (mV)	Maximum Clock Rate (MHz)
7400	9	10	400	35
74L00	33	1	400	3
74H00	6	23	400	50
74S00	3	23	300 (V_{NL})	125
74LS00	9.5	2	300 (V_{NL})	45
ECL	2	25	250	200

8.10   MOS DIGITAL INTEGRATED CIRCUITS

MOS (metal-oxide-semiconductor) technology derives its name from the basic MOS structure of a metal electrode over an oxide insulator over a semiconductor substrate. The transistors of MOS technology are field-effect transistors called MOSFETs. Most of the MOS digital ICs are constructed entirely of MOSFETs and no other components.

The chief advantages of the MOSFET are that it is relatively simple and inexpensive to fabricate, it is small in size, and it consumes very little power. The fabrication of MOS ICs is approximately one third as complex as the fabrication of bipolar ICs (TTL, ECL, etc.). In addition, MOS devices occupy much less space on a chip than bipolar transistors; typically, a MOSFET requires 1 square mil of chip area while a bipolar transistor requires about 50 square mils. More importantly, MOS digital ICs normally do not use the IC resistor elements, which take up so much of the chip area of bipolar ICs.

All of this means that MOS ICs can accommodate a much larger number of circuit elements on a single chip than bipolar ICs. This advantage is evidenced by the fact that MOS ICs are surpassing bipolar ICs in the area of large-scale integration (LSI). The high packing density of MOS ICs results in a greater system reliability because of the reduction in the number of necessary external connections.

The principal disadvantage of MOS ICs is their relatively slow operating speed when compared to the bipolar IC families. In many applications this is not a prime consideration, so MOS logic offers an often superior alternative to bipolar logic. We will examine the MOS logic families after a brief discussion of MOSFETs.

8.11 THE MOSFET

There are presently two general categories of MOSFETs: *depletion* type and *enhancement* type. MOS digital ICs use enhancement MOSFETs exclusively, so only this type will be considered in the following discussion. Furthermore, we will concern ourselves only with the operation of these MOSFETs as ON/OFF switches.

Figure 8.25 shows the schematic symbols for the *N*-channel and *P*-channel enhancement MOSFETs, where the direction of the arrow indicates either *P*- or *N*-channel. The symbols show a broken line between the *source, substrate,* and *drain* to indicate that there is *normally* no conducting channel among these electrodes. The symbol also shows a separation between the *gate* and the other terminals to indicate the very high resistance (typically greater than 10,000 MΩ) between the gate and channel.

Basic MOSFET Switch

Figure 8.26 shows the switching operation of an *N*-channel MOSFET. For the *N*-channel device the drain is always biased positive relative to the source. Note also that the substrate is connected to the source, which is most often the case. The gate-to-source voltage V_{GS} is the input voltage, which is used to control the resistance between drain and source (i.e., the channel resistance) and therefore determines whether the device is ON or OFF.

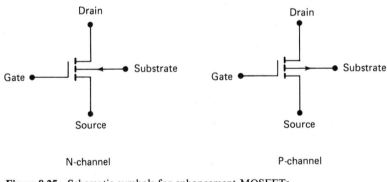

Figure 8.25 Schematic symbols for enhancement MOSFETs.

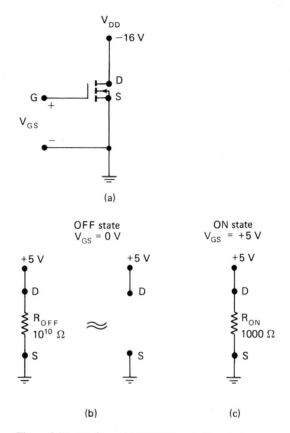

Figure 8.26 *N*-channel MOSFET switching states.

When $V_{GS} = 0$ V, there is no conductive channel between source and drain and the device is OFF. Typically the channel resistance in this OFF state is 10^{10} Ω, which for most purposes is an *open circuit*. The MOSFET will remain OFF as long as V_{GS} is zero or negative. As V_{GS} is made positive (gate positive relative to source), a threshold voltage (V_T) is reached, at which point a conductive channel begins to form between source and drain. Typically $V_T = +1.5$ V for N-MOS-FETs,* so any $V_{GS} = 1.5$ V or more will cause the MOSFET to conduct. Generally, a value of V_{GS} which is much larger than V_T is used to turn ON the MOSFET more completely. As shown in Figure 8.26b, when $V_{GS} = +5$ V, the channel resistance between source and drain has dropped to a value of $R_{ON} = 1000$ Ω.

---

*The newer N-MOSFETs use a special silicon gate to produce lower V_T values. These devices are called low-threshold MOSFETs.

In essence, then, the N-MOSFET will switch from a very high resistance to a low resistance as the gate voltage switches from a LOW voltage to a HIGH voltage. It is helpful simply to think of the MOSFET as a switch that is either opened or closed between source and drain.

The *P*-channel MOSFET operates in exactly the same manner as the *N*-channel except that it uses voltages of the opposite polarity. For P-MOSFETs the drain is connected to a voltage source V_{DD}, which is negative relative to the source. To turn the P-MOSFET ON, a negative voltage that exceeds V_T must be applied to the gate. Table 8.6 summarizes the *P*- and *N*-channel switching characteristics.

Table 8.6

	Drain-to-Source Bias	Gate-to-Source Voltage (V_{GS}) Needed for Conduction	R_{ON}	R_{OFF}
P-channel	Negative	Negative and greater than V_T	1000 Ω (typical)	10^{10} Ω
N-channel	Positive	Positive and greater than V_T	1000 Ω (typical)	10^{10} Ω

8.12 DIGITAL MOSFET CIRCUITS

Digital circuits employing MOSFETs are broken down into three categories: (1) P-MOS, which uses *only* *P*-channel enhancement MOSFETs; (2) N-MOS, which uses *only* *N*-channel enhancement MOSFETs; and (3) CMOS (complementary MOS), which uses both *P*- and *N*-channel devices.

P-MOS and N-MOS digital ICs have a greater packing density (more transistors per chip) and are therefore more economical than CMOS. N-MOS has about twice the packing density of P-MOS. In addition to its greater packing density, N-MOS is also about twice as fast as P-MOS, owing to the fact that free electrons are the current carrier in N-MOS while holes (slower-moving positive charges) are the current carriers for P-MOS. CMOS has the greatest complexity and lowest packing density of the MOS families, but it possesses the important advantages of higher speed and much lower power dissipation.

In this section we will look at some of the basic N-MOS logic circuits, keeping in mind that the P-MOS circuits would be the same except for the voltage polarities. Since P-MOS and N-MOS find their widest applications in LSI (microprocessors, memories, ROMs, etc.), we will defer any applications of these families until later. CMOS, which, like TTL, is widely used in MSI applications, will be covered in more detail beginning with Section 8.14.

Figure 8.27 shows the basic N-MOS logic circuit, which is an inverter. It contains two *N*-channel MOSFETs: Q_1 is called the *load* MOSFET and Q_2 is called the *switching* MOSFET. Q_1 has its gate *permanently* connected to $+5$ V, so it is *always* in the ON state and essentially acts as a load resistor of value R_{ON}. Q_2 will switch from ON to OFF in response to V_{IN}. The Q_1 MOSFET is designed to have a narrower channel than Q_2, so Q_1's R_{ON} is much greater than Q_2's. Typically, R_{ON} for Q_1 is 100 kΩ and R_{ON} for Q_2 is 1 kΩ. R_{OFF} for Q_2 is usually around 10^{10} Ω.

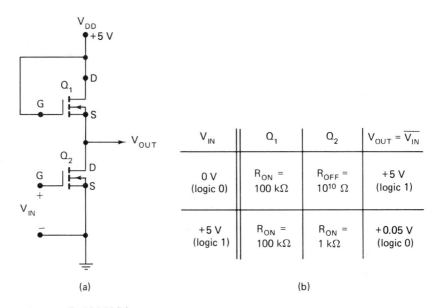

V_{IN}	Q_1	Q_2	$V_{OUT} = \overline{V_{IN}}$
0 V (logic 0)	$R_{ON} =$ 100 kΩ	$R_{OFF} =$ 10^{10} Ω	+5 V (logic 1)
+5 V (logic 1)	$R_{ON} =$ 100 kΩ	$R_{ON} =$ 1 kΩ	+0.05 V (logic 0)

(a)   (b)

Figure 8.27   N-MOS inverter.

The two states of the inverter are summarized in Figure 8.27(b). The best way to analyze this circuit is to consider each MOSFET as a resistance so that the output voltage is taken from a voltage divider formed by the two resistances. With $V_{IN} = 0$ V, transistor Q_2 is OFF, with a very large resistance of 10^{10} Ω. Since Q_1 has $R_{ON} = 100$ kΩ, the voltage-divider output will be essentially $+5$ V. With $V_{IN} = +5$ V, Q_2 is ON, with $R_{ON} = 1$ kΩ. The voltage divider is now 100 kΩ and 1 kΩ, so $V_{OUT} = 1/101 \times (+5$ V$) \approx 0.05$ V.

The circuit functions as an inverter since a LOW input produces a HIGH output and vice versa. This basic inverter can be modified to form NAND and NOR logic gates.

N-MOS NAND Gate

The NAND operation is performed by the circuit of Figure 8.28(a), where Q_1 is again acting as a load resistance while Q_2 and Q_3 are switches controlled by input levels A and B. If either A or B is at 0 V (logical 0), the corresponding FET is OFF, thereby presenting a high resistance from the output terminal to ground so that output X is HIGH ($+5$ V). When both A and B are $+5$ V (logical 1),

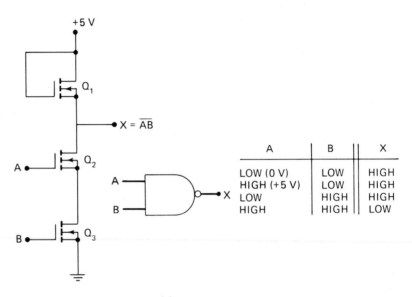

A	B	X
LOW (0 V)	LOW	HIGH
HIGH (+5 V)	LOW	HIGH
LOW	HIGH	HIGH
HIGH	HIGH	LOW

(a)

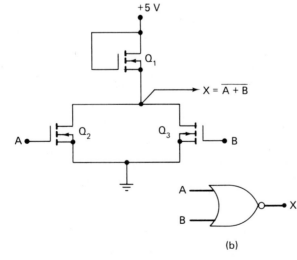

A	B	X
LOW	LOW	HIGH
LOW	HIGH	LOW
HIGH	LOW	LOW
HIGH	HIGH	LOW

(b)

Figure 8.28 (a) N-MOS NAND gate; (b) NOR gate.

both Q_2 and Q_3 are ON, so output X is LOW. Clearly, the output equals the NAND of the inputs ($x = \overline{AB}$).

N-MOS NOR Gate

The NOR gate of Figure 8.28(b) uses Q_2 and Q_3 as parallel switches with Q_1 again acting as a load resistance. When either input A or B is at $+5$ V, the corresponding MOSFET is ON, forcing the output to be LOW. When both inputs are at 0 V, both Q_2 and Q_3 are OFF, so the output goes HIGH. Clearly, this is the NOR operation with $X = \overline{A + B}$.

N-MOS OR gates and AND gates are easily formed by combining the NOR or NAND with inverters.

N-MOS Flip-Flops

Flip-flops are formed by using two cross-coupled NOR gates or NAND gates (Chapter 4). The MOS FF is very important in MOS memories, which will be discussed later.

8.13 CHARACTERISTICS OF MOS LOGIC

Compared to the bipolar logic families the MOS logic families are slower in operating speed, require much less power, have a better noise margin and a higher fan-out, and, as was mentioned earlier, require much less "real estate" (chip area).

Operating Speed

A typical N-MOS NAND gate has a propagation delay time of 50 ns. This is due to *two* factors: the relatively high output resistance (100 kΩ) in the HIGH state and the capacitive loading presented by the inputs of the logic circuits being driven. MOS logic inputs have very high input resistance ($> 10^{12}$ Ω), but they have a reasonably high gate capacitance (MOS capacitor), typically 2-5 picofarads. This combination of large R_{OUT} and large C_{LOAD} serves to increase switching time.

Noise Margin

Typically N-MOS noise margins are around 1 V, which is substantially higher than TTL or ECL.

Fan-Out

Because of the extremely high input resistance at each MOSFET input, one would expect that the fan-out capabilities of MOS logic would be virtually unlimited. This is essentially true for dc or low-frequency operation. However, for frequencies greater than around 100 kHz, the gate input capacitances cause a deterioration in switching time which increases in proportion to the number of loads being driven. Even so, MOS logic can easily operate at a fan-out of 50, which is somewhat better than the bipolar families.

Power Drain

MOS logic circuits draw small amounts of power because of the relatively large resistances being used. To illustrate, we can calculate the power dissipation of the inverter of Figure 8.27 for its two operating states.

1. $V_{IN} = 0$ V: $R_{ON(Q_1)} = 100$ kΩ; $R_{OFF(Q_2)} = 10^{10}$ Ω. Therefore, I_D, current from V_{DD} supply, ≈ 0.05 nA, and $P_D = 5$ V $\times$ 0.05 nA $= 0.25$ nW.
2. $V_{IN} = +5$ V: $R_{ON(Q_1)} = 100$ kΩ; $R_{ON(Q_2)} = 1$ kΩ. Therefore, $I_D = 5$ V/ 101 kΩ ≈ 50 μA and $P_D = 5$ V $\times$ 50 μA $= 0.25$ mW.

This gives an *average* P_D of a little over 0.1 mW for the inverter. This P_D figure is higher for MOS families, which use high-*threshold* MOSFETs that operate at higher supply voltages. The low power drain of MOS logic makes it suitable for LSI, where many gates, FFs, and so on, can be on one chip without causing overheating, which will damage the chip.

Process Complexity

MOS logic is the simplest logic family to fabricate since it uses only one basic element, a N-MOS (or P-MOS) transistor. It requires no other elements, such as resistors, diodes, etc. This characteristic, together with its lower P_D, makes it ideally suited for LSI (large memories, calculator chips, etc.) and this is where MOS logic has made its greatest impact in the digital field. The operating speed of P-MOS and N-MOS is not comparable with TTL, so very little has been done with them in SSI and MSI applications. In fact, there are very few MOS logic circuits in the SSI or MSI categories (gates, FFs, counters, etc.) CMOS, however, is rapidly invading the MSI area, which was heretofore dominated by TTL.

8.14   COMPLEMENTARY MOS LOGIC

The *complementary MOS* (CMOS) logic family uses *both* P- and *N*-channel MOSFETs in the same circuit to realize several advantages over the P-MOS and N-MOS families. Generally speaking, CMOS is faster and consumes even less power than the other MOS families. These advantages are offset somewhat by the increased complexity of the IC fabrication process and a lower packing density. Thus, CMOS cannot yet hope to compete with MOS in applications requiring the utmost in LSI.

However, CMOS logic has undergone a constant growth in the MSI area, mostly at the expense of TTL, with which it is directly competitive. The CMOS fabrication process is simpler than TTL and has a greater packing density, therefore permitting more circuitry in a given area and reducing the cost per function. CMOS uses only a fraction of the power needed for even the low-power TTL series (74L00) and is thus ideally suited for applications using battery power or battery back-up power. CMOS operating speed is not comparable to the faster TTL series yet, but it is expected to continue to improve.

CMOS Inverter

The circuitry for the basic CMOS inverter is shown in Figure 8.29. For this diagram and those that follow, the standard symbols for the MOSFETs have been replaced by blocks labeled P and N to denote a P-MOSFET and N-MOSFET, respectively. This is done simply for convenience in analyzing the circuits. The CMOS inverter has two MOSFETs in series such that the P-channel device has its source connected to $+V_{DD}$ (a positive voltage) and the N-channel device has its source connected to ground. The gates of the two devices are connected together as a common input. The drains of the two devices are connected together as the common output.

The logic levels for CMOS are essentially $+V_{DD}$ for logical 1 and 0 V for logical 0. Consider, first, the case where $V_{IN} = +V_{DD}$. In this situation the gate of Q_1 (P-channel) is at 0 V relative to the source of Q_1. Thus, Q_1 will be in the

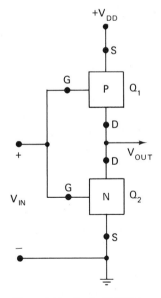

V_{IN}	Q_1	Q_2	V_{OUT}
$+V_{DD}$ (logic 1)	OFF $R_{OFF} = 10^{10}\ \Omega$	ON $R_{ON} = 1\ k\Omega$	$\simeq 0\ V$
0 V (logic 0)	OFF $R_{ON} = 1\ K\Omega$	OFF $R_{OFF} = 10^{10}\ \Omega$	$\simeq +V_{DD}$

$$V_{OUT} = \overline{V_{IN}}$$

Figure 8.29 Basic CMOS inverter.

OFF state with $R_{OFF} \simeq 10^{10}\ \Omega$. The gate of Q_2 (*N*-channel) will be at $+V_{DD}$ relative to its source. Thus, Q_2 will be ON with typically $R_{ON} = 1\ k\Omega$. The voltage divider between Q_1's R_{OFF} and Q_2's R_{ON} will produce $V_{OUT} \simeq 0\ V$.

Next, consider the case where $V_{IN} = 0\ V$. Q_1 now has its gate at a negative potential relative to its source while Q_2 has $V_{GS} = 0\ V$. Thus, Q_1 will be ON with $R_{ON} = 1\ k\Omega$ and Q_2 OFF with $R_{OFF} = 10^{10}\ \Omega$, producing a V_{OUT} of approximately $+V_{DD}$. These two operating states are summarized in Figure 8.29(b), showing that the circuit does act as a logic inverter.

CMOS NAND Gate

Any logic function can be constructed by modifying the basic inverter. Figure 8.30 shows a NAND gate formed by adding a parallel *P*-channel MOSFET and a series *N*-channel MOSFET to the basic inverter. To analyze this circuit it helps to realize that a 0-V input turns ON its corresponding P-MOSFET and turns OFF its corresponding N-MOSFET and vice versa for a $+V_{DD}$ input. Thus, it can be seen that the only time a LOW output will occur is when inputs *A* and *B* are both HIGH ($+V_{DD}$) to turn ON both N-MOSFETs. For any other combination, at

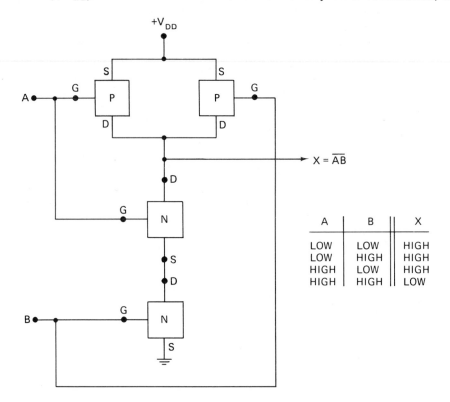

A	B	X
LOW	LOW	HIGH
LOW	HIGH	HIGH
HIGH	LOW	HIGH
HIGH	HIGH	LOW

Figure 8.30   CMOS NAND gate.

least one P-MOSFET will be ON and one N-MOSFET will be OFF, producing a HIGH output.

CMOS NOR Gate

A CMOS NOR gate is formed by adding a series P-MOSFET and a parallel N-MOSFET to the basic inverter as shown in Figure 8.31. Once again this circuit can be analyzed by realizing that a LOW at any input turns ON its corresponding P-MOSFET and turns OFF its corresponding N-MOSFET, and vice versa for a HIGH input. It is left to the reader to verify that this circuit operates as a NOR gate.

CMOS AND and OR gates can be formed by combining NANDs and NORs with inverters.

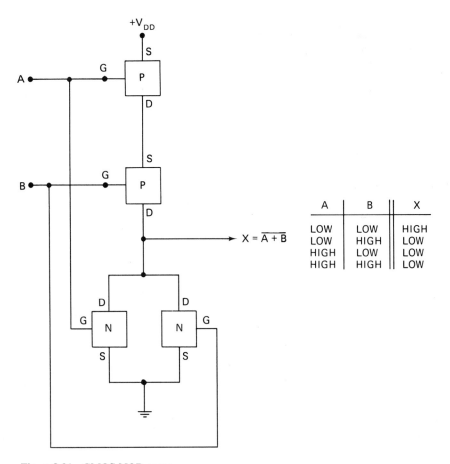

A	B	X
LOW	LOW	HIGH
LOW	HIGH	LOW
HIGH	LOW	LOW
HIGH	HIGH	LOW

Figure 8.31 CMOS NOR gate.

CMOS SET-CLEAR FF

Two CMOS NOR gates or NAND gates can be cross-coupled to form a simple SET-CLEAR FF in the manner discussed in Chapter 4. Additional gating circuitry is used to convert the basic SET-CLEAR FF to clocked D and J-K FFs.

8.15 CMOS SERIES CHARACTERISTICS

The first CMOS logic series was produced by RCA and is known as the 4000 series, which was later developed by other manufacturers. More recently, several manufacturers have developed a CMOS series which is pin-for-pin compatible with TTL. This is the 74C00 series and it contains devices that have the same pin assignments and logic operations as their TTL counterparts. For example, the 74C04 is a hex-inverter chip which is logically the same as the 7404 TTL hex-inverter chip.

Some of the more important characteristics of the CMOS logic family will now be investigated.

Power Dissipation

The quiescent (dc) power dissipation of CMOS logic circuits is extremely low. The reason for this can be seen by reexamining the circuits of Figures 8.29–8.31. For these circuits in either output state, there is *never* a low-resistance path from V_{DD} to ground; that is, for any input condition, there is always an OFF MOSFET in the current path. This fact results in typical CMOS dc power dissipations of 10–20 nW per gate using $V_{DD} = 10$ V, and even lower for $V_{DD} = 5$ V.

Voltage Levels

The CMOS logic levels are 0 V for logical 0 and $+V_{DD}$ for logical 1. The $+V_{DD}$ supply can range from 3 V to 15 V which means that power-supply regulation is not a serious consideration for CMOS. When CMOS is being used with TTL, the V_{DD} supply voltage is made 5 V, so the voltage levels of the two families are the same. The choice of supply voltage often depends on other considerations, which will be discussed shortly.

The required CMOS input levels depend on V_{DD} as follows:

$$V_{IL}(\text{max}) = 30\% \times V_{DD}$$

$$V_{IH}(\text{min}) = 70\% \times V_{DD}$$

For example, with $V_{DD} = 5$ V, $V_{IL}(\text{max})$ is 1.5 V, which is the highest input voltage that is accepted as a LOW, and $V_{IH}(\text{min}) = 3.5$ V, which is the smallest voltage accepted as a HIGH input.

Switching Speed

CMOS, like P-MOS and N-MOS, suffers from the relatively large load capacitances caused by the CMOS inputs being driven. Each CMOS input typically is a 5-pF load. CMOS circuits, however, have a faster switching rate than MOS because of the lower output resistance in the HIGH state. Recall that for N-MOS any load capacitance is charged up in the HIGH state through the load MOSFET, which is about 100 kΩ. In the CMOS circuits the output resistance in the HIGH state is typically 1 kΩ (R_{ON} of a P-MOSFET) so that the load capacitance is charged up more rapidly. A CMOS NAND gate typically has a t_{PD} of around 25 ns when a V_{DD} supply of 10 V is used, and 50 ns for $V_{DD} = 5$ V.

The switching speed of the CMOS family will vary with supply voltage. A large V_{DD} produces lower values of R_{ON}, which produces faster switching, owing to faster charging of load capacitances. This means that for higher-frequency applications, it is best to use a large value of V_{DD} (up to 15 V). Of course, with a larger V_{DD}, the power dissipation will increase, although it will still be extremely low compared to other logic families.

Effect of Frequency on P_D

When CMOS logic circuits are in a stable state for long periods of time or switching at very low frequencies, then the power dissipation will be extremely low. As the switching frequency of the CMOS circuits increases. however, the average power dissipation will increase proportionally. This is because each time the CMOS output switches HIGH, a transient charging current must be supplied to any load capacitance. These momentary pulses of current come from the V_{DD} supply. Clearly, as frequency increases, the average current and therefore average P_D drawn from the V_{DD} supply will also increase (see Figure 8.32). For example, a CMOS inverter gate with a dc dissipation of 10 nW will have an average dissipa-

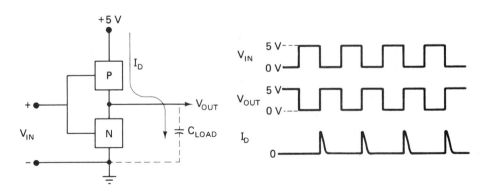

Figure 8.32 Current spikes are drawn from the V_{DD} supply each time the output switches from LOW to HIGH. This is due to charging current of load capacitance.

tion of 0.1 mW at a frequency of 100 kHz. This increases to 1 mW at 1 MHz. Thus, CMOS loses some of its advantages at higher frequencies.

Fan-Out

Because of the high dc input resistance of CMOS, there seems to be no practical limit on the number of CMOS inputs that can be driven by a CMOS output. However, the input capacitance of CMOS becomes a limiting factor when the total load capacitance becomes high enough to limit the switching speed of the circuit below that required for the application. Thus, the fan-out of CMOS is limited by capacitance of the inputs.

To illustrate, a CMOS gate has $t_{pd} = 30$ ns. The value of t_{pd} will increase by approximately 3 ns for each CMOS input (5 pF) being driving by the output. Thus, if a certain application permits a propagation delay of 180 ns, a fan-out of 50 is possible. Because of its lower output resistance, CMOS outputs have a larger fan-out than P-MOS or N-MOS for the same acceptable t_{pd}.

Noise Margin

The CMOS series has the same noise margin in both the HIGH and LOW states. The values of V_{NL} and V_{NH} are guaranteed to be 30 per cent of the V_{DD} supply voltage. Thus, for $V_{DD} = 10$ V, the guaranteed noise margin is 3 V for either state. A V_{DD} of 5 V guarantees a 1.5-V noise margin. In practice, a typical noise margin will be greater than 30 per cent of V_{DD} (more like 45 per cent of V_{DD}), but 30 percent is what the manufacturer guarantees in worst cases. These noise margins are higher than any of the TTL series.

Unused Inputs

All CMOS inputs *must* be tied to some voltage level, preferably ground or V_{DD}. Unused inputs cannot be left floating (disconnected), because these inputs would be susceptible to noise which could bias both the *P*- and *N*-channel MOSFETs in the conducting state, resulting in excessive power dissipation. Unused gate inputs can be tied directly to ground or $+V_{DD}$, whichever is appropriate for the particular logic function. Unused inputs may also be tied to one of the used inputs provided that the fan-out of the signal source is not exceeded. This is highly unlikely for CMOS because of its high fan-out.

Static Charge Susceptibility

The high input resistance of CMOS inputs makes them especially susceptible to static charge buildup, and voltages sufficient to cause electrical breakdown can result from simply handling the devices. Some of the newer CMOS devices have protective diodes on each input to guard against breakdown due to static charges.

8.16 SILICON-ON-SAPPHIRE (SOS)

The SOS family is a modification of the CMOS family. It uses sapphire as an insulating material to reduce the capacitances associated with each MOSFET. SOS works just like CMOS except that it operates at a faster speed due to the reduction of these capacitances. However, it has a more complex fabrication process than the other MOS families and is therefore more expensive.

8.17 INTERFACING CMOS WITH TTL

Many digital systems achieve optimum performance by using more than one logic family, taking advantage of the superior characteristics of each family for different parts of the system. For example, CMOS logic can be used in those parts of the system where high speed is not required, thereby reducing power requirements, while TTL is used in those portions of the system requiring high-speed operation. It becomes necessary, then, to examine the *interface* between CMOS and TTL circuits to see what, if anything, needs to be done to obtain reliable operation when CMOS drives TTL, and vice versa. In the discussions to follow it is assumed that V_{DD} for the CMOS circuits is at $+5$ V, to make it compatible with TTL.

CMOS Driving TTL

Figure 8.33 shows a CMOS gate driving a TTL gate. When the CMOS output is HIGH, there is no problem since $V_{OH} \simeq V_{DD} = +5$ V, which is an acceptable HIGH input for the TTL gate. The TTL input current in the HIGH-state I_{IH} is a maximum of 40 μA, which is easily supplied by the CMOS output through the R_{ON} of the P-channel device.

When the CMOS output is LOW, a problem occurs because of the I_{IL} of the

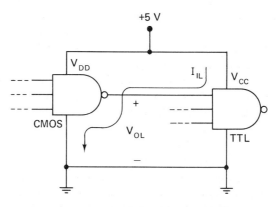

Figure 8.33 CMOS output driving a TTL input must sink I_{IL} without causing V_{OL} to exceed 0.4 V.

TTL. This current, typically 1.6 mA, has to flow back through the CMOS output. In other words, the CMOS output must sink this 1.6 mA to ground through R_{ON} of the N-channel device. The value of R_{ON} will vary for different CMOS circuits and is in the 100 Ω–5 kΩ range. In many cases the 1.6 mA flowing through this CMOS output resistance will produce a voltage at the CMOS output which is too high to satisfy the V_{IL} requirement of the TTL gate.

Recall that $V_{IL} = 0.8$ V for TTL, but also recall that it is necessary to limit the LOW input voltage to 0.4 V to maintain the 0.4 V TTL noise margin. Thus, the CMOS output voltage must remain at or below 0.4 V while sinking 1.6 mA. Some CMOS outputs can meet this requirement while others cannot. In fact, certain CMOS circuits are designed to sink up to 6 mA, which allows them to drive three or four TTL loads. These special CMOS circuits are called *buffers* and can be used in between a conventional CMOS output and several TTL loads.

CMOS outputs can easily drive TTL circuits in the 74L00 and 74LS00 series, since these devices have much lower input current requirements.

EXAMPLE 8.12 Refer to the data sheet for the SCL4001A quad NOR gate in Appendix III. Determine how many standard 7400 loads each NOR gate can reliably drive at 25°C. Also, determine how many 74LS00 loads it can reliably drive at 25°C.

Solution: Look under the "static electrical characteristics" at the I_{DN} entry for "output drive current N-channel," which is the same as *sinking* current in the LOW state. At $V_{DD} = 5$ V and $V_0 = 0.4$ V at 25°C, it is seen that this current is *typically* 1.5 mA. This is enough to barely drive *one* standard TTL load (1.6 mA). However, this is only a *typical* value. The minimum I_{DN} is specified as only 0.4 mA, which is all the manufacturer can guarantee. Thus, the CMOS output *cannot* reliably drive any standard TTL loads.

A 74LS00 has an input loading factor of 0.25 UL in the LOW state (Table 8.4). This represents an I_{IL} of $0.25 \times 1.6 = 0.4$ mA. Thus, the 4001A can reliably drive one 74LS00 load.

TTL Driving CMOS

When a TTL gate output drives a CMOS input, there is no problem in the LOW state since $V_{OL}(\text{max}) = 0.4$ V for the TTL output and the CMOS input will accept anywhere up to 1.5 V for a LOW. However, in the HIGH state a problem can occur because the TTL no-load output voltage is not equal to $+5$ V but is typically around 3.6 V (owing to diode drops across Q_3 and D_1 in the TTL output circuit). This 3.6 V is actually large enough for the CMOS input, which requires 3.5 V or more for a HIGH, but it has essentially cut the noise margin to 0.1 V, which is highly undesirable. Therefore, it is recommended that an *external* pull-up resistor R_p be used as shown in Figure 8.34. The effect of R_p is to raise V_{OH} from the TTL circuit to approximately $+5$ V. The value of R_p is chosen in the same manner as the external resistor R_c for the open-collector TTL circuits (Section 8.7).

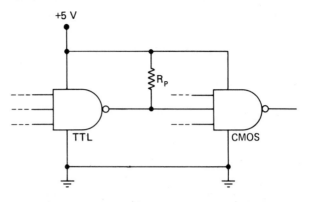

Figure 8.34 External pull-up resistor is used when TTL drives CMOS.

8.18 CMOS TRANSMISSION GATE

A special CMOS circuit that has no TTL or ECL counterpart is the *transmission gate* or *bilateral switch*, which acts essentially as a single-pole, single-throw switch controlled by an input logic level. This transmission gate will pass signals in both directions and is useful for digital and analog applications.

Figure 8.35 shows the basic arrangement for the bilateral switch. It consists of a P-MOSFET and N-MOSFET in parallel so that both polarities of input voltage can be switched. The CONTROL input and its inverse are used to turn the switch ON (closed) and OFF (open). When the CONTROL is HIGH, both

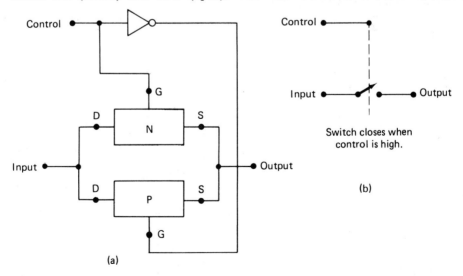

Figure 8.35 CMOS bilateral switch (transmission gate).

MOSFETs are turned ON and the switch is closed. When the CONTROL is LOW, both MOSFETs are OFF and the switch is open. Ideally, this circuit acts like a relay [see Figure 8.35(b)]. However, it is not a perfect short circuit when closed but typically has $R_{ON} \approx 100\ \Omega$. In the OFF state it typically has $R_{OFF} \geq 10^{10}\ \Omega$, which is for most purposes an open circuit.

The CMOS bilateral switch will be examined further in Problem 8.43, and it will also be referred to in subsequent applications.

8.19 INTEGRATED-INJECTION LOGIC (I²L)

This is the newest *bipolar* family and was developed specifically to compete with MOS in LSI applications. It has already made inroads in important applications in games and watches, in chips for television tuning and control, and in memory and microprocessor chips. I²L technology allows greater component densities on a chip (much greater than TTL and in some cases greater than MOS) and offers a variety of speed-power trade-offs. When operated at slow speeds (delays of 100 ns), I²L dissipates less power (5 nW) than any logic family including CMOS. At high speeds (5 ns) it only dissipates 5 mW per gate.

The basic I²L circuit is shown in Figure 8.36(a). Transistor Q_1 is connected as a constant-current source which produces a current I dependent on the value of R_{ext}. This resistor is normally external to the IC chip and is chosen to produce the desired value of I typically between 1 nA and 1 mA. Transistor Q_2 acts as a switching transistor and has multiple collectors similar to the multiple emitters of TTL. Figure 8.36(b) shows the equivalent circuit model for this basic I²L circuit with Q_1 replaced by a constant-current source.

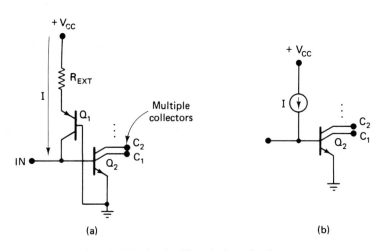

(a)   (b)

Figure 8.36   (a) Basic I²L circuit; (b) equivalent circuit.

The basic circuit operates as follows. If the input terminal is open-circuited, the current I flows into the base of Q_2 and turns it ON so that each collector is a low-resistance path to ground. If the input terminal is shorted to ground, the current I will be shunted away from Q_2's base and will flow through the shorted input terminal to ground. This will turn OFF Q_2 so that each of its collectors will be open circuits.

In practice, the input to this circuit is being driven by one or more collector outputs from similar circuits in order to produce the various logic operations. An example is shown in Figure 8.37. This circuit functions as an OR gate if we define our 0 and 1 logic levels as short circuit and open circuit, respectively. The output will be a 1 (open) when either or both inputs are a 1 (open). When both inputs are 0 (shorted to ground), Q_{2x} and Q_{2y} will be OFF so that Q_{2z} will be turned ON by I_z, thereby producing an output of 0 (short).

Looking at these I²L circuits we can see that a principal reason why they can achieve high component densities is the absence of resistors as part of the IC chip. Resistors represent a significant portion of the chip area, typically requiring *ten* times more space than a transistor. There are also several other facets of the I²L fabrication process which contribute to its superiority in achieving high packing densities, but we will not discuss them here.

In summary, the I²L family is one of the more promising of the bipolar families. Because it is still in the development stage, the cost is still higher than other logic families, but this is gradually changing. It is reasonable to predict that I²L will have a significant impact in the LSI field.

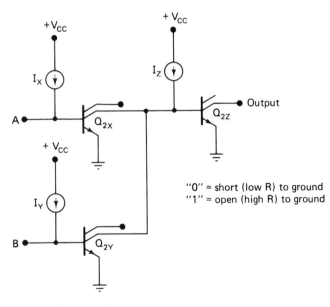

Figure 8.37 I²L OR gate.

QUESTIONS AND PROBLEMS

(Data sheets for the ICs referred to in these problems may be found in Appendix III.)

8.1 Two different logic circuits have the following characteristics:

	Circuit A	Circuit B
V_{supply}	6 V	5 V
V_{IH}	1.6 V	1.8 V
V_{IL}	0.9 V	0.7 V
V_{OH}	2.2 V	2.5 V
V_{OL}	0.4 V	0.3 V
t_{PLH}	10 ns	18 ns
t_{PHL}	8 ns	14 ns
P_D	16 mW	10 mW

(a) Which circuit has the best LOW-state dc noise immunity? The best HIGH-state dc noise immunity?
(b) Which circuit can operate at higher frequencies?
(c) Which circuit draws the most supply current?
(d) Which circuit is more likely to have a better ac noise margin? Explain.

8.2 Explain the difference between current-sinking and current-sourcing logic circuits.

8.3 In the basic TTL NAND gate of Figure 8.5 the function of D_1 is to ensure that Q_3 will not turn ON when the A and B inputs are both HIGH. Assume that Q_2 saturates for this condition and determine the approximate voltage at the base of Q_3. Then show that this amount of voltage is not enough to forward-bias Q_3.

8.4 In which state does the TTL output act as a current sink?

8.5 Give two advantages and one disadvantage of the totem-pole output arrangement.

8.6 Under normal conditions what is the largest LOW-state voltage that should appear at the output of any 7400 series logic circuit?

8.7 Refer to the data sheets in Appendix III. Determine and compare the values of I_{IL}, I_{IH}, I_{OL} and I_{OH} for the 7400, 74S00, and 74LS00 Quad NAND chips.

8.8 (a) Which TTL series (7400, 74L00, 74H00, 74S00, 74LS00) has the lowest power dissipation?
(b) Which TTL series has the longest propagation delay?
(c) Which TTL series can operate at the highest frequency?
(d) Which TTL series use diode clamping to reduce saturation levels?

8.9 Refer to the data sheets in Appendix III. Determine and compare the average P_D and t_{pd} for the 7404, 74H04, and 74S04 ICs.

8.10 A certain TTL logic circuit has a fan-out of 20. What is the most current it can *supply* to another logic circuit in the HIGH state? How much current can it *sink* in the LOW state?

8.11 Refer to the data sheet for the 7473 dual J-K FF.
(a) Determine the input loading factor at the J and K inputs.
(b) Determine the input loading factor at the clock and clear inputs.

(c) How many other 7473 FFs can the output of one 7473 drive at the clock input?

8.12 Figure 8.38(a) shows a 7473 J-K FF whose output is required to drive a total of 14 UL. Since this exceeds the fan-out of the 7473, a buffer of some type is needed. Figure 8.38(b) shows one possibility using one of the NAND gates from the 7437 quad NAND buffer, which has a much higher fan-out than the 7473. Note that $\bar{Q}$ from the 7473 is used since the NAND is acting as an inverter. Refer to the data sheet for the 7437 and determine

(a) Its fan-out.

(b) Its maximum sink current in the LOW state.

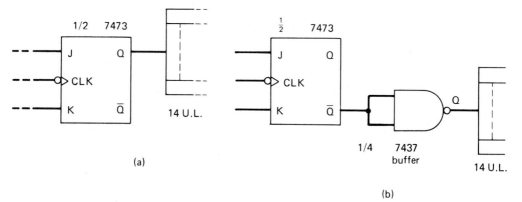

Figure 8.38

8.13 Try to come up with another way of solving the loading problem of Figure 8.38(a) without using a 7437 quad NAND buffer but using a 7400 quad NAND chip.

8.14 Refer to the logic diagram of Figure 8.39, where the 7486 exclusive-OR output is driving several 7420 inputs. Determine whether the fan-out of the 7486 is being exceeded and explain.

8.15 For the circuit of Figure 8.39 determine the longest time it will take for a change in the *A* input to be felt at output *W*. Use all worst-case conditions and maximum values of gate propagation delays. (*Hint:* Remember that NAND gates are inverting gates.)

8.16 Which of the following are acceptable ways of handling unused inputs of a NAND or AND gate?

(a) Leave disconnected.

(b) Connect directly to ground.

(c) Connect to a used input.

(d) Connect directly to V_{CC}.

(e) Connect to V_{CC} through a 1-kΩ resistor.

8.17 Repeat Question 8.16 for NOR and OR gates.

8.18 Figure 8.40 shows a 74121 one-shot being triggered by the closing of the switch. What is the maximum value of *R* which should be used to ensure that the *B* input is biased LOW while the switch is open?

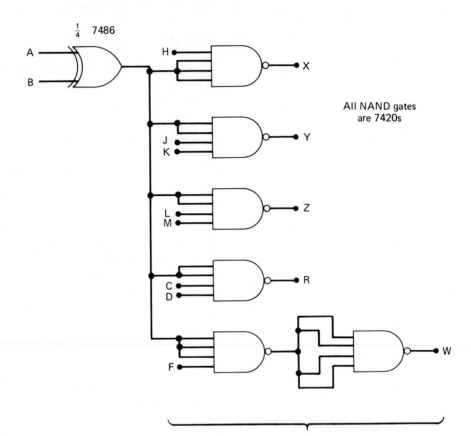

All NAND gates
are 7420s

all 7420's

Figure 8.39

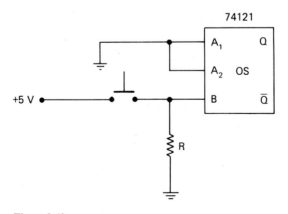

Figure 8.40

8.19 (a)   Will the signal of Figure 8.41 reliably trigger the 74121 one-shot at its A_1 input? Explain.

(b)   Repeat for the B input.

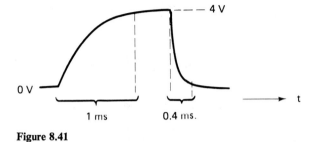

Figure 8.41

8.20 A 74121 OS is being triggered by a 1-MHz pulse waveform. If the OS is using $R_T = 10 \text{ k}\Omega$ and $C_T = 70 \text{ pF}$, determine the approximate value of the average current being supplied to the OS by the V_{CC} source. Use typical I_{CC} values.

8.21 For each waveform in Figure 8.42, determine *why* it will *not* reliably trigger a 7473 FF at its *CLK* input. (*Note:* The t_p of a clock pulse is the positive portion.)

8.22 A certain printed-circuit board contains the following TTL ICs: four 7473s, six

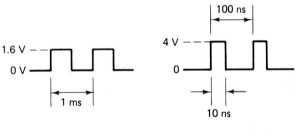

(a)   (b)

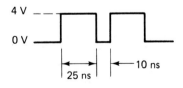

(c)

Figure 8.42

7400s, two 7404s, and two 74121s. How much capacitance should be distributed on the board to filter power-supply spikes due to TTL switching transients?

8.23   Why is it unwise to form the wired-AND connection with conventional TTL NAND gates such as the 7400 quad NAND?

8.24   What are the two advantages of using the wired-AND connection?

8.25   The 7409 TTL IC is a quad two-input AND with open collector outputs. Show how 7409s can be used to implement the operation $x = A \cdot B \cdot C \cdot D \cdot E \cdot F \cdot G \cdot H \cdot I \cdot J \cdot K \cdot M$.

8.26   Determine the logic expression for output X in Figure 8.43.

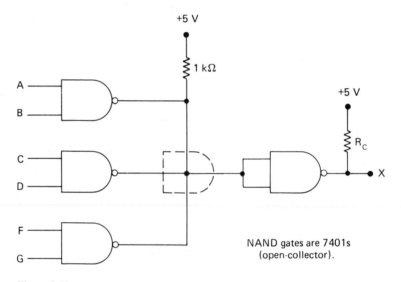

Figure 8.43

8.27   The logic circuit of Figure 8.43 is implemented using one 7401 chip. Can the same function be implemented using a single 7400 chip?

8.28   Determine a value for R_c in Figure 8.43 if output X is driving the clear inputs of four 7473 FFs.

8.29   Figure 8.44 shows a TRI-STATE NAND gate driving a conventional NAND gate. Determine the output X for the following conditions:
(a) $A = 0, B = 1$, ENABLE $= 0$

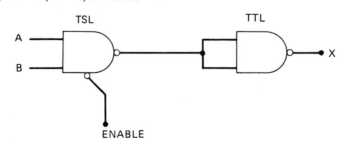

Figure 8.44

(b) $A = 1, B = 1$, ENABLE $= 0$
(c) $A = B = 1$; ENABLE $= 1$
(d) $A = B = 0$; ENABLE $= 1$

8.30   In the TRI-STATE circuit of Figure 8.20 it was assumed that the control inputs will go HIGH one at a time in sequence. What kind of circuit can be used to generate the control inputs?

8.31   Refer to the data sheet for the 7490 decade counter.
(a)   Show all the necessary pin connections for using the 7490 as a BCD counter.
(b)   Show how to connect two 7490s in cascade to count from 00 to 99.
(c)   Modify this arrangement to count from 00 to 59 (then recycle to 00).

8.32   What is the principal reason why ECL logic is faster than TTL?

8.33   What functions do the emitter followers perform in the ECL circuits?

8.34   Which of the following are advantages that ECL has over conventional TTL (7400 series)?
(a)   Lower power dissipation.
(b)   Shorter t_{pd}.
(c)   Greater fan-out.
(d)   Greater noise immunity.
(e)   Complementary outputs.
(f)   No current spikes during switching.

8.35   Show the bias voltages needed to turn ON an *N*-channel MOSFET.

8.36   Draw the circuit diagram of an N-MOS inverter using a supply voltage of 12 V. Determine the output voltages for $V_{\text{IN}} = 0$ V and $+12$ V.

8.37   The circuit of Figure 8.45 is an N-MOS logic gate. Determine what type of gate it is. Assume that $+16$ V $=$ logic 1, 0 V $=$ logic 0.

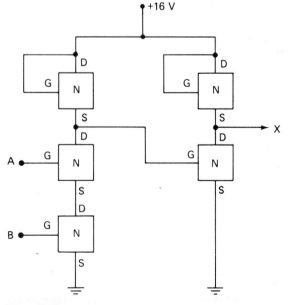

Figure 8.45

8.38 Which of the following are advantages that P-MOS and N-MOS logic have over TTL?
   (a) Greater packing density.
   (b) Greater operating speed.
   (c) Greater fan-out.
   (d) More suitable for LSI.
   (e) Lower P_D.
   (f) Complementary outputs.
   (g) Greater noise immunity.
   (h) Simpler fabrication process.
   (i) More SSI and MSI functions.
   (j) Uses lower supply voltage.

8.39 Which of the following are advantages that CMOS has over P-MOS/N-MOS?
   (a) Greater packing density.
   (b) Higher speed.
   (c) Greater fan-out.
   (d) Lower output impedance.
   (e) Simpler fabrication process.
   (f) More suited for LSI.
   (g) Lower P_D.
   (h) Uses transistors as only circuit element.
   (i) Lower input capacitance.

8.40 Repeat Question 8.39 for CMOS advantages over TTL.

8.41 Which of the following operating conditions will probably result in the lowest average P_D for a CMOS logic system? Explain.
   (a) $V_{DD} = 10$ V, switching frequency $f_{max} = 1$ MHz
   (b) $V_{DD} = 5$ V, $f_{max} = 10$ kHz
   (c) $V_{DD} = 10$ V, $f_{max} = 10$ kHz

8.42 Why is the CMOS switching speed greater than P-MOS/N-MOS?

8.43 Refer to the data sheet for the SCL 4016A, which is a quad bilateral switch (transmission gate). As the block diagram shows, there are four SPST switches (symbolized by circles with X inside) whose state (open or closed) is controlled by its corresponding control input. Each switch will conduct current in both directions, although one end is arbitrarily labeled "input" and the other "output."
   (a) Determine typical switch resistance in the ON (closed) state.
   (b) For what state of control input is the corresponding switch closed?
   (c) What is the maximum amplitude and frequency sine wave that can be switched from input to output?

8.44 The CMOS SCL 4009 is an example of a buffer circuit that can be used when CMOS is driving TTL. Determine how many 7400 series loads can be reliably driven by the 4009 when $V_{DD} = 5$ V. Repeat for 74LS00 loads.

8.45 Refer to the circuit diagram of Figure 8.46. The circuit is logically correct, but it contains at least seven design errors, where the specifications and characteristics of the ICs have not been used properly. Try to find all the errors.

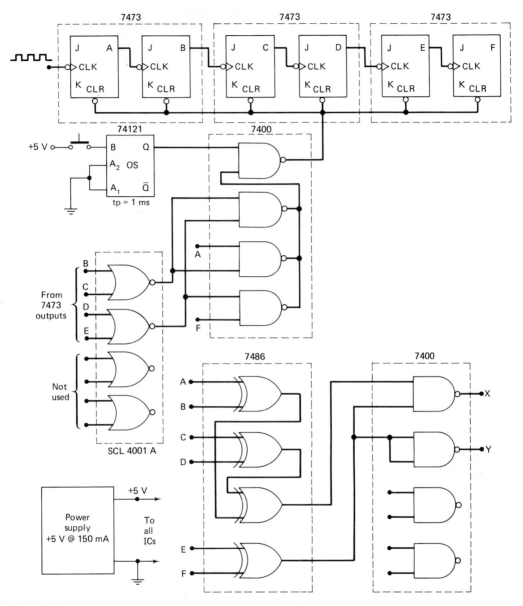

Figure 8.46

8.46 Which *bipolar* family is best suited for LSI?

8.47 Determine the logic function performed by the I²L circuit in Figure 8.47.

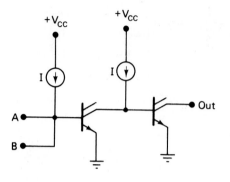

Figure 8.47

DATA-HANDLING LOGIC CIRCUITS

Digital systems always contain data or information that is in some form of binary code, and these data are continuously being operated on in some manner. In this chapter we will examine various types of digital circuits whose applications include (1) changing data from one form to another, (2) selecting one out of several groups of data, and (3) distributing data to one of several destinations. Many of the logic circuits that perform these functions are now available as ICs in the MSI category. For this reason we shall not concentrate on the *design* of these circuits but will investigate how they are used alone or in combination to perform various operations on digital data. Some of the operations that are discussed are decoding, encoding, code conversion, multiplexing and demultiplexing, and data busing.

9.1 DECODERS

A *decoder* is a logic circuit that converts an N-bit binary input code into M* output lines such that each output line will be activated for only one of the possible combinations of inputs. Figure 9.1 shows the general decoder diagram with N inputs and M outputs. Since each of the N inputs can be 0 or 1, there are 2^N possible input combinations or codes. For each of these input combinations only one of the M outputs will be active HIGH; all the other outputs are LOW. Many decoders are designed to produce active LOW outputs, where the selected output is LOW while all others are HIGH. This is always indicated by the presence of small circles on the output lines in the decoder diagram.

*N can be any integer and M is an integer that is less than or equal to 2^N.

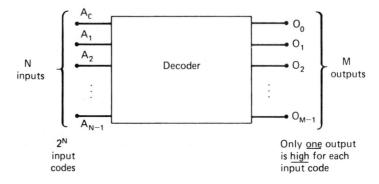

Figure 9.1   General decoder diagram.

Some decoders do not utilize all of the 2^N possible input codes but only certain ones. For example, a BCD-to-decimal decoder has a 4-bit input code which uses only the *ten* BCD code groups 0000 through 1001. Decoders of this type are often designed such that if any of the unused codes are applied to the input, *none* of the outputs will be activated.

In Chapter 7 we saw how decoders are used in conjunction with counters to detect the various states of the counter. In that application it was the FFs in the counter that provided the binary code inputs for the decoder. The same basic decoder circuitry is used no matter where the inputs come from. Figure 9.2 shows the circuitry for a decoder with three inputs and $2^3 = 8$ outputs. It uses all AND gates, so the outputs are active HIGH. For active LOW outputs, NAND gates would be used. Note that for a given input code, the only output which is active (HIGH) is the one corresponding to the decimal equivalent of the binary input code (e.g., output O_6 goes HIGH when CBA = $110_2 = 6_{10}$).

This decoder can be referred to in several ways. It can be called a *3-line-to-8-line decoder*, because it has three input lines and eight output lines. It could also be called a *binary-to-octal decoder or convertor* because it takes a 3-bit binary input code and produces a HIGH at one of the eight (octal) outputs corresponding to that code. It is also sometimes referred to as a *1-of-8 decoder*, because only 1 of the 8 outputs is activated at one time.

EXAMPLE 9.1   How many logic gates would be needed for a 4-line-to-16-line decoder with active LOW outputs, and what type of gate is required?

Solution: The answer is 16 NAND gates, as indicated by the 16 output lines and the requirement for active LOW outputs. All 16 outputs will be normally HIGH, and one will go LOW when its binary code is present at the inputs. This decoder can also be called a *1-of-16 decoder*.

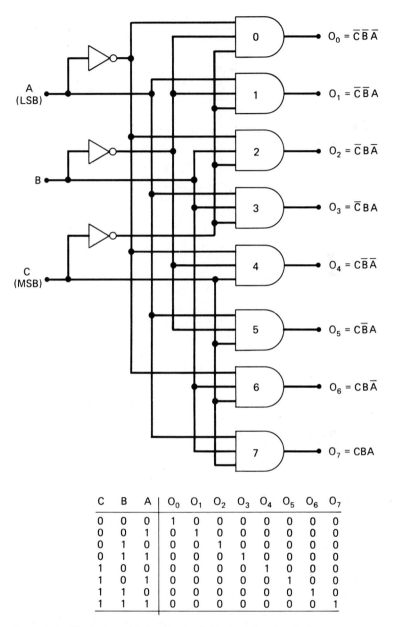

C	B	A	O_0	O_1	O_2	O_3	O_4	O_5	O_6	O_7
0	0	0	1	0	0	0	0	0	0	0
0	0	1	0	1	0	0	0	0	0	0
0	1	0	0	0	1	0	0	0	0	0
0	1	1	0	0	0	1	0	0	0	0
1	0	0	0	0	0	0	1	0	0	0
1	0	1	0	0	0	0	0	1	0	0
1	1	0	0	0	0	0	0	0	1	0
1	1	1	0	0	0	0	0	0	0	1

Figure 9.2 Binary-to-octal decoder (or 3-line-to-8-line decoder).

Open-Collector Outputs

Some decoders such as the TTL 7445 have open-collector outputs. For these types of decoders each output is normally OFF, providing a high-resistance path to ground; the output becomes a low-resistance path to ground when its corresponding input code is applied to the decoder inputs. These open-collector outputs are usually designed to operate at higher current and voltage limits than a normal TTL device. For example, the 7445 outputs can sink up to 80 mA in the LOW state and can tolerate voltages up to 30 V in the HIGH state. This makes them suitable for directly driving loads such as indicator lights or relays. This is illustrated in Figure 9.3, where the open-collector output is connected to the 12-V supply through a 12-V 500-Ω relay. When the transistor is OFF, the relay is de-energized and the transistor collector is at 12 V. When the output transistor is ON, almost the full 12 V is across the relay, producing 24 mA of collector current.

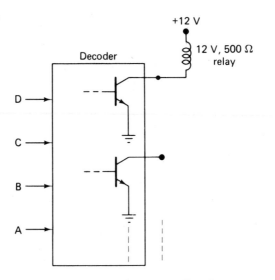

Figure 9.3 Open-collector outputs allow direct connection to high-voltage, high-current loads.

ENABLE Inputs

Some decoders have one or more ENABLE inputs that are used to control the operation of the decoder. For example, refer to the decoder in Figure 9.2 and visualize having a common ENABLE line connected to a fourth input of each gate. With this ENABLE line held HIGH the decoder will function normally and the A, B, C input code will determine which output is HIGH. With ENABLE held LOW, however, the outputs will be forced to the LOW state regardless of the levels at the A, B, C inputs. Thus, the decoder is ENABLED only if ENABLE is HIGH.

Figure 9.4 shows the logic symbol for a 3-line-to-8-line decoder that has two ENABLE inputs, *E1* and *E2*. They are both active-LOW, as indicated by the circles. This means that *both* E1 and E2 must be LOW for the decoder outputs to be operative. If either (or both) is HIGH, the outputs will all be in their inactive HIGH state, regardless of the *A, B, C* inputs.

It might be worth noting at this time that active-LOW inputs and outputs are often represented on a logic diagram with an inversion overbar. For example, the ENABLE inputs in Figure 9.4 could be represented as $\overline{E1}$ and $\overline{E2}$. Likewise, the outputs could be represented as $\bar{O}_0, \bar{O}_1, \bar{O}_2$, and so on. The inversion bar is just another means of indicating that these logic signals are active-LOW.

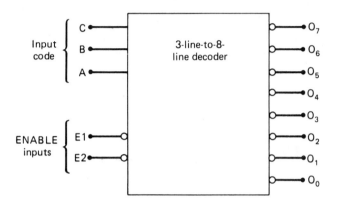

Figure 9.4 Decoder with two ENABLE inputs.

There are several reasons for including ENABLE inputs on decoder circuits. One is to allow the selection of one of several decoders in a system by means of signals applied to the various ENABLE inputs. Another is to allow the decoder to be used as a *demultiplexer*, which we will discuss shortly. Still another is to provide a means for turning off the decoder while the input code is in a transition state, thereby eliminating the possibility of decoding glitches at the decoder outputs. This latter application was described in Section 7.11 in conjunction with decoding a ripple counter. There we called it a "strobe" input rather than ENABLE. Both terms are used for this particular type of input.

9.2 BCD-TO-DECIMAL DECODERS

A decoder that takes a 4-bit BCD input code and produces 10 outputs corresponding to the decimal digits is called a *BCD-to-decimal decoder* (or *converter*). Figure 9.5 shows the basic logic arrangement using AND gates. Each output goes HIGH when its corresponding BCD code group occurs. For example, O_5 will go HIGH only when 0101 (BCD for 5) occurs at the DCBA inputs, respectively. This decoder is also called a *4-line-to-10-line decoder* or a *1-of-10 decoder*.

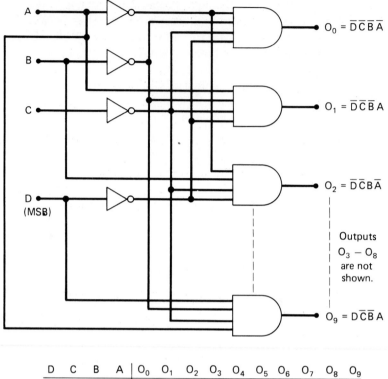

D	C	B	A	O_0	O_1	O_2	O_3	O_4	O_5	O_6	O_7	O_8	O_9
0	0	0	0	1	0	0	0	0	0	0	0	0	0
0	0	0	1	0	1	0	0	0	0	0	0	0	0
0	0	1	0	0	0	1	0	0	0	0	0	0	0
0	0	1	1	0	0	0	1	0	0	0	0	0	0
0	1	0	0	0	0	0	0	1	0	0	0	0	0
0	1	0	1	0	0	0	0	0	1	0	0	0	0
0	1	1	0	0	0	0	0	0	0	1	0	0	0
0	1	1	1	0	0	0	0	0	0	0	1	0	0
1	0	0	0	0	0	0	0	0	0	0	0	1	0
1	0	0	1	0	0	0	0	0	0	0	0	0	1
1	0	1	0										
1	0	1	1										
1	1	0	0				(All outputs = 0)						
1	1	0	1										
1	1	1	0										
1	1	1	1										

Figure 9.5 Logic diagram for BCD-to-decimal decoder.

This decoder is an example of one that does not use all of the input combinations. The code groups 1010 through 1111 are illegal for BCD and do not produce any active outputs. In the TTL MSI family the 7442 is a BCD-to-decimal decoder with active LOW outputs.

BCD-to-Decimal Decoder/Driver

The decoder of Figure 9.5 can be modified for use with decimal displays by including a driver transistor on each output. This is illustrated in Figure 9.6, where the AND gate for the O_7 output feeds a driver transistor. This is an open-collector arrangement, where the transistor is able to handle relatively high current and voltage. The dashed lines in the figure show a typical connection for driving a 24-V incandescent display lamp. Note that the new output becomes $\bar{O}_7$, since the transistor acts as an inverter.

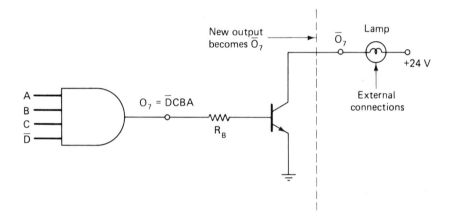

Figure 9.6 Addition of driver transistor to decoder output.

A BCD-to-decimal decoder/driver has a high-voltage transistor on each output so that it can drive lamps, relays, etc. One of its major applications is in driving cold-cathode display tubes (nixie tubes), which display the decimal numerals corresponding to the BCD input code. Figure 9.7 shows a typical arrangement where a BCD-to-decimal decoder/driver is driving a nixie-tube readout. The DCBA inputs usually come from a counter or a storage register. For a given input BCD code, one of the outputs $\bar{O}_0 - \bar{O}_9$ will be LOW (saturated driver transistor). This LOW essentially *grounds* the appropriate filament of the nixie tube so that current flows from the 170-V supply, through the tube, and to this filament, causing it to glow in the shape of the decimal digit corresponding to the input code. In this way the tube displays the decimal digit that is presented at the decoder inputs in BCD form. For any input codes greater than 1001, all the decoder outputs will be HIGH (open driver transistor) and none of the tube filaments will glow.

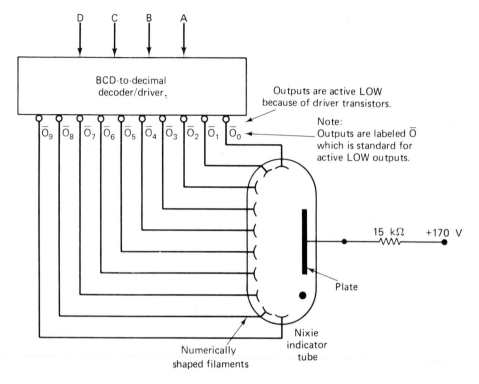

Figure 9.7   Typical application of a BCD-to-decimal decoder/driver.

In the TTL/MSI family, the 74141 is one of the several available BCD-to-decimal decoder/drivers.

9.3   BCD-TO-7-SEGMENT DECODER/DRIVERS

Many numerical displays use a 7-segment configuration [Figure 9.8(a)] to produce the decimal characters 0–9 and sometimes the hex characters A–F. Each segment is made up of a material that emits light when current is passed through it. Most commonly used materials include light-emitting diodes (LEDs) and incandescent filaments. Figure 9.8(b) shows the patterns of segments which are used to display the various digits. For example, to display "6," segments *c, d, e, f,* and *g* are bright while segments *a* and *b* are dark.

A *BCD-to-7-segment decoder/driver* is used to take a 4-bit BCD input and provide the outputs that will pass current through the appropriate segments to display the decimal digit. The logic for this decoder is more complicated than those we looked at previously because each output is activated for more than one combination of inputs. For example, the *e* segment must be activated for any of the

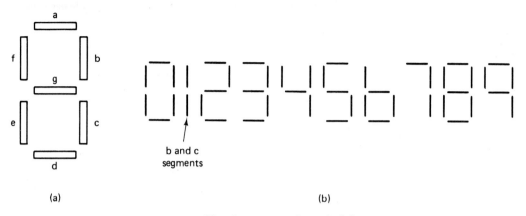

(a)   (b)

Figure 9.8 (a) 7-segment arrangement; (b) active segments for each digit.

digits 0, 2, 6, and 8, which means whenever any of the codes 0000, 0010, 0110, or 1000 occurs. In fact, the following expressions can be used to define the active cases for each segment:

$$a = 0 + 2 + 3 + 5 + 7 + 8 + 9$$
$$b = 0 + 1 + 2 + 3 + 4 + 7 + 8 + 9$$
$$c = 0 + 1 + 3 + 4 + 5 + 6 + 7 + 8 + 9$$
$$d = 0 + 2 + 3 + 5 + 6 + 8$$
$$e = 0 + 2 + 6 + 8$$
$$f = 0 + 4 + 5 + 6 + 8 + 9$$
$$g = 2 + 3 + 4 + 5 + 6 + 8 + 9$$

A truth table can be set up with four inputs (the 4 bits of the BCD input code) and seven outputs (a, b, c, d, e, f, g). Then the expressions above can be used to determine the cases in the table where each output should be HIGH or LOW. The unused input cases 1010 through 1111 can be treated so as to produce no active output segments (blank display).* Once the truth table is set up, the Boolean expressions for each output can be obtained, simplified, and implemented using the techniques we learned in Chapters 2 and 3. This procedure will be left as an exercise at the end of the chapter.

Figure 9.9 shows a BCD-to-7-segment decoder/driver (TTL 7446 or 7447) being used to drive a 7-segment LED readout. Each segment consists of one or two LEDs. The anodes of the LEDs are all tied to $V_{\rm cc}$ (+5 V). The cathodes of the LEDs are connected through current-limiting resistors to the appropriate outputs of the decoder/driver. The decoder/driver has active LOW outputs which

*Many BCD-to-7-segment decoder/drivers do not do this but rather have some specific segment patterns occur for input codes from 1010 to 1111.

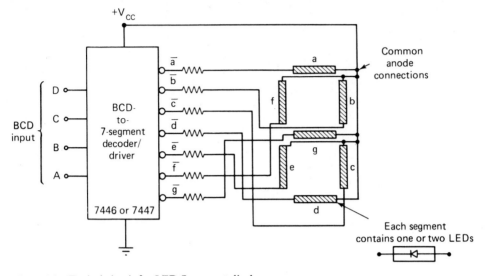

Figure 9.9   Typical circuit for LED 7-segment display.

are open-collector driver transistors that can sink a fairly large current. This is because LED readouts may require 10 mA to 40 mA per segment, depending on their type and size.

To illustrate the operation of this circuit, let us suppose that the BCD input is $D = 0$, $C = 1$, $B = 0$, $A = 1$, which is BCD for 5. With these inputs the decoder/driver outputs $\bar{a}, \bar{f}, \bar{g}, \bar{c}$, and $\bar{d}$ will be driven LOW (connected to ground), allowing current to flow through the a, f, g, c, and d LED segments and thereby displaying the numeral 5. The $\bar{b}$ and $\bar{e}$ outputs will be HIGH (open), so LED segments b and e cannot conduct.

This same arrangement can be used for incandescent 7-segment readouts except that the current limiting resistors are not required. However, the lifetime of these incandescent readouts is somewhat shorter than LEDs. The LED readouts can operate without the resistors if a lower voltage supply is used for the anodes. This supply will typically have to be 2–3 V, depending on the type of LED readout.

9.4 ENCODERS

A decoder accepts an *N*-bit input code and produces a HIGH (or LOW) at *one and only one* output line. In other words, we can say that a decoder identifies, recognizes, or detects a particular code. The opposite of this decoding process is called *encoding* and is performed by a logic circuit called an *encoder*. An encoder has a number of input lines, only one of which is activated at a given time, and produces an *N*-bit *output* code, depending on which input is activated. Figure 9.10 is the general diagram for an encoder with *M* inputs and *N* outputs. Here the inputs are active-HIGH which means they are normally LOW.

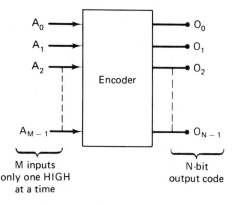

Figure 9.10 General encoder diagram.

We saw that a *binary-to-octal decoder* accepts a 3-bit binary input code and activates one of eight output lines. An *octal-to-binary encoder* operates in the opposite manner. It accepts eight input lines and produces a 3-bit binary output code. Its circuitry is shown in Figure 9.11. It is assumed that only *one* of the input lines is made HIGH at one time, so there are only *eight* possible input conditions. The circuit is designed so that when A_0 is HIGH, the binary code 000 is generated at the output; when A_1 is HIGH, the binary code 001 is generated; when A_2 is HIGH, the code 010 is generated; and so on (see the accompanying truth table). The circuit design is very simple since it only involves looking at each output bit and determining for which input cases that bit is HIGH and then ORing the results. For example, the truth table shows that O_0 (LSB of output code) must be 1 whenever inputs A_1, A_3, A_5, or A_7 are HIGH. Thus, we have

$$O_0 = A_1 + A_3 + A_5 + A_7$$

The other outputs are designed accordingly. This design is made simple by the fact that only *eight* out of the total of 2^8 possible input conditions are used. If more than one input is made HIGH at a given time, the output results will be erroneous. If none of the inputs are HIGH, the outputs will read 000. The encoder of Figure 9.11 is also called an *8-line-to-3-line encoder*.

EXAMPLE 9.2 Describe the structure and operation of a decimal-to-BCD encoder with active LOW inputs.

Solution: This encoder takes 10 input lines, only one of which will be LOW and produces a 4-bit BCD output code. Since there are four outputs, the circuitry contains four gates. The gates used are NAND gates because the outputs are to be normally LOW

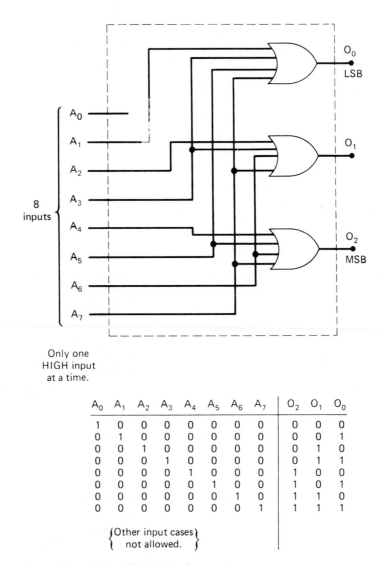

Only one
HIGH input
at a time.

A_0	A_1	A_2	A_3	A_4	A_5	A_6	A_7	O_2	O_1	O_0
1	0	0	0	0	0	0	0	0	0	0
0	1	0	0	0	0	0	0	0	0	1
0	0	1	0	0	0	0	0	0	1	0
0	0	0	1	0	0	0	0	0	1	1
0	0	0	0	1	0	0	0	1	0	0
0	0	0	0	0	1	0	0	1	0	1
0	0	0	0	0	0	1	0	1	1	0
0	0	0	0	0	0	0	1	1	1	1

$\left\{ \begin{array}{c} \text{Other input cases} \\ \text{not allowed.} \end{array} \right\}$

Figure 9.11   Octal-to-binary encoder.

and go HIGH when any of their inputs is made LOW. Figure 9.12 shows the block diagram for this encoder. When one of the inputs is made LOW, the 4-bit code corresponding to this input will appear at the outputs. For example, if $\bar{A}_5$ is held LOW, the outputs will be $O_3 O_2 O_1 C_0 = 0101$.

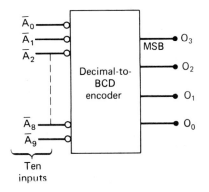

Figure 9.12 Decimal-to-BCD encoder.

9.5 SWITCH ENCODERS

Many digital systems use switches or keys to enter data into the system. One of the most common examples of this are electronic calculators which use keyboard entry. Another is the teletype keyboard entry used in many computer systems. In these and similar cases the actuation of the switches or keys must be encoded in the proper code required by the system. Figure 9.13 shows an example of a decimal-to-BCD encoder (TTL 74147) used with 10 switches that might be keyboard switches, each representing one of the decimal digits 0–9. All the switches are normally open (N. O.), so all the NAND*-gate inputs are HIGH thereby producing all LOW outputs. When one of the switches is depressed, it sends a LOW to the appropriate NAND gates so that the correct BCD output code is produced. For example, when SW7 is depressed, it puts a LOW on NAND gates 0, 1, and 2, so the outputs O_0, O_1, and O_2 all go HIGH while O_3 remains LOW; thus, the output code is 0111. The reader can verify that the other switches produce their proper BCD code when activated.

The switch encoder of Figure 9.13 can be used whenever BCD data have to be manually entered into a digital system. A prime example of this would be in an electronic calculator, where the operator depresses several keyboard switches in succession to enter a decimal number. In a simple, basic calculator the BCD code for each decimal digit is entered into a 4-bit storage register. In other words, when the first key is depressed, the BCD code for that digit is sent to a 4-bit FF register; when the second switch is depressed, the BCD code for that digit is sent to *another* 4-bit FF register, and so on. Thus, a calculator that can handle eight digits will have eight 4-bit registers to store the BCD codes for these digits. Each 4-bit register

*We are using the OR-equivalent representation for NAND gates here because the outputs are active-HIGH and any LOW input will produce the HIGH output state.

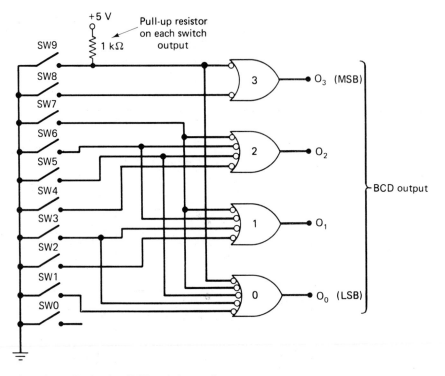

Figure 9.13 Decimal-to-BCD switch encoder.

drives a decoder/driver and numerical display so that the eight-digit number can be displayed.

The operation described above can be accomplished with the circuit in Figure 9.14. This circuit will take three decimal digits entered from the keyboard in sequence, encode them in BCD, and store the BCD in FF output registers. The encoder being used is the one from Figure 9.13, which was previously described. The 12 D-type FFs Q_0–Q_{11} are used to receive and store the BCD codes for the digits. Q_8–Q_{11} stores the BCD code for the most significant digit (MSD), which is the first one entered on the keyboard. Q_4–Q_7 stores the second entered digit and Q_0–Q_3 stores the third entered digit. The X, Y, and Z FFs form a ring counter (Chapter 7) which controls the transfer of data from the encoder outputs to the appropriate output register. The OR gate produces a HIGH output any time one of the keys is depressed. This output may be affected by switch contact bounce, which would produce several pulses before settling down to the HIGH state. The OS is used to neutralize the switch bounce by triggering on the first positive transition from the OR gate and remaining HIGH for 20 ms, well past the time duration of the switch bounce. The OS output clocks the ring counter.

The circuit operation is described as follows for the case where the decimal number 309 is being entered:

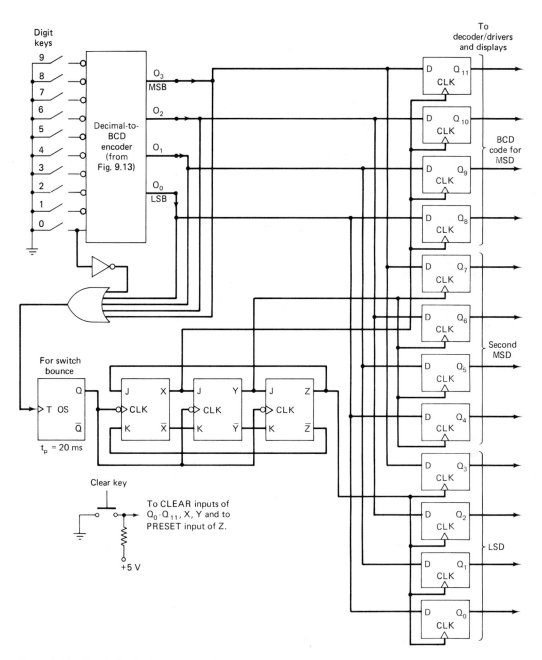

Figure 9.14 Circuit for keyboard entry of three-digit number into storage register.

1. The CLEAR key is depressed. This clears all the Q_0–Q_{11} storage FFs to 0. It also clears FFs X and Y and presets FF Z to 1, so the ring counter begins in the 001 state.

2. The CLEAR key is released and the "3" key is depressed. The encoder outputs now become $O_3 = 0, O_2 = 0, O_1 = 1$, and $O_0 = 1$; that is, 0011, the BCD code for 3. These binary values are sent to the D inputs of the three 4-bit output registers.

3. The OR output goes HIGH (since O_1 and O_0 have gone HIGH) and triggers the OS output $Q = 1$ for 20 ms. After 20 ms Q returns LOW and clocks the ring counter to the 100 state (X goes HIGH). The positive transition at X is fed to the CLK inputs of FFs Q_8–Q_{11}, so the encoder outputs are transferred to these FFs. That is, $Q_{11} = 0, Q_{10} = 0, Q_9 = 1$, and $Q_8 = 1$. Note that FFs Q_0–Q_7 are not affected because their *CLK* inputs have not received a positive transition.

4. The "3" key is released and the OR gate output returns LOW. The "0" key is then depressed. This produces encoder outputs of 0000, which are fed to the inputs of the storage FFs.

5. The OR output goes HIGH in response to the "0" key (note the inverter) and triggers the OS for 20 ms. After 20 ms the ring counter shifts to the 010 state (Y goes HIGH). The positive transition at Y is fed to the *CLK* inputs of Q_4–Q_7 and transfers the encoder outputs 0000 to these FFs. Note that FFs Q_0–Q_3 and Q_8–Q_{11} are not affected by the Y transition.

6. The "0" key is released and the OR output returns LOW. The "9" key is depressed, producing encoder outputs 1001 which are fed to the storage FFs.

7. The OR output goes HIGH again, triggering the OS, which in turn clocks the ring counter to the 001 state (Z goes HIGH). The positive transition at Z is fed to the *CLK* inputs of Q_0–Q_3 and transfers the 1001 from the encoder into these FFs. The other storage FFs are unaffected.

8. At this point the storage register contains 0011 0000 1001, beginning with Q_{11}. This is the BCD code for 309. These register outputs feed decoder/drivers which drive appropriate displays for indicating the decimal digits 309.

9. The storage FF outputs are also fed to other circuits in the system. In a calculator, for example, these outputs would be sent to the arithmetic section to be processed.

Several problems at the end of the chapter will deal with some other aspects of this circuit.

9.6 MULTIPLEXERS (DATA SELECTORS)

A *multiplexer* or *data selector* is a logic circuit that accepts several data inputs and allows only *one* of them at a time to get through to the output. The routing of the desired data input to the output is controlled by SELECT inputs (sometimes referred to as ADDRESS inputs). Figure 9.15 shows the symbol for a general multiplexer (MUX). In this diagram the inputs and outputs are drawn as large arrows to indicate that they may be one or more lines.

The multiplexer acts like a digitally controlled multiposition switch* where the digital code applied to the SELECT inputs controls which data inputs will be switched to the output. For example, output Z will equal data input I_0 for some particular SELECT input code; Z will equal I_1 for another particular SELECT input code; and so on. Stated another way, a multiplexer selects 1 out of N input data sources and transmits the selected data to a single output channel. This is called *multiplexing*.

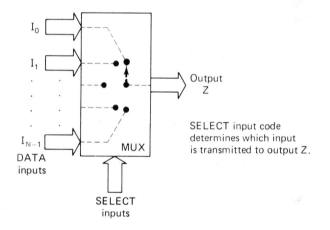

Figure 9.15 Symbol for a digital multiplexer (MUX).

Basic Two-Input Multiplexer

Figure 9.16 shows the logic circuitry for a two-input (or two-channel) multiplexer with data inputs A and B and SELECT input S. The logic level applied to the S input determines which AND gate is enabled so that its data input passes through the OR gate to output Z. Looking at it another way, the Boolean expression for the output is

$$Z = AS + B\bar{S}$$

*This is indicated by the dotted lines in Figure 9.15. These lines are not part of the MUX symbol, but are shown just for illustration of the multiplexing function.

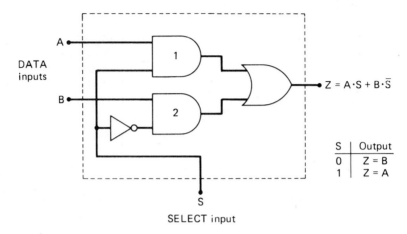

Figure 9.16   Two-input multiplexer.

With $S = 0$, this expression becomes

$$Z = A \cdot 0 + B \cdot 1$$
$$= B$$

which indicates that Z will be identical to input signal B which can be a fixed logic level or a time-varying logic signal. With $S = 1$, the expression becomes

$$Z = A \cdot 1 + B \cdot 0 = A$$

showing that output Z will be identical to input signal A.

EXAMPLE 9.3   Show how multiplexers of the type shown in Figure 9.16 can be used to take two 3-bit binary numbers (X_2, X_1, X_0 and Y_2, Y_1, Y_0) and transmit one or the other number to outputs Z_2, Z_1, and Z_0, depending on an input select level.

Solution: Figure 9.17 shows 3 two-input multiplexers being used to perform the desired operation. Note that the S inputs of each multiplexer are connected together as one common select input. When $S = 1$, the X inputs to each individual multiplexer are routed through to the Z outputs. When $S = 0$, the Y inputs are routed through to the outputs.

Four-Channel Multiplexer

The same basic idea can be used to form the four-input multiplexer shown in Figure 9.18. Here there are four inputs, which are selectively transmitted to the output based on the four possible combinations of the $S_1 S_0$ select inputs. Each

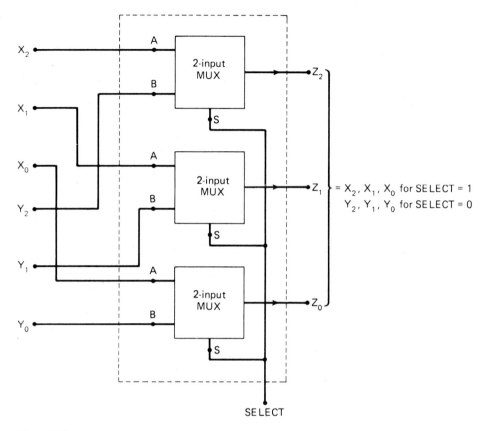

SELECT

Figure 9.17

data input is gated with a different combination of select input levels. I_0 is gated with $\bar{S}_1\bar{S}_0$ so that I_0 will pass through its AND gate to output Z only when $S_1 = 0$ and $S_0 = 0$. The table in the figure gives the outputs for the other three input select codes.

Two, 4, 8, and 16 input multiplexers are readily available in the TTL and CMOS logic families. These basic ICs can be combined for multiplexing a larger number of inputs.

EXAMPLE 9.4 Draw the block-diagram symbol and truth table for an eight-input multiplexer showing all inputs.

Solution: Figure 9.19 shows the diagram for this multiplexer, which has eight data inputs I_0–I_7. To select one of eight input channels, a 3-bit select code is needed ($2^3 = 8$). Thus, S_2, S_1, and S_0 are the select inputs that control the output according to the truth table shown. The TTL 74151 is a multiplexer of this type.

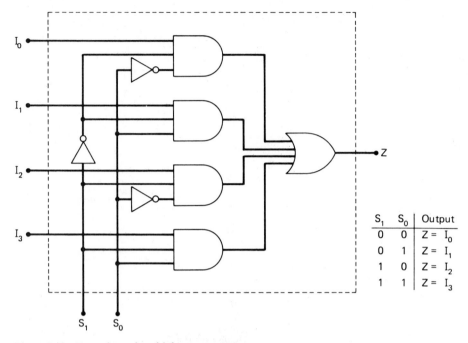

S_1	S_0	Output
0	0	$Z = I_0$
0	1	$Z = I_1$
1	0	$Z = I_2$
1	1	$Z = I_3$

Figure 9.18   Four-channel multiplexer.

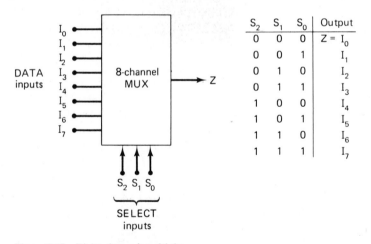

S_2	S_1	S_0	Output
0	0	0	$Z = I_0$
0	0	1	I_1
0	1	0	I_2
0	1	1	I_3
1	0	0	I_4
1	0	1	I_5
1	1	0	I_6
1	1	1	I_7

Figure 9.19   Eight-channel multiplexer.

Two-Channel, 4-Bit Multiplexer

A very useful type of multiplexer is the two-channel, 4-bit multiplexer shown in Figure 9.20. This multiplexer operates basically the same as the multiplexer of Figure 9.16 except that the A and B inputs and the Z output are 4-bit data groups.

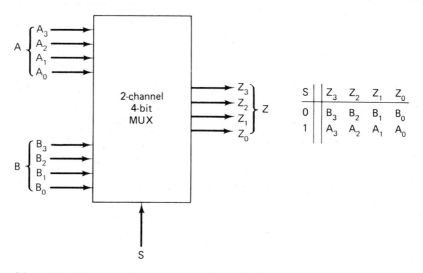

Figure 9.20 Diagram of a two-channel, 4-bit multiplexer.

The four output bits Z_3, Z_2, Z_1, and Z_0 will match either the four A inputs or the four B inputs, depending on the select input S. The TTL version of this multiplexer is the 74157.

9.7 MULTIPLEXER APPLICATIONS

Multiplexer circuits find numerous and varied applications in digital systems of all types. These applications include data selection, data routing, operation sequencing, parallel-to-serial conversion, waveform generation, and logic-function generation. We shall look at some of these applications here and several more in the problems at the end of the chapter.

Data Routing

Multiplexers usually route data from one of several sources to one destination. One typical application uses the multiplexer of Figure 9.20 to select and display the contents of either of two BCD counters using a *single* set of decoder/drivers and LED displays. The circuit arrangement is shown in Figure 9.21.

When the COUNTER SELECT line is held HIGH, the BCD output states of counter 2 are transmitted through to the multiplexer outputs and onto the readouts to be displayed. When the COUNTER SELECT is held LOW, the states of counter 1 are displayed. Thus, the decimal contents of either counter can be displayed on the same readouts under the control of the COUNTER SELECT input.

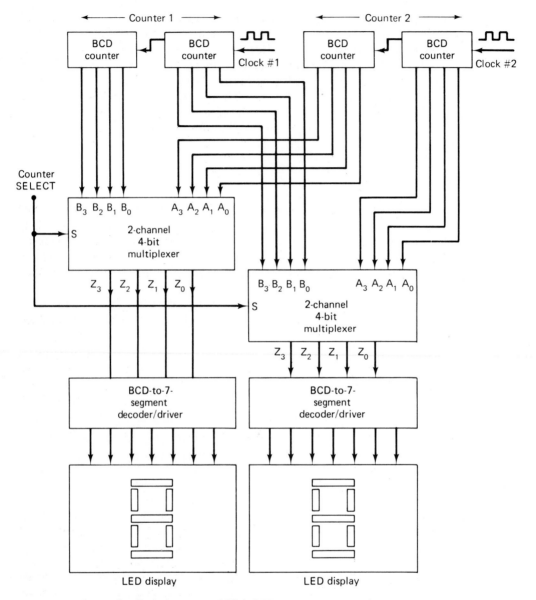

Figure 9.21   System for displaying two multidigit BCD counters one at a time.

The purpose of the multiplexing technique, as it is used here, is to *time share* the decoder/drivers and display circuits between the two counters rather than have a separate set of decoder/drivers and displays for each counter. This results in a significant savings in the number of wiring connections, especially when more BCD stages are added to each counter. Even more importantly, it represents a significant decrease in power consumption, because decoder/drivers and LED readouts typically draw relatively large amounts of current from the V_{CC} supply. Of course, this technique has the limitation that only one counter contents can be displayed at a time. However, in many applications this is not a drawback. A mechanical switching arrangement could have been used to perform the function of switching first one counter and then the other to the decoder/drivers and displays, but the number of required switch contacts, the complexity of wiring, and the physical size could all be disadvantages over the completely logic method of Figure 9.21.

Parallel-to-Serial Conversion

Many digital systems process binary data in parallel form (all bits simultaneously) because it is faster. When these data are to be transmitted over relatively long distances, however, the parallel arrangement is undesirable because it requires a large number of transmission lines. For this reason, binary data or information that are in parallel form are often converted to serial form before being transmitted to a remote destination. One method for performing this *parallel-to-serial conversion* uses a multiplexer, as illustrated in Figure 9.22.

The data are present in parallel form at the outputs of the X register and are fed to the eight-channel multiplexer. A 3-bit (MOD-8) counter is used to provide the select code bits $S_2 S_1 S_0$ so that they cycle through from 000 to 111 as clock pulses are applied. In this way, the output of the multiplexer will be X_0 during the first clock period, X_1 during the second clock period, and so on. The output Z is a waveform which is a serial representation of the parallel input data. The waveforms in the figure are for the case where $X_7 X_6 X_5 X_4 X_3 X_2 X_1 X_0 = 10110101$. This conversion process takes a total of eight clock cycles. Note that X_0 (the LSB) is transmitted first and the X_7 (MSB) is transmitted last.

Operation Sequencing

The circuit of Figure 9.23 uses an eight-channel multiplexer as part of a control sequencer that steps through seven steps, each of which actuates some portion of the physical process being controlled. This process could, for example, be a large high-temperature oven which is energized by seven different heaters, which must be activated one at a time. The circuit also uses a 3-line-to-8-line decoder (from Figure 9.2) and a MOD-8 binary counter. The operation is described as follows:

1. Initially, the counter is in the 000 state. The counter outputs are fed to the select inputs of the multiplexer and to the inputs of the decoder. Thus, the decoder output $O_0 = 1$ and the others are all 0, so all the ACTUATOR inputs of the process are LOW. The SENSOR outputs of the process all start out LOW. The multiplexer output $Z = I_0 = 0$ since the S inputs are 000.

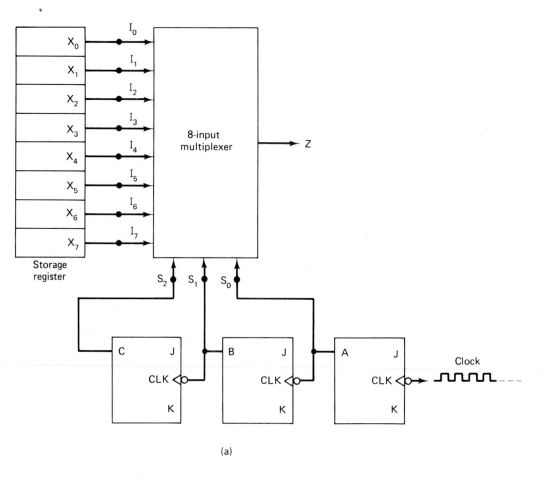

(a)

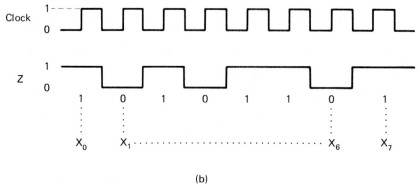

(b)

Figure 9.22 (a) Parallel-to-serial converter; (b) waveforms for $X_7 X_6 X_5 X_4 X_3 X_2 X_1 X_0 = 10110101$.

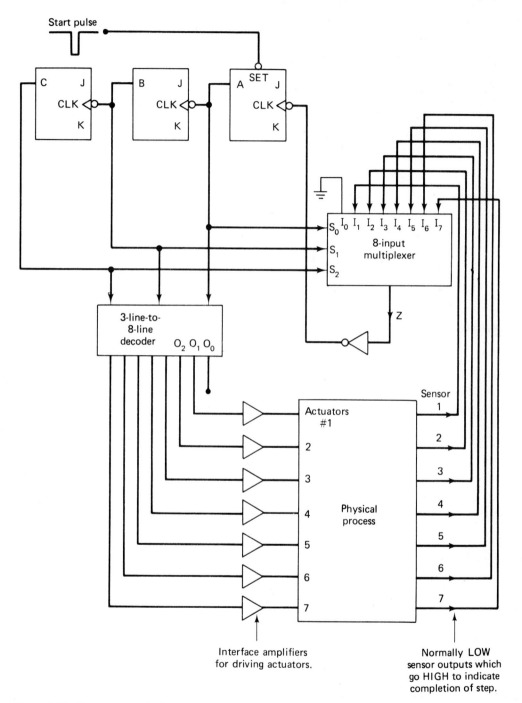

Figure 9.23 Seven-step control sequencer.

2. The start pulse initiates the sequencing operation by setting FF *A* HIGH, bringing the counter to the 001 state. This causes decoder output O_1 to go HIGH, thereby activating actuator 1, which is the first step in the process (it could be the turning on of a heater).

3. Some time later SENSOR output 1 goes HIGH, indicating the completion of the first step (it could be the reaching of a certain temperature level). This HIGH is now present at the I_1 input of the multiplexer and thereby reaches the *Z* output since the select code from the counter is 001.

4. The HIGH at *Z* is inverted to a LOW and fed to the *CLK* of FF *A*. This negative transition advances the counter to the 010 state.

5. Decoder output O_2 now goes HIGH, activating actuator 2, which is the second step in the process. *Z* now equals I_2 (select code is 010), which is SENSOR output 2, which is still LOW.

6. When the second process step is complete, SENSOR output 2 goes HIGH, producing a HIGH at *Z* and advancing the counter to 011.

7. This same action is repeated for each of the other steps. When the seventh step is completed, SENSOR output 7 goes HIGH, causing the counter to go from 111 to 000, where it will remain until another start pulse reinitiates the sequence.

9.8 DEMULTIPLEXERS (DATA DISTRIBUTORS)

A multiplexer takes several inputs and transmits *one* of them to the output. A *demultiplexer* performs the reverse operation; it takes a single input and distributes it over several outputs. Figure 9.24 shows the general diagram for a demultiplexer

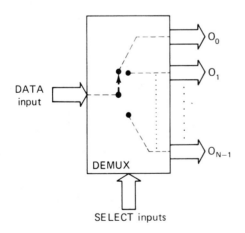

DATA input is transmitted only to one of the outputs as determined by select input code

Figure 9.24 General demultiplexer symbol.

(DEMUX). The large arrows for inputs and outputs can represent one or more lines. The SELECT input code determines to which output the DATA input will be transmitted. In other words, the demultiplexer takes one input data source and selectively distributes it to 1 of N output channels just like a multi-position switch.

1-Line-to-8-Line Demultiplexer

Figure 9.25 shows the logic diagram for a demultiplexer that distributes one input line to eight output lines. The single data input line I is connected to all eight AND gates, but only one of these gates will be enabled by the SELECT input lines. For example, with $S_2S_1S_0 = 000$, only AND gate 0 will be enabled, and data input I will appear at output O_0. Other SELECT codes cause input I to reach the other outputs. The truth table summarizes the operation.

The demultiplexer circuit of Figure 9.25 is very similar to the 3-line-to-8-line decoder circuit in Figure 9.2 except that a fourth input (I) has been added to each gate. It was pointed out earlier that many IC decoders have an ENABLE input, which is an extra input added to the decoder gates. This type of decoder chip can therefore be used as a demultiplexer, with the binary code inputs (e.g., A, B, C in Figure 9.2) serving as the SELECT inputs and the ENABLE input serving as the data input I. For this reason, IC manufacturers often call this type of device a *decoder/demultiplexer*.

Clock Demultiplexer

Many applications of the demultiplexing principle are possible. Figure 9.26 shows the demultiplexer of Figure 9.25 being used as a *clock demultiplexer*. Under control of the SELECT lines, the clock signal is routed to the desired destination. For example, with $S_2S_1S_0 = 000$, the clock signal applied to I will appear at output O_0. With $S_2S_1S_0 = 101$, the clock will appear at O_5.

Synchronous Data Transmission System

Multiplexing and demultiplexing are often used together in systems where digital data are being transmitted over relatively long distances. As we know, it is prohibitive to transmit large numbers of *bits* of data in parallel. It is much more feasible to transmit the data in serial one bit at a time. Figure 9.27 shows a system used for the synchronous transmission of four 4-bit data words from a transmitter to a remote receiver. The four data words are stored in registers A, B, C, and D of the transmitter; these registers are connected as 4-bit *circulating* shift registers* with a common SHIFT (clock) input that causes shifting from left to right on

*Like the ring counter of Figure 7.19.

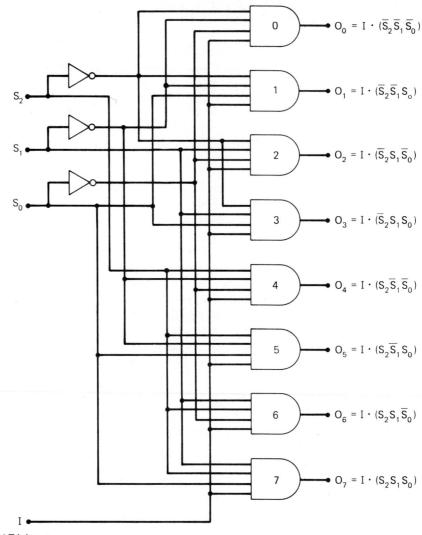

$O_0 = I \cdot (\overline{S_2}\,\overline{S_1}\,\overline{S_0})$

$O_1 = I \cdot (\overline{S_2}\,\overline{S_1}\,S_0)$

$O_2 = I \cdot (\overline{S_2}\,S_1\,\overline{S_0})$

$O_3 = I \cdot (\overline{S_2}\,S_1\,S_0)$

$O_4 = I \cdot (S_2\,\overline{S_1}\,\overline{S_0})$

$O_5 = I \cdot (S_2\,\overline{S_1}\,S_0)$

$O_6 = I \cdot (S_2\,S_1\,\overline{S_0})$

$O_7 = I \cdot (S_2\,S_1\,S_0)$

I •

DATA input

SELECT code			Outputs							
S_2	S_1	S_0	O_7	O_6	O_5	O_4	O_3	O_2	O_1	O_0
0	0	0	0	0	0	0	0	0	0	I
0	0	1	0	0	0	0	0	0	I	0
0	1	0	0	0	0	0	0	I	0	0
0	1	1	0	0	0	0	I	0	0	0
1	0	0	0	0	0	I	0	0	0	0
1	0	1	0	0	I	0	0	0	0	0
1	1	0	0	I	0	0	0	0	0	0
1	1	1	I	0	0	0	0	0	0	0

Figure 9.25 One-line-to-8-line demultiplexer.

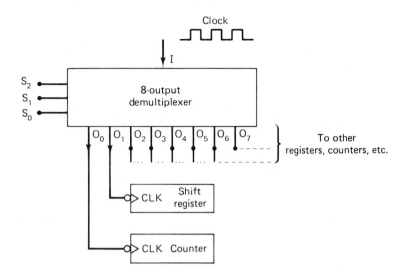

Figure 9.26 Clock demultiplexer.

a positive transition. The last FF of each of these shift registers is connected to one of the inputs of the four-input multiplexer.

The two MOD-4 counters in the transmitter control the multiplexer output; the *word counter* outputs select the input channel; the *bit counter* allows the 4-bits from the corresponding data register to shift into the multiplexer before advancing the word counter to its next state. In other words, every four control pulses out of AND gate 2 causes the bit counter to recycle and trigger the word counter to its next state. In this way, the four data registers have their contents transmitted to multiplexer output Z one bit at a time, starting with register A (for $S_0 = S_1 = 0$) and proceeding through each register as the word counter advances one count. The Z signal thus contains 16 bits of serial data.

The AND gates, FF X, and the OS in the transmitter circuit generate the control pulses for the data shift registers and the bit counter as well as the *transmitted clock* signal. The operation of this portion of the circuit will be explained shortly.

The Receiver

The *receiver* has a four-output demultiplexer which receives the Z output signal of the transmitter multiplexer and demultiplexes (distributes) this signal to the data shift registers at the demultiplexer outputs. The 16 bits of data contained in the transmitted Z signal are distributed in groups of 4 to these registers. The first 4 bits are allowed to pass through to output O_0 and shifted into the A register; the second 4 bits are allowed to pass through to O_1 and shifted into the B register; and

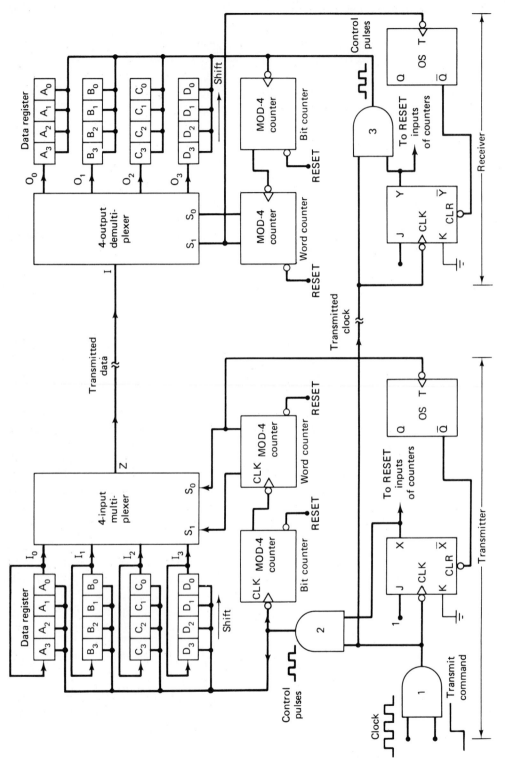

Figure 9.27 Synchronous data transmission system.

so on. Thus, when the operation is complete, the receiver data registers will contain the same data as the corresponding registers in the transmitter.

The two MOD-4 counters in the receiver control the demultiplexer in much the same way as the counters in the transmitter control the multiplexer. The word counter selects the demultiplexer output channel and the bit counter allows four control pulses to the shift registers before advancing the word counter to its next state.

Synchronization

It should be clear that in order for this system to work properly there has to be some means for synchronizing the selection of multiplexer inputs in the transmitter with the selection of demultiplexer outputs in the receiver. This synchronization is accomplished as follows:

1. FF X in the transmitter and FF Y in the receiver are normally LOW. The LOW outputs of these FFs keep both sets of counters in the zero state. Thus, both the multiplexer and demultiplexer initially have select inputs of 00.

2. The CLOCK input signal is applied to the transmitter AND gates 1 and 2. These gate outputs, however, are disabled since $X = 0$ and TRANSMIT COMMAND $= 0$.

3. The data transmission operation is initiated when the TRANSMIT COMMAND signal goes HIGH, allowing clock pulses through AND gate 1. These gated clock pulses also become the transmitted clock sent to the receiver. The first of these pulses sets FF X to the 1 state, thereby disabling the CLEAR inputs of the counters and enabling AND gate 2, which will allow all subsequent clock pulses through to the bit counter and shift registers. The same action occurs at the receiver where the first transmitted clock pulse sets FF Y to enable AND gate 3 to pass all subsequent pulses.

4. The control pulses in the transmitter and receiver are now available to clock the shift registers and counters. The first four control pulses shift the transmitter's A-register contents into and out of the multiplexer, into the demultiplexer, and out of output O_0 into the receiver's A register. The next four control pulses do the same with the B registers; and so on.

5. After 16 control pulses, the transmitter word counter recycles to 00. The negative transition from the MSB FF of the word counter triggers the OS, which produces a narrow pulse to clear FF X. The same operation occurs in the receiver to clear FF Y. With FFs X and Y LOW, the control pulses are cut off and the counters are held in the 0 state.

6. If the TRANSMIT COMMAND level is still HIGH, the operation repeats the sequence beginning with step 3. If not, the transmission is terminated until the TRANSMIT COMMAND goes HIGH again.

Actually, one portion of the logic circuitry for this data-transmission system has been purposely left out. The effect of this omission will be investigated in one of the end-of-chapter problems.

9.9   IC REGISTERS

Many types of flip-flop registers are used in digital systems to temporarily store data or to assist in transmitting data from one location to another. We will now look at some of the numerous types of registers that are available as off-the-shelf ICs.

Data-Latching Registers

This type of register uses the D-type latches discussed earlier. Figure 9.28 shows the representation for a 4-bit latching register. The *CLK* input is common to each latch and causes the data outputs Q_3, Q_2, Q_1, and Q_0 to respond to the data inputs D_3, D_2, D_1, and D_0 as follows:

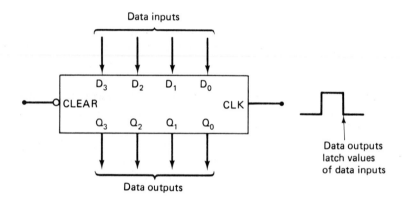

Figure 9.28   Four-bit latching register.

1. While *CLK* is high, each *Q* output follows the logic levels present on its corresponding *D* input (e.g., Q_3 follows D_3).

2. When *CLK* goes low, each *Q* output latches (holds) the last *D* value and cannot change even if the *D* input changes.

The CLEAR input is used to clear each output to 0 simultaneously on a low level.

Edge-Triggered Register

This type of register (Figure 9.29) uses edge-triggered D FFs and operates exactly like the register in Figure 9.28 except that the data inputs only affect the data outputs at the instant when *CLK* makes a *low-to-high* transition. At that time,

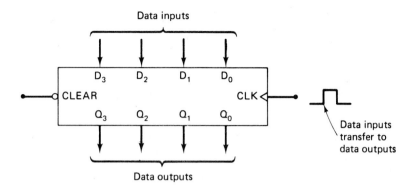

Figure 9.29   Four-bit edge-triggered register.

the levels present on the D inputs transfer to the Q outputs. Variations in the D input levels will cause no change in the Q outputs, but it is necessary that the D inputs be stable when the *CLK* transition occurs in order for proper data transfer to occur.

Shift Registers

There are several integrated variations of the basic shift register circuit. They differ as to how the data can enter the register (serial, parallel, or both) and as to whether the output data are available in parallel as well as serial form. Some also have control inputs to determine the direction of the shift operation (left to right, or vice versa). To illustrate, Figure 9.30 shows the block representation of two common IC shift registers.

The IC in Figure 9.30(a) is called a *parallel-in, serial-out* shift register. It consists of eight FFs arranged to shift data left to right on the low-to-high transition of the clock pulses. Data can be entered into the register in two different ways. *Serial* data can be entered through the serial input D_S. These data are shifted into FF Q_7 on the clock pulses. *Parallel* data can be entered via the data inputs $D_0 - D_7$ under the control of the parallel-load input, P_L. Whenever P_L goes LOW, the levels on $D_0 - D_7$ are immediately loaded (transferred) into the register FFs. With P_L at a high level, the parallel data inputs are disabled (have no effect).

This IC has only one FF output, Q_0, which is the rightmost bit. It represents the serial output that would occur as data are shifted left to right. The reason that the $Q_1 - Q_7$ outputs are not available is because of the limited number of IC pins (usually 14 or 16).

Figure 9.30(b) shows a *serial-in, parallel-out* shift register. Data can only enter this register serially through D_S on the low-to-high edge of the clock pulses. All eight FF outputs are available for external use as data are shifted on the clock pulses. The mode control, M, is used to control the shifting direction. For one logic level on M, the shifting will be left to right. For the other logic level on M, shifting will be right to left.

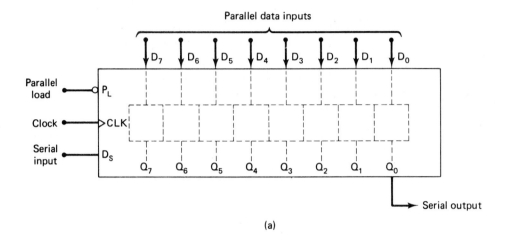

Parallel data inputs

(a)

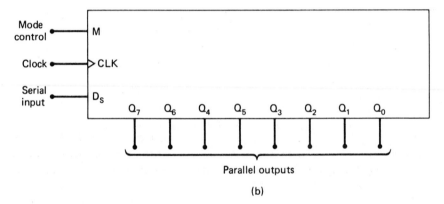

Parallel outputs

(b)

Figure 9.30   IC shift registers.

Tristate Registers

In most modern computers, the transfer of data from register to register takes place over a group of connecting lines called a *bus*. These *bus-organized* computers utilize tristate devices for reasons which will soon become clear. The tristate register is one of these devices. A typical 4-bit tristate register is represented in Figure 9.31. It contains four edge-triggered D FFs with a common clock input. The outputs O_0–O_3 are tristate outputs that operate as normal outputs as long as the *output disable input*, *OUTD*, is kept LOW. A HIGH on *OUTD* places the outputs in the HIGH impedance (Hi-Z) state.

The outputs are all cleared to 0 by a low level on the CLR input. This condition will always clear the register independent of all other inputs. However, if *OUTD* = 1 when *CLR* goes LOW, the outputs will not register as 0s since they are in the Hi-Z state. When *OUTD* returns to 0, the cleared outputs will be 0.

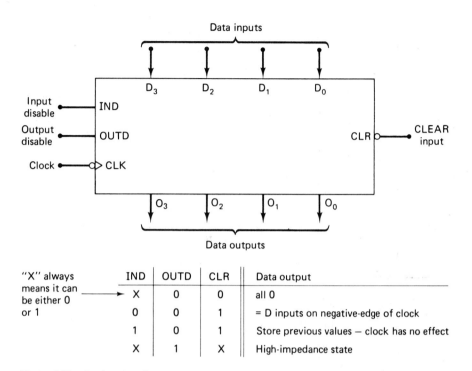

Figure 9.31 A tristate register.

IND	OUTD	CLR	Data output
X	0	0	all 0
0	0	1	= D inputs on negative-edge of clock
1	0	1	Store previous values — clock has no effect
X	1	X	High-impedance state

"X" always means it can be either 0 or 1

The *input disable input*, *IND*, is used to inhibit the effect of the *CLK* and *D* inputs. With $IND = 0$, the data inputs will be transferred to the data outputs on the high-to-low edge of *CLK*. With $IND = 1$, the *D* and *CLK* inputs are essentially disabled, so that the data outputs are not affected.

Although this is a typical tristate register, some of the input designations will vary among the many IC manufacturers. For example, the *IND* input might be called the *LOAD* input and the *OUTD* input might be the ENABLE input. The manufacturers' IC data sheets will always provide information as to the functions of the various inputs.

9.10 DATA BUSING

We mentioned the term *bus* as being very important in computer systems. Its total significance cannot be appreciated until our study of microcomputers. For now, the bus concept can be illustrated for register-to-register transfers.

A *data bus* is a group of wires used as a common path connecting all the inputs and outputs of several registers, such that data can be easily transferred from any one register to any other using appropriate control signals. Figure 9.32 shows a bus-organized system of three registers. Each register is a 4-bit tristate register like the one in Figure 9.31.

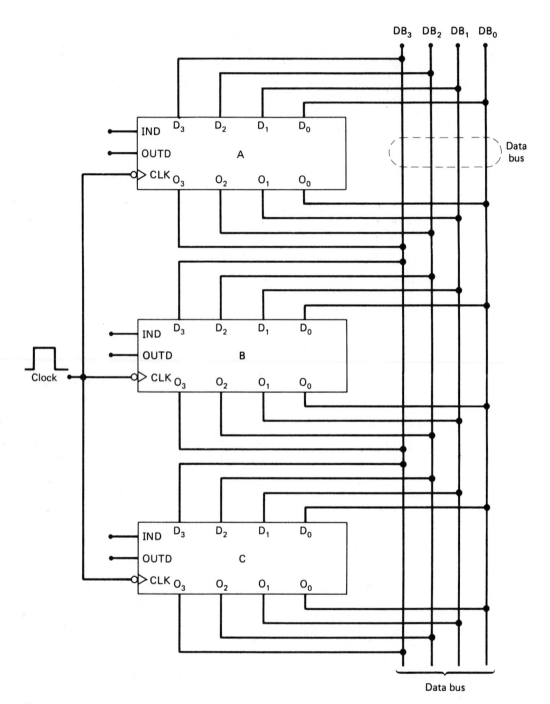

DB$_3$ DB$_2$ DB$_1$ DB$_0$

Data
bus

Clock

Data bus

Figure 9.32 Registers connected to a common bus.

In this arrangement, there are four lines that make up the data bus since the registers are each 4 bits wide. Corresponding outputs of each register are connected to the same one line of the bus (e.g., each O_0 is connected to the DB_0 line). This appears to be an ambiguous situation for determining the logic level on each bus line. The ambiguity, however, is removed by ensuring that only *one* of the three registers has its outputs enabled while the other two register outputs are in the **Hi-Z** state. For example, assume that registers A and B have their $OUTD = 1$ while register C has $OUTD = 0$. This essentially disconnects the register A and B outputs from the data bus so that only the register C outputs have their logic levels present on the data bus lines.

The data inputs to each register are also connected to the corresponding bus lines. Thus, the levels on the bus are always ready to be transferred to any one of the registers. Normally, though, only the register that is to receive the data will have $IND = 0$.

Data Transfer Operation

Consider transferring the contents of register A to register C. In order to perform this transfer, the data outputs of register A must be placed on the data bus by setting $OUTD_A = 0$. Also, register C must have $IND_C = 0$, so that it can receive the data. All other disable inputs must be kept at the 1 level. Summarizing,

$$\left.\begin{array}{l} IND_A = 1, OUTD_A = 0 \\ IND_B = 1, OUTD_B = 1 \\ IND_C = 0, OUTD_C = 1 \end{array}\right\} \quad \begin{array}{c} \text{conditions needed for} \\ [A] \longrightarrow [C] \end{array}$$

With these levels on the disable inputs, a pulse applied to the common *CLK* inputs will cause the levels on the data bus (A outputs) to be transferred to register C.

It should be easy to see how we can add more registers to the data bus by repeating the same connections for each register. Of course, each register adds two more disable inputs that have to be controlled for each data transfer. For the time being, we will not be concerned about what provides the signals to these disable inputs.

Simplified Representation of Bus

In all but the simplest computer systems, there will be many devices, as well as registers, connected to the same bus. On a circuit schematic, this can produce a very confusing array of lines and connections, especially for larger numbers of lines per bus. For this reason, we will often use a simplified representation for bus organization where a group of wires is replaced by one wide line. This is illustrated in Figure 9.33 for a three-register, 8-bit bus arrangement. The bracketed numbers

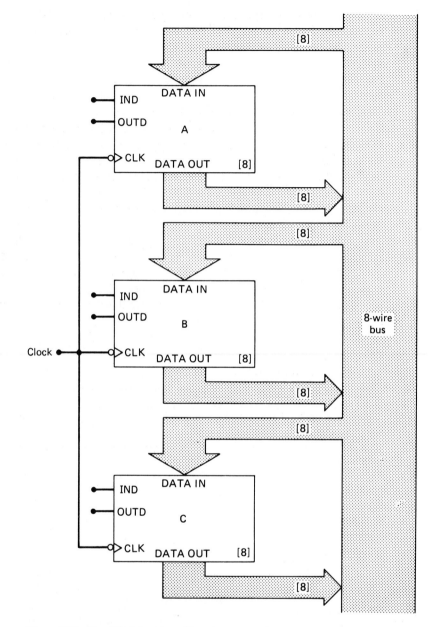

Figure 9.33 Simplified drawing of bus arrangement.

inside the register always indicate the number of bits the register contains. Like-wise, the number inside the wide lines indicates the number of actual wires which they represent.

Bidirectional Busing

Refer again to Figure 9.32. Since each register has both its inputs and its outputs tied to the data bus, it is obvious that corresponding inputs and outputs are shorted together. For example, output O_3 and input D_3 for each register are connected to bus line DB_3, and therefore they are connected to each other. As we have seen, there is nothing wrong with this because in normal operation a register will never have both its data inputs and outputs enabled at the same time.

Since inputs and outputs are going to be tied together anyway, many IC manufacturers *internally* connect them. This cuts down the number of pins that have to be used for the IC. Instead of four separate pins for data input and four for output, there are just four input/output (I/O) pins [see Figure 9.34(a)]. Depend-ing on the status of the disable inputs, the I/O lines function as inputs or outputs and are therefore called *bidirectional data lines*. The simplified representation for these bidirectional registers is shown in Figure 9.34(b), where the double-pointed arrow means that it is *bidirectional*.

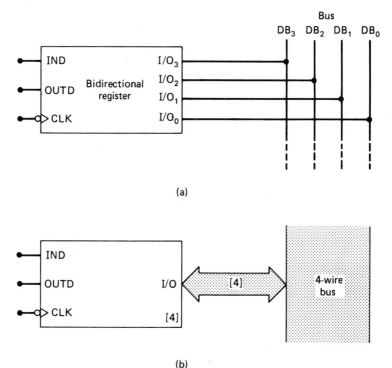

(a)

(b)

Figure 9.34   (a) Bidirectional register; (b) simplified diagram.

QUESTIONS AND PROBLEMS

9.1 What is the number of inputs and number of outputs of a decoder that accepts 32 different input combinations?

9.2 A certain 3-line-to-8-line decoder has active-LOW outputs and two ENABLE inputs E1 and E2. What conditions are necessary to produce a LOW at output O_6?

9.3 Draw the logic circuit for the decoder described in Problem 9.2.

9.4 Design the logic circuitry for the a output of a BCD-to-7-segment decoder with active LOW outputs.

9.5 Show how a 4-line-to-16-line decoder can be used as a 3-line-to-8-line decoder.

9.6 In a certain digitally-controlled process, six solenoid-controlled valves are to be actuated sequentially for intervals of 1 s each; i.e., valve 1 closes for 1 s, valve 1 opens and valve 2 closes for 1 s, and so on. Draw the complete circuit diagram for accomplishing this operation using the decoder/driver of Figure 9.7 and any other necessary logic. Assume that the solenoids operate from +24 V.

9.7 Show how to connect BCD-to-7-segment decoder/drivers (Figure 9.9) and LED 7-segment displays to the clock circuit of Figure 7.27. Assume that the displays require 10 mA per segment at a forward voltage drop of 2.5 V.

9.8 Figure 9.35 shows two 3-to-8 decoders combined to form a 4-to-16 decoder. Construct a truth table showing all 16 possible conditions of inputs w, x, y, and z, and

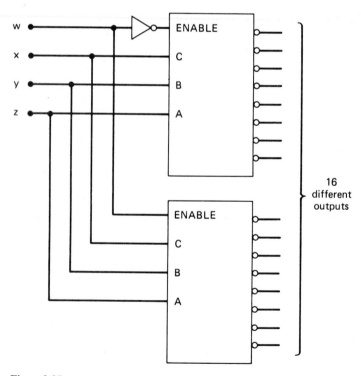

Figure 9.35

the resulting output levels for each case. This example illustrates one of the principal uses of the ENABLE inputs.

9.9 What will be the outputs of the encoder of Figure 9.11 if both A_4 and A_7 inputs are made HIGH?

9.10 Figure 9.36 shows the block diagram of a logic circuit used to control the number of copies made by a copy machine. The machine operator selects the number of desired copies by activating one of the selector switches S_0-S_9. This number is encoded in BCD by the encoder and sent to a comparator circuit. The operator then hits a momentary-contact START switch, which clears the counter and initiates a HIGH operate output that is sent to the machine to signal it to make copies. As the machine makes each copy, a copy pulse is generated and fed to the BCD counter. The counter outputs are continually compared with the switch encoder outputs in the comparator. When the two BCD numbers match, indicating that the desired number of copies have been made, the comparator output X goes LOW; this causes the operate level to return LOW and stop the machine so that no more copies are made. Activating the START switch will cause this process to be repeated. Design the complete logic circuitry for the comparator and control sections of this system.

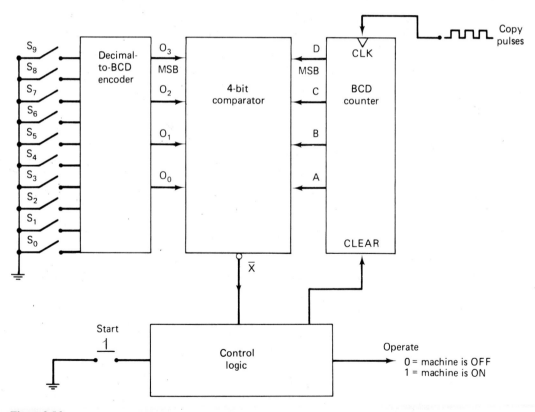

Figure 9.36

9.11 The keyboard circuit of Figure 9.14 is designed to accept a three-digit decimal number. What will happen if *four* digit keys were activated (e.g., 3095)? Design the necessary logic to be added to this circuit so that after three digits have been entered, any further digits will be ignored until the CLEAR key is depressed. In other words, if 3095 is entered on the keyboard, the output registers will display 309 and ignore the 5 and any subsequent digits until the circuit is CLEARED.

9.12 A technician breadboards the keyboard entry circuit of Figure 9.14 and tests it out by depressing various digit keys. He observes that each time he hits a digit key other than 0, the wrong result appears at the corresponding register outputs while the other register outputs are not affected. He also observes that there is no particular pattern to the malfunction—that it is more or less random. He replaces each of the ICs in the circuit but the malfunction persists. Which of the following could be the reasons for the malfunction? Explain your choice.
(a)   The technician has neglected to ground the unused input of the OR gate.
(b)   He has mistakenly used $\bar{Q}$ instead of Q from the OS.
(c)   The switch bounce from the digit keys lasts longer than 20 ms.

9.13 Refer to the data sheet for the Fairchild 9318 *priority* encoder in Appendix III. This encoder is an 8-line-to-3-line encoder, but how does its operation differ from that of the circuit of Figure 9.11? Determine the $\bar{A}_0$, $\bar{A}_1$, and $\bar{A}_2$ output levels if inputs $\bar{3}$, $\bar{7}$, and $\overline{\text{EI}}$ are made LOW while all other inputs are HIGH.

9.14 If the multiplexer of Figure 9.16 were to be constructed from TTL circuits, it would probably use all NAND gates. Convert the circuit to all NAND gates.

9.15 Several two-input multiplexers (Figure 9.16) can be combined in other multiplexing arrangements, such as the one shown in Example 9.3. Show how three of these circuits can be combined to form a four-input multiplexer. Use the block-diagram approach as in Figure 9.17. (*Hint:* Use 2 two-input multiplexers with their outputs feeding another two-input multiplexer.)

9.16 Design the logic circuitry for the multiplexer of Figure 9.19.

9.17 (a)   Expand the system of Figure 9.21 to display the contents of 2 three-stage BCD counters.
(b)   Count the number of connections in your circuit and compare it to the number of connections required if a separate decoder/driver and display were used for each counter.

9.18 Two 12-bit registers are each storing the binary equivalent of a four-digit octal number. Design a system for displaying the four octal digits of each register using the multiplexing technique used in Figure 9.21.

9.19 Design the additional logic needed for the parallel-to-serial convertor (Figure 9.22) so that it operates under the control of an input CONVERT level as follows:
(a)   The CONVERT input starts LOW and the counter holds in the 000 state.
(b)   When the CONVERT input goes HIGH, the counter goes through one complete cycle and stops at 000.
(c)   This operation is repeated when CONVERT returns LOW and then goes HIGH.

9.20 A multiplexer can be used to generate logic waveforms of any desirable pattern. These waveforms are often used as control signals. Figure 9.37 shows a circuit for

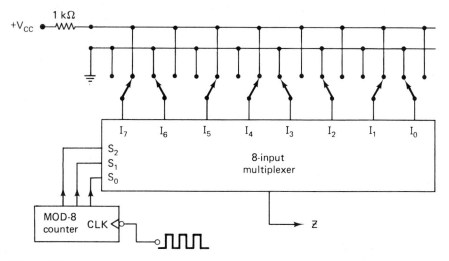

Figure 9.37

generating such waveforms. The pattern can be programmed using the eight SPDT switches. The MOD-8 counter is continuously cycled by the clock pulses. Draw the waveform at Z for the switches in the position shown.

9.21 Multiplexers can be used to implement logic functions directly from a truth table without the need for simplification. When used as a function generator, the multiplexer's select inputs are used as the logic variables while the data inputs are connected HIGH or LOW to satisfy the truth table. To illustrate, Figure 9.38 shows an eight-input multiplexer used to generate a certain logic function. Determine this logic function by filling in the values of Z in the truth table and then writing down the sum-of-products expression for Z in terms of A, B, and C.

9.22 Reconnect the data inputs of Figure 9.38 to produce $Z = AB + AC + BC$. (*Hint:* Construct the truth table for Z.)

9.23 Show how a 16-input multiplexer can be used to generate the function $Z = \bar{A}\bar{B}\bar{C}D + BCD + A\bar{B}\bar{D} + AB\bar{C}D$.

9.24 Add the necessary logic to the circuit of Figure 9.26 so that it operates as follows:

1. Initially the select code is 000. Four clock pulses are transmitted to O_0 and into the counter.

2. The next four clock pulses are transmitted to O_1, the four after that are transmitted to O_2, and so on until O_7 receives four clock pulses.

3. The cycle then repeats itself, beginning with (1).

9.25 The synchronous data-transmission system of Figure 9.27 will not work exactly as it should. Specifically, the problem is caused by the fact that the data registers in the receiver circuit will all shift simultaneously since they receive common SHIFT pulses. What is actually required is that the first four control pulses cause the A register to shift, the second four control pulses cause the B register to shift, and so on. Add the necessary logic so that the desired operation is accomplished.

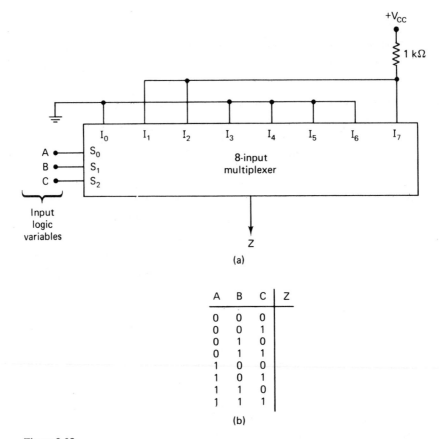

(a)

A	B	C	Z
0	0	0	
0	0	1	
0	1	0	
0	1	1	
1	0	0	
1	0	1	
1	1	0	
1	1	1	

(b)

Figure 9.38

9.26 Assume that the transmitter data registers in Figure 9.27 are storing 0011, 0110, 1001, and 0111, respectively. Construct the waveform diagrams for the following signals in response to a TRANSMIT COMMAND level: (a) AND 1 output, (b) AND 2 output, (c) X, (d) S_0 and S_1 of multiplexer, (e) Z, (f) AND 3 output, (g) Y, (h) S_0 and S_1 of demultiplexer, and (i) O_0, O_1, O_2, and O_3. (Assume that the circuit has been modified as in Problem 9.25 so that it operates properly.)

9.27 What could happen in the bus arrangement of Figure 9.32 if we applied LOW levels to the $OUTD$ inputs of register A and register B simultaneously?

9.28 Assume that the registers in Figure 9.32 are initially at $A = 1011$, $B = 1000$, $C = 0111$. The signals in Figure 9.39 are applied to the register inputs. Determine the contents of each register at times t_1, t_2, and t_3.

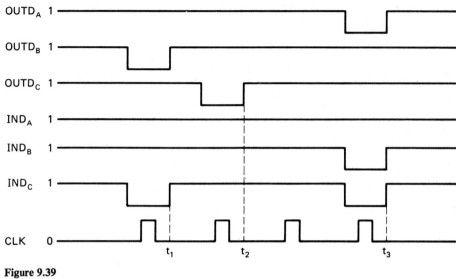

Figure 9.39

INTERFACING
WITH THE ANALOG WORLD

Digital systems perform all their internal operations in binary or some type of binary code. Any information that is to be input to a digital system must be put into binary form before it can be processed by the digital circuits. On the other hand, the outputs of a digital system are in some type of binary code and very often must be converted to a different form depending on how the outputs are to be used. Many devices are used on the input and/or output sides of digital systems to serve as the communications link to the outside world. (Refer to Figure 10.1.) Process-related I/O devices provide the means by which a digital system (i.e., computer) monitors and controls a physical process. On the *input* side, measurements of process parameters that are *analog* in nature are usually transduced (changed to a proportional electronic voltage or current) and sent to an analog-to-digital converter, ADC, which converts the analog quantity to a corresponding digital representation. Some process parameters are already digital in nature but must be changed to the appropriate form for the digital system. For example, digital input circuitry is used to count pulses from magnetic flow and turbine meters or to sense the ON/OFF status of switches or valves within the process.

Process-related *output* devices translate the digital system outputs to appropriate actuating signals needed to control the process. These actuating signals might be simply opening and closing of switch contacts or pulses to a stepping motor. Many times, however, the required actuating signal has to be analog in nature, such as a voltage for controlling the speed of a dc motor. In these cases, a digital-to-analog converter, DAC, is needed to convert the digital system output to the required analog form.

Thus, we see that ADCs and DACs function as the *interface* between a completely digital system or device, such as a computer, and the analog outside world. This function is becoming increasingly more important as inexpensive microcom-

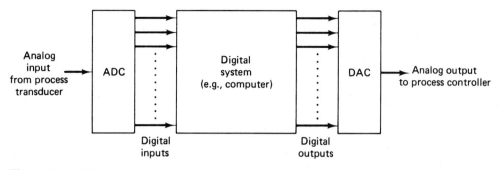

Figure 10.1   ADC and DAC are used to interface a purely digital system with the analog outside world.

puters move into areas of process control where computer control was previously unfeasible.

10.1   DIGITAL-TO-ANALOG CONVERSION

The two most important process-related I/O operations are digital-to-analog (D/A) and analog-to-digital (A/D) conversion. Since many A/D conversion methods utilize the D/A conversion process, we will examine D/A conversion first.

Basically, *D/A conversion* is the process of taking a value represented in *digital* code (such as straight binary or BCD) and converting it to a voltage or current which is proportional to the digital value. This voltage or current is an *analog* quantity, since it can take on many different values over a given range. Figure 10.2(a) shows the block diagram of a typical 4-bit D/A converter. We will not concern ourselves with the internal circuitry until later. For now, we will examine the various input/output relationships.

The digital inputs D, C, B, and A are usually derived from the output register of a digital system. The $2^4 = 16$ different binary numbers represented by these 4 bits are listed in Figure 10.2(b). For each input number, the D/A converter output voltage is a different value. In fact, for this case, the analog output voltage V_{OUT} is equal in volts to the binary number. It could also have been twice the binary number or some other proportionality factor. The same idea would hold true if the D/A output was a current I_{OUT}.

EXAMPLE 10.1   A 5-bit D/A converter has a current output. For a digital input of 10100, an output current of 10 mA is produced. What will I_{OUT} be for a digital input of 11101?

Solution: The digital input 10100 is the binary equivalent of 20_{10}. Since $I_{OUT} = 10$ mA for this case, the proportionality factor is 0.5; that is, $I_{OUT} = 0.5 \times$ binary value. Thus, the binary input 11101 is equivalent to 29_{10}, so $I_{OUT} = 0.5 \times 29 = 14.5$ mA.

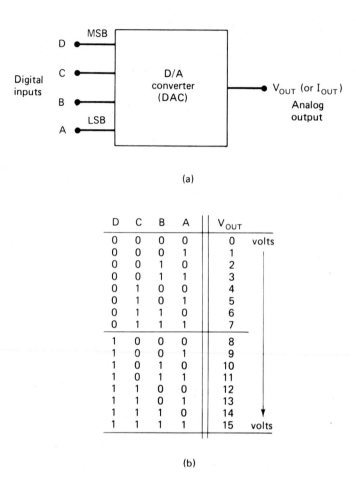

(a)

D	C	B	A	V_{OUT}	
0	0	0	0	0	volts
0	0	0	1	1	
0	0	1	0	2	
0	0	1	1	3	
0	1	0	0	4	
0	1	0	1	5	
0	1	1	0	6	
0	1	1	1	7	
1	0	0	0	8	
1	0	0	1	9	
1	0	1	0	10	
1	0	1	1	11	
1	1	0	0	12	
1	1	0	1	13	
1	1	1	0	14	
1	1	1	1	15	volts

(b)

Figure 10.2   Four-bit D/A convertor with voltage output.

Input Weights

For the DAC of Figure 10.2 it should be noted that each digital input contributes a different amount to the analog output. This is easily seen if we examine the cases where only one input is HIGH:

D	C	B	A		V_{OUT} (V)
0	0	0	1	$\longrightarrow$	1
0	0	1	0	$\longrightarrow$	2
0	1	0	0	$\longrightarrow$	4
1	0	0	0	$\longrightarrow$	8

The contributions of each digital input are *weighted* according to their position in the binary number. Thus, A, which is the LSB, has a *weight* of 1 V, B has a weight of 2 V, C has a weight of 4 V and D, the MSB, has the largest weight, 8 V. The weights are successively doubled for each bit, beginning with the LSB. Thus, we can consider V_{OUT} to be the weighted sum of the digital inputs. For instance, to find V_{OUT} for the digital input 0111 we can add the weights of the C, B, and A bits to obtain 4 V + 2 V + 1 V = 7 V.

EXAMPLE 10.2 A 5-bit D/A converter produces $V_{OUT} = 0.2$ V for a digital input of 00001. Find the value of V_{OUT} for a 11111 input.

Solution: Obviously, 0.2 V is the weight of the LSB. Thus, the weights of the other bits must be 0.4 V, 0.8 V, 1.6 V, and 3.2 V, respectively. For a digital input of 11111, then, the value of V_{OUT} will be 3.2 V + 1.6 V + 0.8 V + 0.4 V + 0.2 V = 6.2 V.

Resolution (Step Size)

Resolution of a D/A converter is defined as the smallest change that can occur in the analog output as a result of a change in the digital input. Referring to the table in Figure 10.2, we can see that the resolution is 1 V, since V_{OUT} can change by no less than 1 V when the input code is changed. The resolution is always equal to the weight of the LSB and is also referred to as the *step size*, since it is the amount V_{OUT} will change as the input code goes from one step to the next. This is illustrated more graphically in Figure 10.3, where the digital inputs are being derived from the outputs of a 4-bit binary counter. The counter is being continuously cycled through its 16 states by the clock input. The waveform at the D/A output is a repetitive staircase which goes up 1 V per step as the counter advances from 0000 to 1111. When the counter returns to 0000, the D/A output returns to 0 V. The resolution or step size is the size of the jumps in the staircase waveform. In this example each step is 1 V.

Although resolution can be expressed as the amount of voltage or current per step, it is more useful to express it as a percentage of the *full-scale output*. To illustrate, the D/A converter of Figure 10.3 has a maximum full-scale output of 15 V (when the digital input is 1111). The step size is 1 V which gives a percentage resolution of

$$\% \text{ resolution} = \frac{\text{step size}}{\text{full scale (F.S.)}} \times 100\% \qquad (10.1)$$

$$= \frac{1 \text{ V}}{15 \text{ V}} \times 100\% = 6.67\%$$

EXAMPLE 10.3 A 10-bit D/A converter has a step size of 10 mV. Determine the full-scale output voltage and the percentage resolution.

Solution: With 10 bits, there will be $2^{10} - 1 = 1023$ steps of 10 mV each. The fullscale output will therefore be 10 mV $\times$ 1023 = 10.23 V and

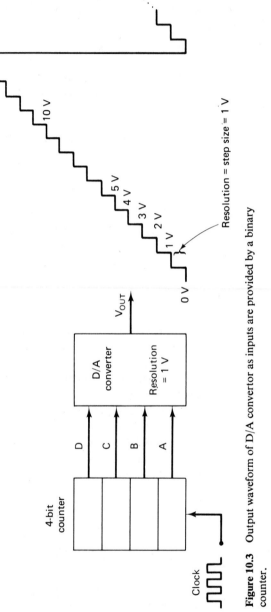

Figure 10.3 Output waveform of D/A convertor as inputs are provided by a binary counter.

$$\% \text{ resolution} = \frac{10 \text{ mV}}{10.23 \text{ V}} \times 100\% \simeq 0.1\%$$

Example 10.3 helps to illustrate the fact that the percentage resolution becomes smaller as the number of input bits is increased. In fact, the percentage resolution can also be calculated from

$$\% \text{ resolution} = \frac{1}{\text{total } \# \text{ steps}} \times 100\% \qquad (10.2)$$

For an N-bit binary input code the total number of steps is $2^N - 1$. Thus, for the previous example

$$\% \text{ resolution} = \frac{1}{2^{10} - 1} \times 100\%$$

$$= 1/1023 \times 100\%$$

$$\approx 0.1\%$$

This means that it is *only the number of bits* which determines the *percentage* resolution. Increasing the number of bits increases the number of steps to reach full scale, so each step is a smaller part of the full-scale voltage. Many DAC manufacturers specify resolution as the number of bits.

BCD Input Code

The D/A converters we have considered thus far have used a binary input code. Many D/A converters use a BCD input code where 4-bit code groups are used for each decimal digit. Figure 10.4 shows the diagram of an 8-bit (two-digit) convertor of this type. Each 4-bit code group can range from 0000 to 1001, so the BCD inputs can represent any decimal number from 00 to 99. *Within* each code group the weights of the different bits are proportioned the same as in the binary code, but the *group* weights are different by a factor of 10. For example, A_0, the LSB of the least-significant digit (LSD), could have a weight of 0.1 V. Thus, B_0, C_0, and D_0

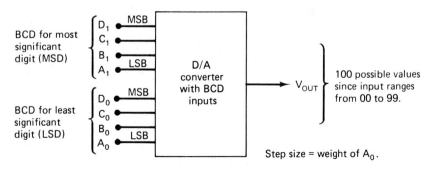

Figure 10.4   D/A convertor using BCD input code.

would be 0.2 V, 0.4 V, and 0.8 V, respectively. The weight of A_1, the LSB of the MSD, would be 1 V (10 times A_0). Similarly, B_1, C_1, and D_1 would be 2 V, 4 V, and 8 V, respectively.

EXAMPLE 10.4   If the weight of A_0 is 0.1 V in Figure 10.4, find:
  (a)   Step size.
  (b)   Full-scale output and percentage resolution.
  (c) V_{OUT} for $D_0 C_0 B_0 A_0 = 1000$ and $D_1 C_1 B_1 A_1 = 0101$.

Solution:
  (a)   Step size is the weight of the LSB of the LSD, 0.1 V.
  (b)   There are 99 steps since there are two BCD digits. Thus, full-scale output is $99 \times 0.1 = 9.9$ V. The resolution is (using equation 10.1)

$$\frac{\text{step size}}{\text{F.S.}} \times 100\% = \frac{0.1}{9.9} \times 100\% \approx 1\%$$

We could also have used equation (10.2) to calculate percentage resolution since the total number of steps is 99.
  (c)   To find V_{OUT}, add the weights of all the bits that are 1s. Thus,

$$V_{OUT} = \overset{D_0}{\overbrace{0.8 \text{ V}}} + \overset{C_1}{\overbrace{4 \text{ V}}} + \overset{A_1}{\overbrace{1 \text{ V}}} = 5.8 \text{ V}$$

Alternatively, $D_1 C_1 B_1 A_1 = 5_{10}$ and $D_0 C_0 B_0 A_0 = 8_{10}$, so the input code is 58, multiplied by 0.1 V per step, to again give 5.8 V.

10.2   D/A-CONVERTER CIRCUITRY*

There are several methods and circuits for producing the D/A operation which has been described. We shall examine only one of the basic schemes, to give an insight into the ideas used. It is not important to be familiar with all the various schemes because D/A converters are available as self-contained, encapsulated packages that do not require any circuit knowledge. Instead, it is important to know the significant performance characteristics of D/A converters, in general, so that they can be used intelligently. These will be covered in Section 10.10.

Figure 10.5(a) shows the basic circuit for one type of 4-bit D/A converter. The inputs A, B, C, and D are binary inputs which are assumed to have values of either 0 V or 5 V. The *operational amplifier* is employed as a summing amplifier, which produces the weighted sum of these input voltages. It may be recalled that the summing amplifier multiplies each input voltage by the ratio of the feedback resistor R_F to the corresponding input resistor R_{IN}. In this circuit $R_F = 1$ kΩ and

---

* This section can be omitted by those not concerned with the internal circuitry of D/A converters.

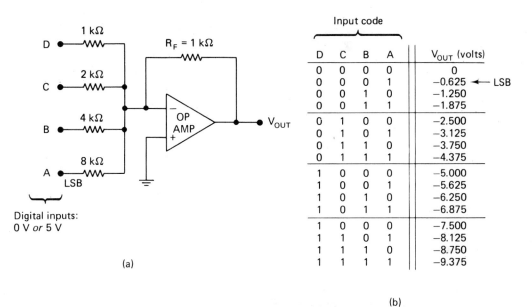

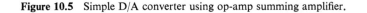

Figure 10.5 Simple D/A converter using op-amp summing amplifier.

the input resistors range from 1 to 8 kΩ. The D input has $R_{IN} = 1$ kΩ, so the summing amplifier passes the voltage at D with no attenuation. The C input has $R_{IN} = 2$ kΩ so it will be attenuated by 0.5. Similarly, the B input will be attenuated by 0.25 and the A input by 0.125. The amplifier output can thus be expressed as

$$V_{OUT} = -(V_D + 0.5V_C + 0.25V_B + 0.125V_A) \tag{10.3}$$

The negative sign is present because the summing amplifier is a polarity-inverting amplifier, but it will not concern us here.

Clearly, the summing amplifier output is an analog voltage which represents a weighted sum of the digital inputs, as shown by the table in Figure 10.5(b). This table lists all of the possible input conditions and the resultant amplifier output voltage. The output is evaluated for any input condition by setting the appropriate inputs to either 0 V or 5 V. For example, if the digital input is 1010, then $V_D = V_B = 5$ V and $V_C = V_A = 0$ V. Thus, using (10.3),

$$V_{OUT} = -(5 \text{ V} + 0 \text{ V} + 0.25 \times 5 \text{ V} + 0 \text{ V})$$
$$= -6.25 \text{ V}$$

The resolution of this D/A converter is equal to the weighting of the LSB, which is 0.125×5 V $= 0.625$ V. As shown in the table, the analog output increases by 0.625 V as the binary input number advances one step.

EXAMPLE 10.5

(a) Determine the weights of each input bit of Figure 10.5(a).

(b) Change R_F to 250 Ω and repeat.

Solution:

(a) The MSB passes with gain = 1, so its weight in the output is 5 V. Thus,

$$\text{MSB} \longrightarrow 5 \text{ V}$$
$$\text{2nd MSB} \longrightarrow 2.5 \text{ V}$$
$$\text{3rd MSB} \longrightarrow 1.25 \text{ V}$$
$$\text{4th MSB} = \text{LSB} \longrightarrow 0.625 \text{ V}$$

(b) If R_F is reduced by a factor of four to 250 Ω, each input weight will be four times *smaller* than the values above.

Conversion Accuracy

The table in Figure 10.5(b) gives the *ideal* values of V_{OUT} for the various input cases. How close the circuit comes to producing these values depends primarily on two factors: (1) the precision of the input and feedback resistors, and (2) the pre-

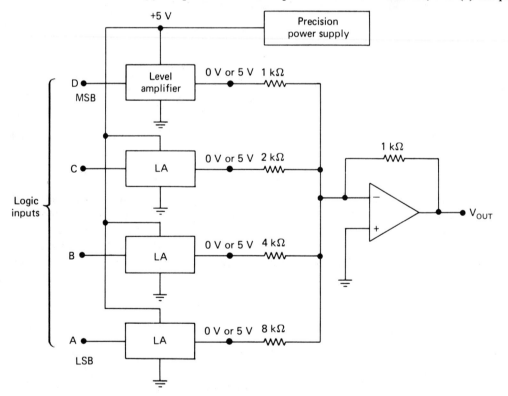

Figure 10.6 Complete 4-bit D/A converter including precision-level amplifiers.

cision of the input voltage levels. The resistors can be made very accurate (within 0.01 per cent of the desired values) by trimming, but the input voltage levels must be handled differently. It should be clear that the digital inputs cannot be taken directly from the outputs of FFs or logic gates because the output logic levels of these devices are not precise values like 0 V and 5 V but vary over a given range. For this reason, it is necessary to insert a *precision-level amplifier* in between each logic input and its input resistor to the summing amplifier. This is shown in Figure 10.6.

The level amplifiers produce precise output levels of 5 V and 0 V, depending on whether the digital inputs are HIGH or LOW. A very stable, precise 5-V reference supply is required to produce the accurate 5-V level.

Other D/A Methods

There are many D/A conversion techniques besides the one we described here using the weighted summing amplifier, but we will not discuss them here. As stated earlier, most DACs are manufactured as complete packages (some in IC form), so it is more important to understand the various DAC specifications than the internal circuitry.

10.3 DAC SPECIFICATIONS

DACs are available in a wide range of specifications and prices. One must be familiar with the manufacturers' specifications in order to intelligently evaluate a DAC for a particular application. One specification, *resolution*, has already been discussed. Some of the other important ones are described below.

Accuracy

DAC manufacturers have several ways of specifying accuracy. The two most common are called *relative accuracy* and *differential linearity*, which are normally expressed as a percentage of the converter's full-scale output (%F.S.).

Relative accuracy is the maximum deviation of the DAC's output from its expected (ideal) value. For example, assume that the DAC of Figure 10.5 has a relative accuracy of $\pm 0.01\%$F.S. Since this converter has a full-scale output of 9.375 V, this percentage converts to

$$\pm 0.01\% \times 9.375 \text{ V} = \pm 0.9375 \text{ mV}$$

This means that the output of this DAC can, at any time, be off by as much as 0.9375 mV from its expected value.

Differential linearity is the maximum deviation in step size from the ideal step size. For example, the DAC of Figure 10.5 has an expected step size of 0.625 V.

If this converter has a differential linearity of $\pm 0.01\%$F.S., this would mean that the actual step size could be off by as much as 0.9375 mV.

Some of the more expensive DACs have relative and differential accuracies as low as 0.001%F.S. General-purpose DACs usually have accuracies in the 0.01–0.1% range.

It is important to understand that accuracy and resolution of a D/A converter must be compatible. It is illogical to have a resolution of, say, 1 per cent and an accuracy of 0.1 per cent, or vice versa. To illustrate, a D/A converter with a resolution of 1 per cent and a F.S. output of 10 V can produce an output analog voltage within 0.1 V of any desired value, assuming perfect accuracy. It makes no sense to have a costly accuracy of 0.01 per cent of F.S. (or 1 mV) if the resolution already limits the closeness to the desired value to 0.1 V. The same can be said for having a resolution that is very small (many bits) while the accuracy is poor; it is a waste of input bits.

Operating Speed

The speed of a DAC is usually specified as *settling time*, which is the maximum time interval required for the output to go from zero to full scale as the input code is changed from all 0s to all 1s. Typical values of settling time fall in the range 1–20 μs. In general, a DAC with a current output has a shorter settling time than does a DAC with a voltage output.

10.4　D/A MULTIPLEXING

In many applications more than one digital input is to be converted to an analog output. For example, a process-control computer may provide several digitally coded control signals which are to be used to drive different actuating devices, such as motors or valves. Basically, there are two methods for performing the D/A conversion of more than one digital input code.

The most obvious method is simply to use a separate DAC for each digital input and is illustrated in Figure 10.7. This method has the advantage that each digital input is continuously applied to its DAC so that the analog output is continually generated with no need for analog storage. On the other hand, separate D/A converters are required, and these are relatively high-cost elements because of the precision components they contain (precision resistors, reference supply, level amplifiers, etc.).

A second method for performing multiple D/A conversions uses a multiplexing technique that involves time-sharing a single DAC among several digital inputs. The basic arrangement is shown in Figure 10.8. The different sets of digital inputs are switched into the input register one at a time, in sequence. For the first digital input, the DAC produces an analog voltage V_A which is transmitted to

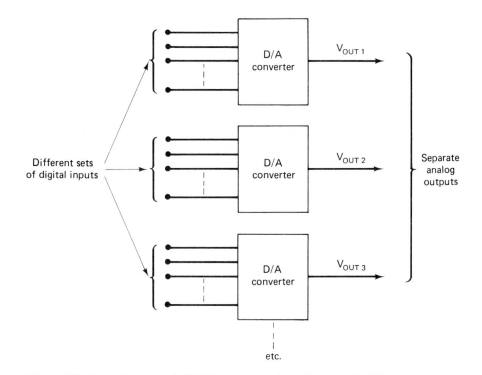

Figure 10.7 Converting several digital inputs to analog using separate D/A converters.

the first *sample-and-hold circuit* via the closed multiplexing switch S_1. The other switches are held open. The capacitor C_1 charges to the voltage V_A while its switch is closed and *holds* this value when the switch is open.

When the second digital input is switched into the input register, its analog equivalent appears at V_A and is transmitted to C_2 via switch S_2, which is now closed (all other switches are open). This sequence continues for all the digital inputs until each has been converted to an analog output. It is then repeated beginning with the first input. The operational amplifier associated with each capacitor has a very high input impedance to prevent the capacitor from discharging while its switch is open. The unity gain amplifier serves to isolate the capacitor from any load being driven.

Multiplexing Rate

The multiplexing rate is the rate at which the various digital inputs are switched sequentially into the D/A converter. One complete cycle of operation involves transferring the new digital value to the input register, converting the new digital

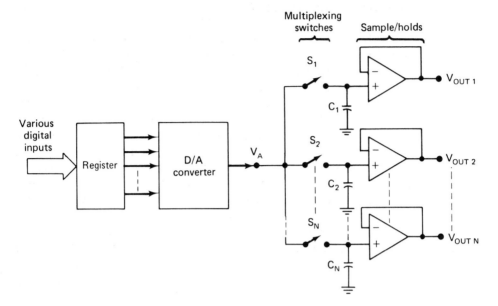

Figure 10.8   Time-sharing a single D/A converter using a multiplexer technique.

input to an analog value V_A, closing the appropriate switch and opening all others, and allowing the corresponding holding capacitor to charge to the value of V_A. Clearly, the *maximum* multiplexing rate that can be used depends on the amount of time needed for each of the above operations. Usually, the response time of the D/A converter is the principal factor limiting the multiplexing rate. An exception to this occurs when the multiplexing switches are electromechanical relay contacts. High-speed relays can respond in about 1 ms, thereby limiting the multiplexing rate to 1 kHz or less. For greater multiplexing rates, the switches must be a solidstate type, such as the CMOS bilateral switch discussed in Chapter 8.

The *minimum* multiplexing rate is determined by the ability of the capacitors to hold a voltage. For example, if the D/A multiplexer has four channels (four switches and four sample-and-hold circuits), then each capacitor is charged to its voltage during *one* cycle and must hold this voltage for *three* cycles while the other channels are being activated. The capacitor voltage must not discharge appreciably during this time.

Further investigation of D/A multiplexing is reserved for some of the end-of-chapter problems. It should be mentioned that although D/A multiplexing uses only one D/A converter, it does have the disadvantage of more complex circuitry and slower conversion speed than the method of Figure 10.7, so it is only used when its advantages outweigh its disadvantages.

10.5 ANALOG-TO-DIGITAL CONVERSION

An *A/D converter* takes an analog input voltage and after a certain amount of time produces a digital output code which represents the analog input. The A/D conversion process is generally more complex and time consuming than the D/A process, and many different methods have been developed and used. We shall examine several of these methods in detail, even though it may never be necessary to design or construct A/D converters (they are available as completely packaged units). However, the techniques that are used provide examples of logic control and give an insight into what factors determine an A/D converter's performance.

Several important types of ADC utilize a D/A converter as part of their circuitry. Figure 10.9 is a general block diagram for this class of ADC. The timing for the operation is provided by the input clock signal. The control unit contains the logic circuitry for generating the proper sequence of operations in response to the START COMMAND, which initiates the conversion process. The comparator is a special circuit that has two *analog* inputs and a *digital* output which switches states, depending on which analog input is greater.

The basic operation of A/D converters of this type consists of the following steps:

1. The START COMMAND goes HIGH, starting the operation.

2. At a rate determined by the clock, the control unit continually modifies the binary number, which is stored in the register.

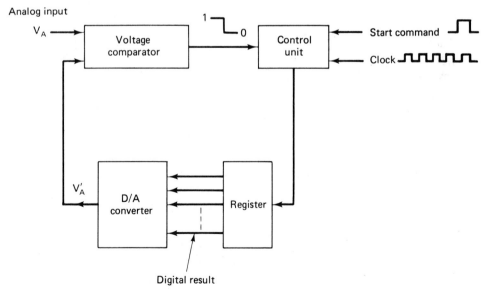

Figure 10.9 General diagram of one class of A/D converters.

3. The binary number in the register is converted to an analog voltage $V_{A'}$, by the D/A converter.

4. The comparator compares $V_{A'}$ with the analog input V_A. As long as $V_{A'} < V_A$, the comparator output stays HIGH. When $V_{A'}$ equals or exceeds V_A, the comparator output goes LOW and stops the process of modifying the register number. At this point, $V_{A'}$ is a close approximation to V_A, and the digital number in the register, which is the digital equivalent of $V_{A'}$, is also the digital equivalent of V_A, within the resolution and accuracy of the system.

The several variations of this A/D conversion scheme differ mainly in the manner in which the control section continually modifies the numbers in the register. Otherwise, the basic idea is the same, with the register holding the required digital output when the conversion process is complete.

Comparator Circuit

A simple implementation of the comparator circuit is shown in Figure 10.10(a). It consists only of an operational amplifier used in its high-gain differential mode. Any difference in the inputs V_1 and V_2 is amplified by the gain G_V of the amplifier (typically 10,000 times). Thus, if V_1 is slightly greater than V_2 by an amount V_T, called the *threshold voltage*, the comparator output saturates at $+10$ V. Similarly, if V_2 is greater than V_1 by an amount V_T, the comparator output saturates at -10 V. For differences in V_1 and V_2 that are less than V_T, the comparator output is G_V times the difference.

To summarize:

$$V_{\text{OUT}} = \begin{cases} G_V(V_1 - V_2) & \text{if } |V_1 - V_2| \leq V_T \\ +10 \text{ V} & \text{if } V_1 - V_2 > V_T \\ -10 \text{ V} & \text{if } V_2 - V_1 > V_T \end{cases}$$

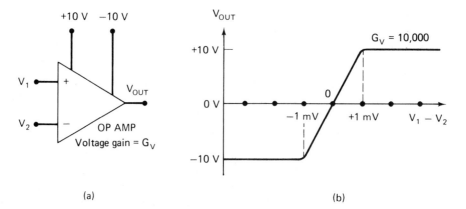

Figure 10.10   (a) Op-amp comparator; (b) output/input graph.

The graph in Figure 10.10(b) displays the relationship between V_{OUT} and the differential input, $V_1 - V_2$ for a $G_V = 10,000$.

The output levels of the comparator of Figure 10.10 are not compatible with TTL logic. The addition of a simple Zener diode circuit on the output will produce the proper levels. A Zener diode of 3–4 V should be used for TTL. Some IC comparators are designed to have TTL compatible output levels.

EXAMPLE 10.6
  (a) Determine V_T for the comparator of Figure 10.10.
  (b) Determine V_{OUT} for $V_1 = 9.342$ V and $V_2 = 9.340$ V.
  (c) Determine V_{OUT} for $V_1 = 2.176$ V and $V_2 = 2.180$ V.
  (d) Determine V_{OUT} for $V_1 = 1.920$ V and $V_2 = 1.920$ V.

Solution:
  (a) From the graph, V_{OUT} saturates at $+10$ V when $V_1 - V_2 \geq 1$ mV. Thus, $V_T = 1$ mV.
  (b) $V_1 - V_2 = 0.002$ V $= 2$ mV, which is greater than V_T. Thus, $V_{OUT} = +10$ V.
  (c) $V_1 - V_2 = -0.004$ V $= -4$ mV. Thus $V_{OUT} = -10$ V.
  (d) $V_1 - V_2 = 0$ V. Thus, $V_{OUT} = G_V \times 0 = 0$ V.

10.6   DIGITAL-RAMP A/D CONVERTER

One of the simplest versions of the general A/D converter of Figure 10.9 uses a binary counter as the register and allows the clock to increment the counter one step at a time until $V_{A'} \geq V_A$. This type of A/D converter is called a *digital-ramp A/D converter* because the waveform at $V_{A'}$ is a step-by-step ramp (actually a staircase) like the one shown in Figure 10.3.

Figure 10.11 shows the complete diagram for a digital-ramp A/D converter. Its operation proceeds as follows:

1. A positive start pulse is applied, which resets the counter to zero. It also inhibits the AND gate so that no clock pulses get through to the counter while the start pulse is HIGH.

2. With the counter at zero, $V_{A'} = 0$, so the comparator output is HIGH (assume that V_A is some positive voltage).

3. When the start pulse returns to LOW, the AND gate is enabled and pulses are allowed into the counter.

4. As the counter advances, the D/A output $V_{A'}$ increases in steps of voltage equal to its resolution.

5. This continues until $V_{A'}$ reaches a step that exceeds V_A by an amount equal to V_T or greater. At this point, the comparator output goes LOW, stopping the pulses to the counter so that the counter has stopped at a count that is the desired digital representation of V_A. The conversion process is now complete.

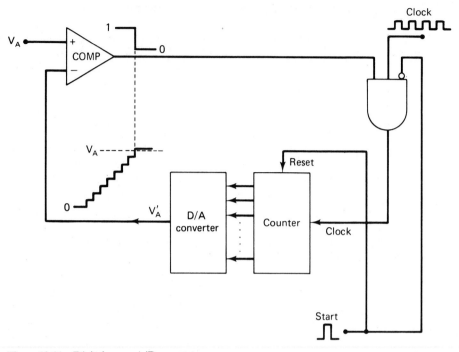

Figure 10.11   Digital-ramp A/D converter.

EXAMPLE 10.7   Assume the following values for the A/D converter of Figure 10.11:
Clock frequency = 1 MHz; V_T = 1 mV; D/A converter has F.S. output = 10.23 V
and a 10-bit input. Determine: (a) the digital equivalent obtained for V_A = 3.728 V;
(b) the conversion time; and (c) the resolution of this converter.

Solution:

(a)   The DAC has a 10-bit input and a 10.23-V F.S. output. Thus, the number of
total possible steps is $2^{10} - 1 = 1023$, so the step size is

$$\frac{10.23 \text{ V}}{1023} = 10 \text{ mV}$$

This means that $V_{A'}$ increases in steps of 10 mV as the counter counts up from zero.
Since V_A = 3.728 V and V_T = 1 mV, then $V_{A'}$ has to reach 3.729 V or more before
the comparator switches low. This will require

$$\frac{3.279 \text{ V}}{10 \text{ mV}} = 372.9 = 373 \text{ steps}$$

At the end of the conversion, then, the counter will hold the binary equivalent of
373, which is 0101110101. This is the desired digital equivalent of V_A = 3.728 V,
as produced by this A/D convertor.

(b)   373 steps were required to complete the conversion. Thus, 373 clock pulses occurred at the rate of one per microsecond. This gives a total conversion time of 373 μs.

(c)   The resolution of this converter is equal to the step size of the D/A converter, which is 10 mV. In percent it is 10 mV/10.23 V $\times$ 100 per cent $\simeq$ 0.1 per cent.

A/D Resolution and Accuracy

As pointed out in Example 10.7, the resolution of the A/D converter is equal to the resolution of the D/A converter which it contains. The D/A output voltage $V_{A'}$ is a staircase waveform that goes up in discrete steps until it exceeds V_A. Thus, $V_{A'}$ is an approximation to the value of V_A and the best we can expect is that $V_{A'}$ is within 10 mV of V_A if the resolution (step size) is 10 mV. We can think of the resolution as being an inherent error which is often referred to as *quantization error*. This quantization error, which can be reduced by increasing the number of bits in the counter and D/A converter, is sometimes specified as an error of $+1$ LSB, indicating that the result could be off by that much due to the step size. In a later problem we will see how this quantization error can be modified so that it is $\pm\frac{1}{2}$ LSB, which is a more common situation.

Looking at it from a different aspect, the input V_A can take on an *infinite* number of values from 0 V to F.S. The approximation $V_{A'}$, however, can only take on a finite number of discrete values. This means that a small range of V_A values will have the same digital representation. To illustrate, in Example 10.7 any value of V_A from 3.720 V to 3.729 V will require 373 steps, thereby resulting in the same digital representation. In other words, V_A must change by 10 mV (the resolution) to produce a change in digital output.

As in the D/A converter, *accuracy* is not related to the resolution but is dependent on the accuracy of the circuit components, such as the comparator, the D/A precision resistors and level amplifiers, the reference supplies, and so on. An accuracy rating of 0.01 per cent F.S. indicates that the A/D converter result may be off by 0.01 per cent of F.S., owing to nonideal components. This error is *in addition* to the quantization error due to the resolution. These two sources of error are usually of the same order of magnitude for a given ADC.

Average Conversion Time

In the digital-ramp A/D converter, the counter begins at zero and counts up until $V_{A'} > V_A$. Obviously, the time it takes to complete the conversion depends on the value of V_A. A larger value of analog input requires more steps and therefore more clock cycles. The longest conversion time would be 2^N clock periods, where N is the number of bits in the counter. For example, for the A/D converter in Example 10.7 the *maximum* conversion time is

$$2^{10} \times 1 \ \mu s = 1024 \ \mu s$$

$$= 1.024 \ ms$$

The *average* conversion time is sometimes specified; it is half of the maximum and is given by $2^N/2 = 2^{N-1}$ clock periods. Clearly, the conversion time for this type of A/D converter increases by a factor of 2 for each bit added to the counter and D/A. Thus, resolution can be improved only at the cost of a much longer conversion time.

The relatively long conversion times of the digital-ramp converter make it unsuitable for high-speed applications. A modification of this converter which reduces the conversion time considerably uses an up/down counter which is not reset to zero before each conversion. Instead, the sequence begins at the level of the previous conversion and goes either up or down, depending on the state of the comparator output. In this way, the A/D converter can better follow a slowly changing analog input with shorter conversion times. This type of converter is called a *continuous digital-ramp converter*.

10.7 THE SUCCESSIVE-APPROXIMATION A/D CONVERTER

The most widely used A/D converter is the *successive-approximation* type. It has more complex circuitry than the digital-ramp types, but it has a much shorter conversion time. In addition, its conversion time is a fixed value independent of the value of the analog input.

This type of A/D converter does not use a counter to provide the input to the D/A converter block. It uses a REGISTER instead and so has the basic arrangement of Figure 10.9. Whereas the counter used in the digital-ramp method progressed through each binary count step by step, the REGISTER has its binary values modified in a much different sequence. The sequence takes place as follows:

1. The MSB is set to 1 and all other bits are 0. This produces a value of $V_{A'}$ at the D/A output equal to the weight of the MSB. If $V_{A'}$ is now greater than V_A, the comparator goes LOW. This causes the MSB to be reset to 0. If $V_{A'}$ is less than V_A, the MSB is kept at 1.

2. The second MSB is set to 1. Again, if the new value of $V_{A'}$ is now greater than V_A, this bit is reset to 0. If not, it is kept at 1.

3. This process is continued for all bits in the register. This trial-and-error process requires *one* clock period per bit.

To better illustrate this process, consider an A/D converter that uses a 4-bit REGISTER and a D/A converter with a step size of 1 V. Assume that the V_A input is 10.4 V. Table 10.1 shows the sequence of operations. This illustration shows that in each step a better and better approximation to the analog input V_A is being generated. It also shows that for this 4-bit A/D converter, only *four* steps are required. In general, an N-bit successive-approximation ADC requires N steps regardless of the value of V_A. Each step requires one clock period.

Table 10.1

Steps		Register	$V_{A'}$ (V)	V_A (V)	Comparator
	Initial status	0000	0	10.4	HIGH
I.	A. Set MSB to 1	1000	8	10.4	HIGH
	B. Leave it at 1 since $V_{A'} < V_A$.				
II.	A. Set second MSB to 1	1100	12	10.4	LOW
	B. Reset it to 0 since $V_{A'} > V_A$.	1000	8	10.4	HIGH
III.	A. Set third MSB to 1	1010	10	10.4	HIGH
	B. Leave it at 1 since $V_{A'} < V_A$.				
IV.	A. Set LSB to 1.	1011	11	10.4	LOW
	B. Reset it to 0 since $V_{A'} > V_A$.	1010	10	10.4	HIGH
	Digital number now in the REGISTER is the final result.	<u>1010</u>			

Circuitry

The successive-approximation converter is much faster than the digital-ramp type but it requires more complex control circuitry. It will be instructive to look at this circuitry because it uses many of the ideas covered earlier. Figure 10.12 shows the complete diagram for a 4-bit successive-approximation A/D converter. The circuitry is easily extended to any number of bits.

FFs A, B, C, and D form the register, which provides the digital input to the D/A converter. At the end of the conversion process the register holds the digital result. A ring counter composed of D-type FFs V, W, X, Y, and Z provides the sequencing action in synchronism with the CLOCK input. The various gates and FF M provide the logic for implementing the successive-approximation method. The circuit operation proceeds as follows:

1. A momentary start pulse is applied to clear the register (A, B, C, D) to zero and set control FF M to the HIGH state. In addition, it clears FFs W, X, Y, Z and sets FF V so that the ring counter begins in the 00001 state. With the register outputs at zero, the value of $V_{A'}$ is 0 V, so the comparator output is HIGH (COMP = 1).

2. With $M = 1$, AND gate 9 transmits the clock pulses to the ring counter. The first clock pulse brings the ring counter to the 10000 state with $Z = 1$.

3. With COMP = 1 and $Z = 1$, NAND gate 1 goes LOW, which sets FF D HIGH. This is the MSB input to the D/A converter and causes $V_{A'}$ to equal 8 V. If this 8 V exceeds V_A, COMP will go LOW and $\overline{\text{COMP}}$ will go HIGH. This produces a negative transition at NAND gate 5 output that clocks FF

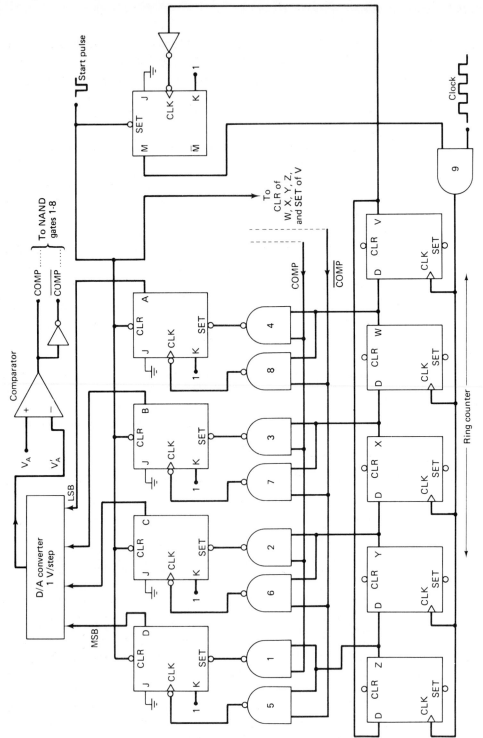

Figure 10.12 Circuitry for a 4-bit successive-approximation converter.

D back LOW (since $J = 0$, $K = 1$).* If the 8 V is less than V_A, COMP will stay HIGH and FF D will be kept in the 1 state.

4. The second clock pulse brings the ring counter to the 01000 state with $Y = 1$.

5. NAND gate 2 now sets FF C HIGH, producing a new value of $V_{A'}$ at the D/A output. If $V_{A'}$ exceeds V_A, $\overline{\text{COMP}}$ goes HIGH, producing a negative transition at NAND gate 6 which clocks FF C back LOW. Otherwise, FF C is kept in the 1 state.

6. The third and fourth clock pulses repeat the above operations on register FFs B and A. After the fourth pulse, the register contains the digital representation of the analog input V_A.

7. The fifth clock pulse brings the ring counter back to 00001 with FF $V = 1$. When V goes HIGH, it produces a negative-going transition to clock FF M to the LOW state. This disables AND gate 9 so that no further clock pulses can affect the ring counter.

8. Although the above operation requires five clock periods, the A/D conversion is actually complete after four periods. The fifth pulse is used to terminate the operation.

EXAMPLE 10.8   Compare the maximum conversion times of a 10-bit digital-ramp A/D converter and a 10-bit successive approximation A/D converter if both utilize a 1-MHz clock frequency.

Solution: For the digital-ramp converter, the maximum conversion time is

$$2^N \times 1 \ \mu s = 2^{10} \times 1 \ \mu s = 1024 \ \mu s$$

For a 10-bit successive approximation converter, the conversion time is always 10 clock periods or

$$10 \times 1 \ \mu s = 10 \ \mu s$$

Thus, it is 100 times faster than the digital-ramp converter.

10.8   DIGITAL VOLTMETER

One of the most common applications of A/D conversion is the *digital voltmeter* (DVM). Any type of A/D converter can be used as part of a DVM. In the DVM, the analog input is converted to a BCD-code representation, which is then decoded and displayed on some type of numerical readout. Figure 10.13 shows a three-digit

---

*It may be necessary to add a small amount of signal delay at the output of NAND gate 5 to ensure that the SET returns HIGH before the negative transition is applied to *CLK*. The same is true for NAND gates 6, 7, and 8.

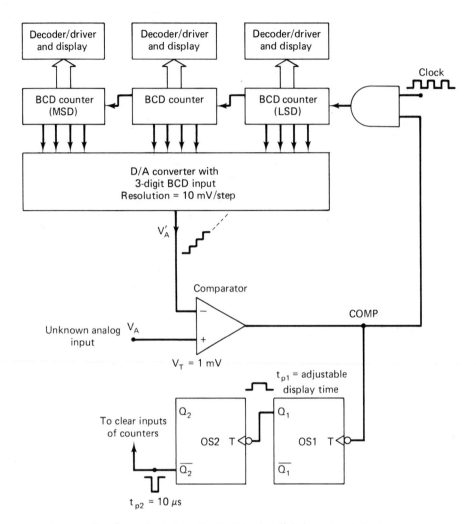

Figure 10.13   Simple continuous-reading DVM using digital-ramp technique.

DVM which uses the digital-ramp technique. The three cascaded BCD counters provide the digital inputs to a BCD-type DAC, which produces a 10 mV/step output $V_{A'}$. Each BCD counter feeds a decoder/driver and display so that each digit of the count is continually being displayed as the counters progress from 000 to 999.

The clock pulses are gated into the counters with the COMPARATOR output. As long as $V_A > V_{A'}$, COMP = 1 and the counter will receive pulses. As the counter advances, the $V_{A'}$ waveform goes up 10 mV per step until $V_{A'}$ exceeds V_A by 1 mV. At that point, COMP goes LOW and disables the AND gate so that the counter will no longer advance. The negative transition at COMP also triggers one-shot OS1, called the *display time* OS, so the Q_1 output goes HIGH for a time t_{p1} determined by the values of timing resistor and capacitor used. After t_{p1}, the negative transition at Q_1 triggers OS2, which provides a short pulse at $\overline{Q_2}$ to CLEAR all the BCD counters to the 0 state. This brings $V_{A'}$ to 0 V and COMP returns HIGH, allowing pulses into the counter and the process is repeated.

It should be clear that the counter holds the final count for a time t_{p1} during which time it is displayed so that it can be easily viewed. After OS1 returns LOW, the counter resets to 000 and a new conversion process begins. The display time t_{p1} is usually made adjustable by using a variable resistor as the timing resistor for OS1.

A numerical example will help illustrate this circuit's operation. Assume that V_A is 6.372 V. In order to cause the COMPARATOR output to switch LOW, $V_{A'}$ must exceed 6.373 V. Since the DAC output increases by 10 mV/step, this requires

$$\frac{6.373 \text{ V}}{10 \text{ mV}} = 637.3 \longrightarrow 638 \text{ steps}$$

Thus, the counter will read and display 638 when the conversion is complete. A small bulb can be used to display the decimal point, so the viewer sees a result of 6.38 V. The last digit is in error because of the 10-mV resolution of the D/A converter. This quantization error can be reduced by adding a fourth BCD counter and using a four-digit (16-bit) D/A converter.

The DVM can be modified to read input voltages over several ranges by using a suitable amplifier or attenuator between V_A and the comparator. For example, if V_A was 63.72 V, it could be attenuated by a factor of 10, so the comparator would receive 6.372 V at its + input and the counters would display 638 at the end of the conversion. The decimal-point indicator would be placed in front of the LSD, so the display would read 63.8 V.

10.9   TRISTATE ADC

When the tristate concept was first discussed, it was shown that many tristate devices could be connected to a common bus as long as only one device is activated at one time. The rapid emergence of microprocessors and their bus-type structure has brought about the development of tristate ADCs whose outputs can be tied to a system bus. Figure 10.14 shows a functional diagram for an ADC of this type.

This ADC converts the analog input voltage V_A to an 8-bit output. It has a full-scale rating of 2.55 V that translates to a resolution (step size) of 10 mV. The conversion process is initiated by a pulse applied to its START input. The BUSY output from the ADC goes LOW while the conversion is taking place and returns HIGH to signal the end of the conversion.

The digital output lines D_7-D_0 come from tristate latches which are part of

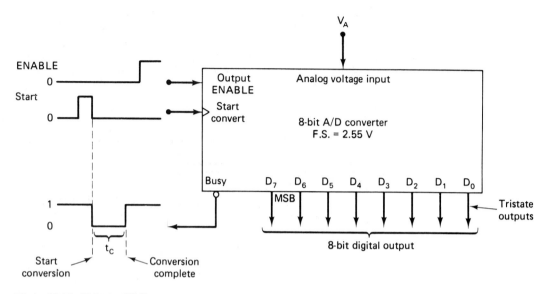

Figure 10.14   Tristate ADC.

the converter. A HIGH on the ENABLE input will enable these outputs so that the digital representation of V_A is present on these lines. A LOW on the ENABLE input puts these output lines in their HIGH-Z state. In most situations, the ENABLE input will be pulsed HIGH only after the BUSY output has indicated that the conversion is complete. If the ENABLE input is made HIGH during the t_c interval, the output lines will indicate the results of the previous A/D conversion.

10.10 TRANSDUCTION OF PROCESS VARIABLES

The monitoring and control of relatively complex processes requires the acquisition of low-level signals from a variety of sources. These signals are present at the outputs of transducers and are often in the millivolt range, perhaps 10 or 100 mV full scale. In addition, there is usually a relatively large amount of noise present in the process environment. Figure 10.15 illustrates what is involved in acquiring these signals and converting them to a digital representation for the digital-control circuitry or computer.

The transducer generates a voltage somewhere between 0 V and 100 mV between its two output terminals, neither of which is at ground (e.g., from a bridge circuit). A long shielded cable carries the transducer output to a cabinet that contains the majority of the electronic circuitry. The noise picked up by the v_1 and v_2 leads has two components: *common-mode noise*, which is the same on both

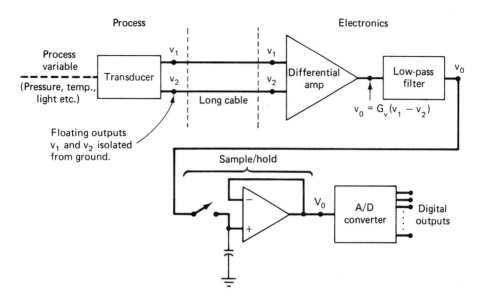

Figure 10.15 System for acquiring and processing low-level signals.

leads (such as 60 Hz hum), and *differential noise*, which is different for each lead and is usually a lot smaller. The differential amplifier eliminates the common-mode noise because it amplifies the *difference* in v_1 and v_2. The low-pass filter hopefully eliminates most of the differential noise. The low-pass output v_0 is ideally a noise-free voltage proportional to v_1-v_2, the transducer output, and scaled upward to a full-scale range of 10 V to be applied to the A/D converter.

The sample-and-hold circuit may or may not be necessary, depending on the circumstances. If the voltage v_0 is changing continuously, then it may produce an erroneous A/D conversion if the A/D conversion time is too long compared to the rate at which v_0 is changing. If such is the case, the sample-and-hold circuit is used to sample the value of v_0 by momentarily closing the switch and allowing C to charge to the present value of v_0. The switch opens and C holds this sampled value of v_0 constant while it is converted to digital by the A/D converter. The sample-and-hold circuit is not needed if the A/D conversion time is short compared to the rate at which v_0 changes.

Multiplexing

When signals from several transducers are to be processed, a multiplexing technique can be used so that one A/D converter may be time-shared. The basic scheme is illustrated in Figure 10.16 for a three-channel acquisition system. Switches $S_1, S_2,$ and S_3 are used to switch each analog signal sequentially to the A/D converter. The control circuitry controls the actuation of these switches, which

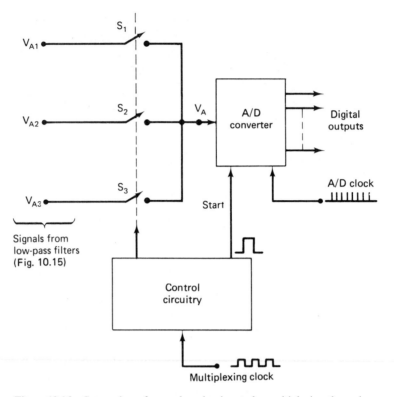

Figure 10.16   Conversion of several analog inputs by multiplexing through one A/D converter.

are usually semiconductor switches, so that only one switch is closed at a time. The control circuitry also generates the START pulse for the ADC. The operation proceeds in the following manner:

1. The control circuit closes S_1, which connects V_{A1} to the ADC input.
2. A START pulse is generated and the ADC converts V_{A1} to its digital equivalent. S_1 stays closed long enough to allow the conversion to be completed.
3. The ADC outputs representing V_{A1} can now be transferred to another location. In many cases this location would be in the memory of a computer.
4. The control circuit opens S_1 and closes S_2 to connect V_{A2} to the ADC input.
5. Steps 2 and 3 are repeated.
6. S_2 is open and S_3 is closed to connect V_{A3} to the ADC input.
7. Steps 2 and 3 are repeated.

The multiplexing clock controls the rate at which the analog signals are sequentially switched into the ADC. The maximum rate is determined by the delay time of the switches and the conversion time of the ADC. The switch delay time can be minimized by using semiconductor switches such as the CMOS bilateral switches described in Chapter 8. It may be necessary to connect a sample-and-hold circuit at the input of the ADC if the analog inputs will change significantly during the ADC conversion time.

QUESTIONS AND PROBLEMS

10.1 An 8-bit D/A converter produces an output voltage of 2.0 V for an input code of 01100100. What will be the value of V_{OUT} for an input code of 10110011?

10.2 Determine the weights of each input bit for the D/A converter of Problem 10.1.

10.3 What is the resolution of the D/A converter of Problem 10.1? Express it in volts and in per cent.

10.4 What is the resolution in volts of a 10-bit D/A converter whose F.S. output is 5 V?

10.5 How many bits are required for a D/A converter so that its F.S. output is 10 V and its resolution is less than 40 mV?

10.6 What is the percentage resolution of the D/A converter of Figure 10.17? What is the step size?

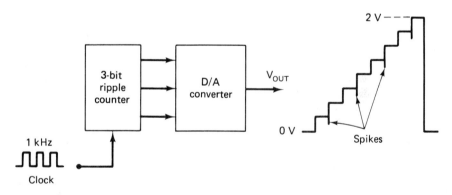

Figure 10.17

10.7 What is the cause of the negative-going spikes on the V_{OUT} waveform of Figure 10.17? (*Hint:* Note that the counter is a ripple-counter and that the spikes occur on every other step.)

10.8 TRUE or FALSE: The percentage resolution of a D/A converter depends *only* on the number of input bits.

10.9 A 12-bit (three-digit) D/A converter which uses the BCD input code has a full-scale output of approximately 2 V. Determine the step size, percentage resolution, and the value of V_{OUT} for an input code of 0110 1001 0101.

10.10 Which will have a higher percentage resolution, a D/A converter using an 8-bit binary input code or one using an 8-bit BCD code?

10.11 The step size of the D/A converter of Figure 10.5 can be changed by changing the value of R_F. Determine the required value of R_F for a step size of 0.5 V. Will the new value of R_F change the percentage resolution?

10.12 Draw the circuit diagram for a 6-bit D/A converter that uses a summing amplifier and has a step size of 50 mV. The input levels are 0 V and 2 V.

10.13 Why are precision-level amplifiers needed in a D/A converter?

10.14 Why is D/A multiplexing used?

10.15 Which of the following factors can affect the *maximum* multiplexing rate in Figure 10.8?
(a) Size of holding capacitors.
(b) Input impedance of amplifiers.
(c) Output resistance of D/A converter.
(d) Resistance of closed switch.
(e) Resistance of open switch.
(f) Settling time of D/A converter.
(g) Response time of switches.

10.16 Repeat Question 10.15 for *minimum* multiplexing rate.

10.17 Figure 10.18 shows the complete block diagram for a four-channel D/A multiplexer which converts four 4-bit inputs to their analog equivalents using a single D/A converter. The four sets of digital inputs are fed into a 4-bit, four-channel digital multiplexer which transmits *one* of the sets of digital inputs to the D/A converter, depending on the values of the S_1 and S_0 select lines.

A 1-kHz clock controls the multiplexing by causing the control circuit to generate the sequence of S_1, S_0 select codes while activating the appropriate multiplexing switch SW1–SW4. For example, when S_1S_0 is 00, SW1 is closed and the *A* digital input group is transmitted through the digital multiplexer to the D/A converter and its analog equivalent is stored on C_1, producing output V_A. The next clock cycle changes S_1S_0 to 01 and closes SW_2 (SW_1 now opens), allowing the analog equivalent of the *B* inputs to appear at V_B, and so on for the 10 and 11 select codes.
(a) Determine the values of V_A, V_B, V_C, and V_D for the following digital inputs: $A_3A_2A_1A_0 = 0101$; $B_3B_2B_1B_0 = 1001$; $C_3C_2C_1C_0 = 1101$; and $D_3D_2D_1D_0 = 0001$.
(b) With a 1-kHz clock, how long does each holding capacitor have to maintain its charge before it is recharged?
(c) A capacitor will lose 10 per cent of its charge in 0.1 time constants. What is the minimum value of C that can be used if the capacitor is to hold its voltage within 10 per cent? Assume that Z_{IN} of each amplifier is 1 MΩ and the switches have 10^{10} Ω OFF resistance.
(d) If a fifth channel is added, what minimum clock frequency must be used, assuming the value of capacitor calculated in part (c)?
(e) Use the CMOS SCL4016A bilateral switches for SW1–SW4 (Appendix III) and design the complete control circuitry for generating S_0 and S_1 and activating the multiplexing switches.

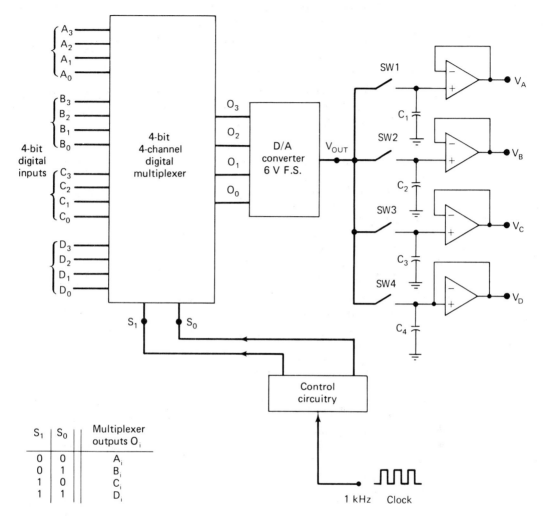

Figure 10.18

10.18   The waveform of Figure 10.19 is observed at the V_A output of the D/A multiplexer of Figure 10.18. What causes the tilt in the waveform? How can it be decreased?

10.19   An 8-bit D/A converter has a relative accuracy of 0.2 per cent F.S. If the DAC has a full-scale output of 10 mA, what is the most that it can be in error for any digital input? If the D/A output reads 50 μA for a digital input of 00000001, would this be within the specified range of accuracy?

10.20   The control of a positioning device may be achieved using a *servomotor*, which is a motor designed to drive a mechanical device as long as an error signal exists. Figure 10.20 shows a simple servo-controlled system which is controlled by a digital input that could be coming directly from a computer or from an output

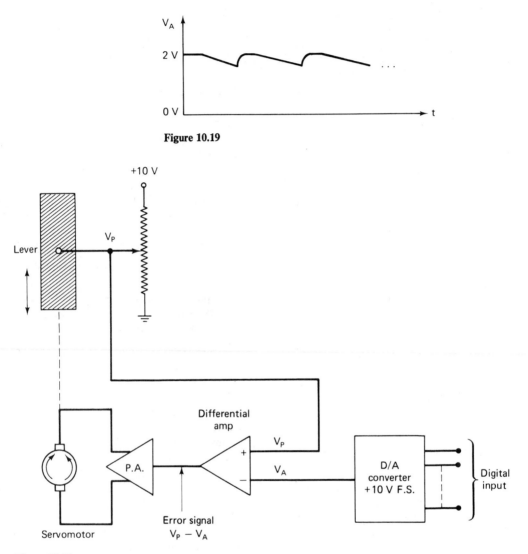

Figure 10.19

Figure 10.20

medium such as magnetic tape. The lever arm is moved vertically by the servo-motor. The motor rotates clockwise or counterclockwise, depending on whether the voltage from the power amplifier (PA) is positive or negative. The motor stops when the PA output is zero.

The mechanical position of the lever is converted to a dc voltage by the poten-tiometer arrangement shown. When the lever is at its zero reference point, $V_P =$ 0 V. The value of V_P increases at the rate of 1 V/inch until the lever is at its highest point (10 inches) and $V_P = 10$ V. The desired position of the lever is provided as a digital code from the computer and is then fed to a D/A converter, producing

V_A. The *difference* between V_P and V_A (called *error*) is produced by the *differential* amplifier and amplified by the PA to drive the motor in the direction that causes the error signal to decrease to 0, that is, moves the lever until $V_P = V_A$.

(a)   If it is desired to position the lever within a resolution of 0.1 inch, what is the number of bits needed in the digital input code?

(b)   In actual operation, the lever arm might oscillate slightly around the desired position, especially if a *wirewound* potentiometer is used. Can you explain why?

10.21   Outline the operation of the basic A/D converter of Figure 10.9.

10.22   A certain operational amplifier comparator has $G_V = 5000$ and full-scale outputs of ± 12 V. With $V_1 = +6.372$ V, what value of V_2 is needed to cause the output to switch from $+12$ V to -12 V?

10.23   An 8-bit digital-ramp A/D converter with a 40-mV resolution uses a clock frequency of 2.5 MHz and a comparator with $V_T = 1$ mV. Determine:

(a)   The digital output for $V_A = 6.000$ V.

(b)   The digital output for 6.035 V.

(c)   The maximum and average conversion times for this ADC.

10.24   Why were the digital outputs the same for parts (a) and (b) of Problem 10.23?

10.25   What would happen in the A/D converter of Problem 10.23 if an analog voltage of $V_A = 10.853$ V were applied to the input? What waveform would appear at the D/A output? Incorporate the necessary logic in this A/D converter so that an "over-scale" indication will be generated whenever V_A is too large.

10.26   An A/D converter has the following characteristics: resolution, 12 bits; relative accuracy, 0.03 per cent F.S.; full-scale output, $+5$ V.

(a)   What is the quantization error in volts?

(b)   What is the total possible error in volts?

10.27   The quantization error of an A/D converter such as the one in Figure 10.11 is always positive since the $V_{A'}$ value must exceed V_A in order for the COMPARATOR output to switch states. This means the value of $V_{A'}$ could be as much as 1 LSB greater than V_A. This quantization error can be modified so that $V_{A'}$ would be within $\pm\frac{1}{2}$ LSB of V_A. This can be done by adding a fixed voltage equal to $\pm\frac{1}{2}$ LSB ($\frac{1}{2}$ of a step) to the value of $V_{A'}$. Figure 10.21 shows this symbolically for a converter that has a resolution of 10 mV/step. A fixed voltage of $+5$ mV is added to the D/A output in the summing amplifier and the result, $V_{A''}$, is fed to the comparator which has $V_T = 1$ mV.

For this modified converter, determine the digital output for:

(a) $V_A = 5.022$ V

(b) $V_A = 5.028$ V

Determine the quantization error in each case by comparing $V_{A'}$ and V_A. Note that the error is positive in one case and negative in the other.

10.28   Explain how the successive-approximation converter improves the conversion speed over a digital ramp converter. How many clock periods are needed for an 8-bit successive-approximation converter? For a 10-bit converter?

10.29   Construct a table such as Table 10.1 showing the sequence of operations for a 6-bit successive-approximation converter which has a resolution of 0.1 V per step and an analog input of $V_A = 6.25$ V. Assume that $V_T = 1$ mV for the comparator.

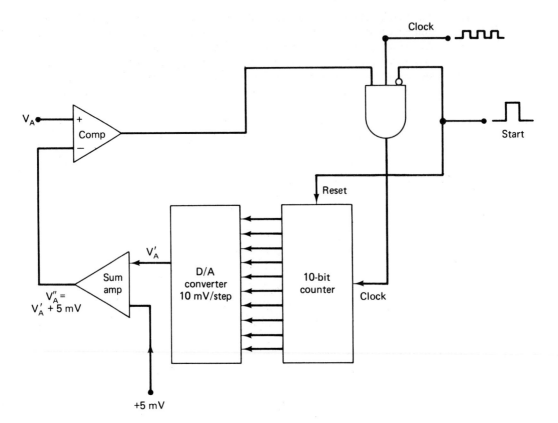

Figure 10.21

10.30 In the circuit of Figure 10.12 why is it necessary to gate $\overline{COMP}$ with Z, before feeding it to the CLK input of FF D?

10.31 When the START pulse is applied in Figure 10.12, FF V is set HIGH. Why won't the resulting negative-going transition at the CLK input of FF M cause it to go to the 0 state?

10.32 In the circuit of Figure 10.12 there is a possible race problem between the gates driving the S inputs of the REGISTER FFs and those driving the CLK inputs. Explain the race problem and suggest a means for eliminating it.

10.33 A voltage $V_A = 3.853$ V is applied to the DVM of Figure 10.13. Assume that the counters are initially reset to zero and sketch the waveforms at $V_{A'}$, COMP, Q_1, $\overline{Q_2}$, and the AND gate output for *two* complete cycles. Use a 100-kHz clock and $t_{p1} = 1$ s.

10.34 The DVM of Figure 10.13 has the disadvantage of a blinking display like the frequency counter discussed in Chapter 7. Modify the circuitry using D-type FFs such as was done for the frequency counter (Problem 7.45) so that only the final digital result is displayed. (*Hint:* The Q_1 waveform can be used to transfer the counter contents to the D FFs.)

10.35   Use the CMOS SCL4016A bilateral switches for S_1, S_2, and S_3 in Figure 10.16 and design the necessary control logic so that each analog input is converted to its digital equivalent in sequence. The A/D converter is a 10-bit successive-approximation type using a 50-kHz clock signal and requires a 10-μs duration start pulse to begin each conversion. The digital outputs are to remain stable for 100 μs after the conversion is complete before switching to the next channel.

11

MEMORY DEVICES

One of the major advantages that digital systems have over analog systems is the ability to easily store large quantities of digital information and data for short or long periods of time. This memory capability is what makes digital systems so versatile and adaptable to many situations. For example, in a digital computer the internal main memory stores instructions that tell the computer what to do under *all* possible circumstances so that the computer will do its job with a minimum amount of human intervention.

This chapter is devoted to a review of the most commonly used types of memory devices and systems. We have already become very familiar with the flip-flop, which is an electronic memory device. We have also seen how groups of FFs called registers can be used to store information and how this information can be transferred to other locations. FF registers are high-speed memory elements which are used extensively in the internal operations of a digital computer where digital information is continually being moved from one location to another. Recent advances in LSI technology have made it possible to obtain large numbers of FFs on a single chip arranged in various memory-array formats. These bipolar and MOS semiconductor memories are the fastest memory devices available and their cost has been continuously decreasing as LSI technology improves.

Because of their high speed, semiconductor memories can be used as the *internal* memory of a computer, where fast operation is important. The internal memory of a computer is in constant communication with the rest of the computer as it executes a program of instructions. In fact, the instructions and data are usually stored in this internal memory.

A reasonably good compromise between speed and cost is provided by magnetic core memories. This type of memory has been widely used as the internal working memory of digital computers since the later 1950s. Its popularity is due mainly to its high speed at relatively low cost. Recently, MOS memories have begun to replace magnetic cores in many computers. The MOS memories are smaller, faster, and require less power than magnetic cores, and their cost is approaching that of magnetic cores. The IBM 370, a very large computer, uses MOS memory for internal storage. Many minicomputers and almost all microcomputers use MOS (or bipolar) memories for internal storage.

Although semiconductor and magnetic core memories are well suited for high-speed internal memory, their cost per bit of storage prohibits their use as *bulk* memory devices. Bulk memory refers to memory *external* to the main computer, which is capable of storing large quantities of information (millions of bits). External bulk memory, which is normally much slower than internal memory, is used to transfer data and information to a computer's internal memory; likewise, data from internal memory is often transferred to bulk memory for long-term storage. Magnetic tape, magnetic disc, and magnetic drum are popular bulk memory devices that are much less expensive in cost per bit than internal memory devices. Two newer entries into the bulk memory field are *charge-coupled devices* (CCD) and *magnetic bubble memory* (MBM). These are semiconductor devices which are still in the developmental stage and it is too early to predict their impact in the area of bulk storage.

11.1 APPLICATION OF VARIOUS MEMORY DEVICES

Each memory device is better suited for certain applications than others. For example, bulk storage of very large quantities of data is usually done on magnetic tape. Payroll information and records are stored on magnetic tape to be used each time a business computer computes the employee payroll. We can appreciate the different types of memory functions by taking a look at a typical large computer system such as the one illustrated in Figure 11.1.

This diagram shows the interfacing of the internal memory with external storage and the input/output devices. The card reader is used as the input device and it converts the information punched on cards to binary signals that are recorded on the magnetic tape. The cards fed to the card reader are prepared by a programmer who determines the instructions and data (the program) which the computer requires to solve a problem or perform some other task. The central processor can only work on one program at a time. However, many programs are run through the card reader and placed on magnetic tape well ahead of time. This is done for a very good reason: binary information can be transferred from the magnetic tape to the central processor much faster than it can from the card reader. When the central processor finishes a program, it sends a control signal to the magnetic tape unit requesting the next program stored on the tape. New programs are continuously being fed into the card reader to be placed on tape

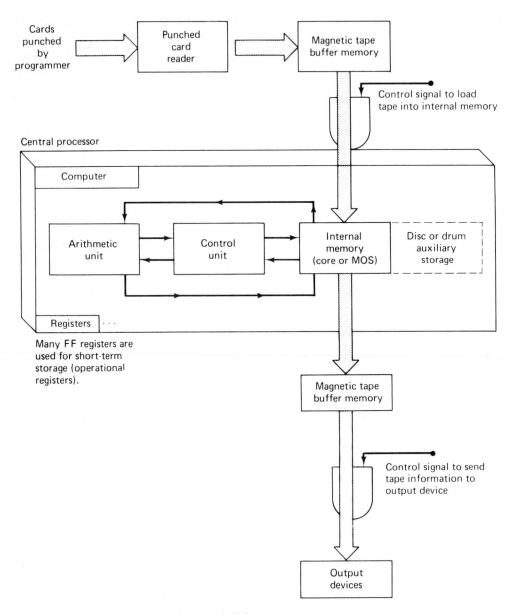

Figure 11.1   Various memory devices in a digital computer system.

so that the central processor rarely has to wait long between jobs. In essence, the magnetic tape unit acts as a buffer between the very slow input unit (card reader) and the very fast central processor.

    The instructions and data from the magnetic tape are loaded into the internal main memory unit of the central processor. This internal memory then stores one complete program of instructions and associated data. It works in conjunction with the control and arithmetic units to perform the program execution. Instructions and data are fetched from the internal memory and sent to the control and arithmetic units, respectively, and data are sent from the arithmetic unit to the internal memory.

In addition to the internal memory, some computers use *auxiliary storage* in situations where it is desired to extend the capacity of the internal memory. Magnetic discs and drums are often used for this purpose. This auxiliary storage handles instructions or data that exceed the capacity of the internal memory. The information stored in the slower auxiliary memory is transferred to the faster internal memory when it is needed. This technique can be used to increase the *virtual* capacity of the internal memory to up to 20 million bits.

Once the central processor has completed a program, the output data are transferred from the internal memory to an output buffer memory, which is usually a magnetic tape unit. The magnetic tape stores the output results until the output unit (printer, CRT, card puncher, etc.) is ready for it. This is done because the output units are usually slow devices which cannot keep up with the central processor. In essence, the magnetic tape unit acts as a buffer between the very fast central processor and the very slow output units.

The memory operations illustrated in Figure 11.1 are typical for a larger computer installation. Minicomputers and microcomputers employ memory devices in a more or less similar manner. One major difference in mini- and microcomputers is that the slower central processor communicates directly with the input/output devices without the need for a buffer memory. In these smaller computers magnetic tape and disc are used to store programs and data that are transferred into the internal memory when the computer is ready for them.

11.2   MEMORY TERMINOLOGY

Before looking at the various memory devices, we will define some of the more important terms that are common to most memory systems.

Memory Cell   This is a device or electrical circuit used to store a single bit (0 or 1). Examples of memory cells include a flip-flop, a single magnetic core, and a single spot on magnetic tape.

Memory Word   A group of bits (cells) in a memory that represents information or data of some type. For example, a register consisting of eight FFs can be considered to be a memory that is storing an 8-bit word. Word sizes in modern computers typically range from 4 to 64 bits, depending on the size of the computer.

Byte   This is a special term used for an 8-bit word. A byte always consists of 8 bits, which is the most common word size in microcomputers.

Capacity   This is a way of specifying how many bits can be stored in a particular memory device or complete memory system. To illustrate, suppose that we have a memory which can store 4096 20-bit words. This represents a total capacity of 81,920 bits. We could also express this memory's capacity as 4096 × 20. When expressed this way, the first number (4096) is the number of words and the second number (20) is the number of bits per word (word size). The number of words in a memory is often a multiple of 1024. It is common to use the designation "1K" to represent 1024 when referring to memory capacity. Thus, a memory that has a storage capacity of 4K × 20 is actually a 4096 × 20 memory.

EXAMPLE 11.1   A certain semiconductor memory chip is specified as 2K × 8. How many words can be stored on this chip? What is the word size? How many total bits can this chip store?

Solution:

$$2K = 2 \times 1024 = \textbf{2048} \text{ words}$$

Each word is 8 bits (one byte). The total number of bits is therefore

$$2048 \times 8 = \textbf{16,384} \text{ bits}$$

Address   This is a number that identifies the location of a word in memory. Each word stored in a memory device or system has a unique address. Addresses are always specified as a binary number, although octal, hexadecimal, and decimal numbers are often used for convenience. Figure 11.2 illustrates a small memory consisting of eight words. Each of these eight words has a specific address represented as a 3-bit number ranging from 000 to 111. Whenever we refer to a specific word location in memory, we use its address code to identify it.

Read Operation   This is the operation whereby the binary word stored in a specific memory location (address) is sensed and then transferred to another location. For example, if we want to use word 4 of the memory of Figure 11.2 for

Addresses

0 0 0	Word 0
0 0 1	Word 1
0 1 0	Word 2
0 1 1	Word 3
1 0 0	Word 4
1 0 1	Word 5
1 1 0	Word 6
1 1 1	Word 7

Figure 11.2 Each word location has a specific binary address.

some purpose, we have to perform a read operation on address 100. The read operation is often called a *fetch* operation, since a word is being fetched from memory. We will use both terms interchangeably.

Write Operation The operation whereby a new word is placed into a particular memory location. It is also referred to as a *store* operation. Whenever a new word is written into a memory location, it replaces the word that was previously stored there.

Access Time This is a measure of a memory device's operating speed. It is the amount of time required to perform a read operation. More specifically, it is the time between the memory receiving a read command signal and the data becoming available at the memory output. The symbol t_{ACC} is used for access time.

Cycle Time Another measure of a memory device's speed. It is the amount of time required for the memory to perform a read or write operation and then return to its original state ready for the next command. Cycle time is normally longer than access time.

Volatile Memory This refers to any type of memory that requires the application of electrical power in order to store information. If the electrical power is removed, all information stored in the memory will be lost. Many semiconductor memories are volatile, while all magnetic memories are nonvolatile.

Random Access Memory (RAM) This refers to memories in which the actual physical location of a memory word has no effect on how long it takes to read from or write into that location. In other words, the access time is the same for any address in memory. Most semiconductor memories and magnetic core memories are RAMs.

Sequential Access Memory (SAM) A type of memory in which the access time is not constant, but varies depending on the address location. A particular stored word is found by sequencing through all address locations until the desired address is reached. This produces access times which are much longer than those of random access memories. Examples of sequential access memory devices include magnetic tape, disc, and drum; charge-coupled devices; and magnetic bubble memory.

Read/Write Memory (RWM) Any memory that can be read from or written into with equal ease.

Read-Only Memory (ROM) Refers to a broad class of semiconductor memories designed for applications where the ratio of read operations to write operations is very high. Technically, a ROM can be written into (programmed) only once, and this operation is normally performed at the factory. Thereafter information can only be read from the memory. Other types of ROM are actually read-mostly memories (RMM) which can be written into more than once, but the write operation requires special circuitry and is not done while the memory is part of a working system.

Static Memory Devices   Semiconductor memory devices in which the stored data will remain permanently stored as long as power is applied, without the need for periodically rewriting the data into memory.

Dynamic Memory Devices   Semiconductor memory devices in which the stored data *will not* remain permanently stored, even with power applied, unless the data are periodically rewritten into memory. The latter operation is called a *refresh* operation.

11.3   GENERAL MEMORY OPERATION

Although each type of memory is different in its internal operation, there are certain basic operating principles that are the same for all memory systems. An understanding of these basic ideas will help in our study of individual memory devices.

Every memory system requires several different types of input and output lines to perform the following functions:

1.   Select the address in memory that is being accessed for a read or write operation.

2.   Select either a read or write operation to be performed.

3.   Supply the input data to be stored in memory during a write operation.

4.   Hold the output data coming from memory during a read operation.

5.   Enable (or disable) the memory so that it will (or will not) respond to the address inputs and read/write command.

Figure 11.3 illustrates these basic functions in a simplified diagram of a 32×4 memory that stores 32 4-bit words. Since the word size is 4 bits, there are four data input lines $I_0 - I_3$ and four data output lines $O_0 - O_3$. During a write operation the data to be stored into memory have to be applied to the data input lines. During a read operation the word being read from memory appears at the data output lines.

This memory has 32 different storage locations and therefore 32 different addresses, ranging from 00000_2 to 11111_2 (0 to 31 in decimal). Thus, we need *five* address inputs, $A_0 - A_4$, to specify one of the 32 address locations. To access one of the memory locations for a read or write operation, the 5-bit address code for that particular location has to be applied to the address inputs.

The read/write (R/W) input line determines which memory operation is to take place. Some memory systems use two separate inputs, one for read and one for write. When a single R/W input is used, it is common to have $R/W = 1$ commanding a read operation and $R/W = 0$ a write operation.

Many memory systems have some means for completely disabling all or part of the memory so that it will not respond to the other inputs. This is represented

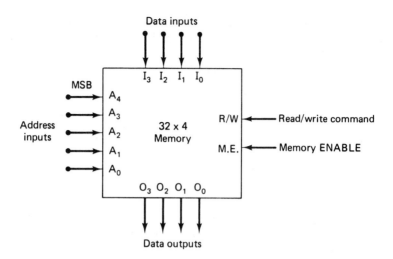

Figure 11.3   General diagram for a 32×4 memory.

in Figure 11.3 as the MEMORY ENABLE input, although it can have different names in the various memory systems. It is an active-HIGH input that enables the memory to operate normally when it is kept HIGH. A LOW on this input disables the memory so that it will not respond to the address and R/W inputs. This type of input is useful when several memory modules are combined to form a larger memory. We will examine this idea later.

EXAMPLE 11.2   A certain memory has a capacity of 4K $\times$ 8.
   (a)   How many data input and data output lines does it have?
   (b)   How many address lines does it have?
   (c)   What is its capacity in bytes?

Solution:
   (a)   Eight of each, since the word size is eight.
   (b)   The memory stores 4K $= 4 \times 1024 = 4096$ words. Thus, there are 4096 memory addresses. Since $4096 = 2^{12}$, it requires a 12-bit address code to specify one of 4096 addresses.
   (c)   A byte is 8 bits. This memory has a capacity of 4096 bytes.

The example memory in Figure 11.3 illustrates the important input and output functions common to most memory systems. Of course, each type of memory will often have other input and output lines that are peculiar to that memory. These will be described as we discuss the individual memory types.

11.4 MAGNETIC CORES

The basic memory cell in magnetic-core memories is a small toroidal (ring-shaped) piece of ferromagnetic material called a *ferrite core*. These cores are typically 50 mils in diameter, although more recently 18-mil cores have been introduced.

Ferrite cores have a very low reluctance, which means that they are easily magnetized, and they have a high retentivity, which means that they will stay magnetized indefinitely when not disturbed.

Figure 11.4(a) shows a ferrite core, many times its actual size, with a wire shown threaded through the core. If a current is passed through the wire, magnetic flux will be produced around the wire in a direction dependent on the direction

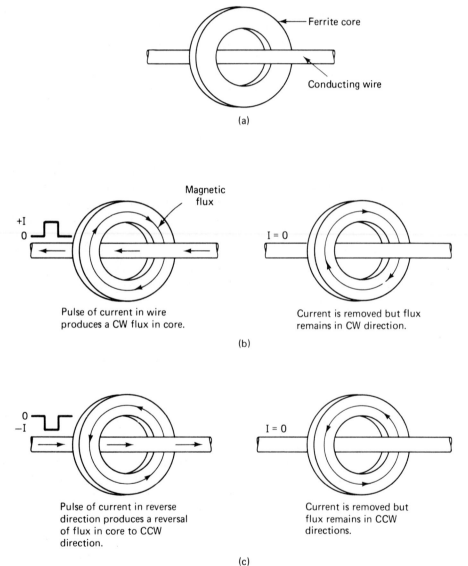

Figure 11.4   (a) Ferrite core; (b) & (c) magnetizing the core in different directions.

of current.* Because of the ferrite core's low magnetic reluctance, these lines of flux will enter the core and magnetize the core in either a clockwise or counterclockwise direction. When the current in the wire is reduced to zero, the retentivity of the ferrite core results in the core retaining a large portion of the flux, *provided that the wire current exceeded the threshold magnetizing current value, I_M.*

To illustrate, consider Figure 11.4(b), which shows a pulse of current I flowing right to left. This establishes a clockwise (CW) flux pattern in the core. If the value of I is *greater than* I_M, then when the current is reduced to zero, the core will remain saturated (fully magnetized) in the CW direction. This current pulse need only be applied for a short duration to produce the required magnetization. Typically, current pulses of 0.1–1.0 μs pulse width are used to magnetize the ferrite cores.

Now consider Figure 11.4(c), where the direction of current has been reversed. Assume that the core has been previously magnetized in the CW direction. If the reversed current pulse has an amplitude greater than $-I_M$, the core will become magnetized in the CCW direction and will remain that way when the current is reduced to zero. This is called *switching the core* from one magnetic state to the other. If the reversed current pulse is smaller than $-I_M$, the core will not switch but will remain in its CW state when the current returns to zero.

Once the core is magnetized in the CCW direction, it can be switched back to the CW direction by applying a current from right to left as in Figure 11.2(b), provided that it is larger than I_M. In essence, then, the magnetic core acts like a magnetic flip-flop which can be switched from one state to another by applying appropriate electrical pulses. It is important to remember that the core will switch from one flux direction to the other only if the current pulse is greater than I_M. The value of I_M depends on core size and material but is typically 50–100 mA.

Magnetic Core Logic

The two possible states of magnetization of the ferrite core act just like the logic levels of a flip-flop and can be assigned logic values. The usual convention is:

$$\text{CW flux} = \text{logical 1 state}$$

$$\text{CCW flux} = \text{logical 0 state}$$

We have seen how a given core can be pulsed to the 0 state (CCW) and the 1 state (CW). As with FFs, the terms *clearing* and *setting* are used to denote these respective operations. In addition, the terms *writing* and *storing* are used to denote the operation of placing a 0 or 1 into the core.

---

*The right-hand rule can be used to determine the flux direction as follows: Grasp the wire in the right hand with the thumb pointing in the direction of *conventional* current. The fingers then wrap around the wire in the direction of the flux lines.

Sensing a Core

Once a core has been pulsed to a given logic state, it can retain that logic state indefinitely without the need for any electrical power. Each core is a memory cell storing 0 or 1. Since a core's logic state is magnetic rather than electrical, a somewhat unusual technique is required to convert the core's stored information to an electrical signal to determine whether the core contains a 0 or 1. This technique of *sensing* or *reading* the core requires a second wire, called a *sense line*, threaded through the core as illustrated in Figure 11.5(a).

To sense or read the state of the core, a negative current pulse is applied to the input winding to bring the core to the 0 state. A small voltage pulse will be induced in the sense line with an amplitude and duration that will depend on what the state of the core was prior to its being reset to 0. If the core was already magnetized in the 0 state, then a negative current pulse would cause only a small change in the already established CCW flux. This small change in flux would induce a small voltage in the sense line (Faraday's law) as shown in Figure 11.5(b). If the core was in the 1 state, the negative current pulse would produce a much larger change

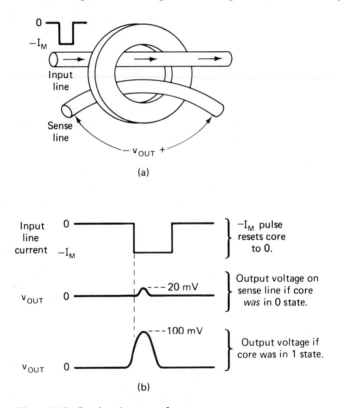

(a)

(b)

Figure 11.5 Sensing the state of a core.

in flux, since the core would go from CW saturation to CCW saturation. This larger flux change would induce a larger voltage in the sense line. Thus, a small output on the sense line indicates that the core originally contained a 0, and a large output on the sense line indicates that it originally contained a 1.

As Figure 11.5(b) shows, the 1-state output waveform has a much larger amplitude and pulse width than the 0-state output waveform. Thus, a 1 can be detected by using an amplitude-discriminating amplifier. This amplifier will amplify the 1 output pulse to several volts so that it can drive logic circuitry, but it will not amplify the 0 pulse.

Destructive Readout

The main drawback of this technique for sensing the states of magnetic cores is that all the cores end up reset to the 0 state. This means that the information previously contained in the cores is no longer there. This sensing technique is therefore referred to as *destructive readout*. Generally speaking, it is necessary to retain the information in the core memory after it has been read. This requires that the information be written back into the memory after each read operation.

11.5 MAGNETIC CORE PLANES

We have seen how a single bit of information can be written into a core for storage and how it can be read out. In digital computer systems a large number of cores are used to store information, so a technique is needed for selecting a certain core or group of cores to be written into or read out of. The selection technique most commonly used is called the *coincident-current technique*, which will now be described.

Figure 11.6 shows 16 cores arranged in a 4×4 matrix array called a *memory plane*. The four horizontal rows in the plane are designated X_0 through X_3 and the four vertical columns are designated Y_0 through Y_3. The *address* or location of any one core is specified by giving its X–Y coordinates. For example, the core in the upper-left-hand corner is located at address X_0–Y_0. Similarly, the core in the lower-left-hand corner is at address X_3–Y_0, and so on.

Each core in the plane has *four* wires threaded through its center, two input *select* lines, one *sense* line and one *inhibit* line. The X and Y lines are the select lines that are used to select the one core in the plane that is to be set or cleared. Each core has a unique X–Y pair threaded through its center. Note that the X and Y lines are shown as going to other core planes. We will see why this is done when we look at a complete core memory. A single sense line is threaded through each core and fed to a sense amplifier. A single inhibit line is also threaded through each core. We will ignore the function of this inhibit line for the moment.

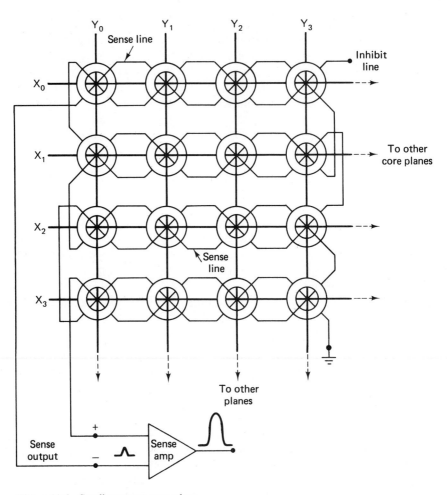

Figure 11.6   Small memory core plane.

Setting or Clearing a Core

The reason for using two select lines per core is to provide an efficient means for selecting any given core to be switched to the 0 or 1 state. For example, Figure 11.7 shows how the X_0–Y_0 core is set to the 1 state. Current pulses with an amplitude of $+I_M/2$ are simultaneously applied to both the X_0 and Y_0 lines. All the cores in the X_0 row will receive the $+I_M/2$ current from the X_0 line; similarly, all the cores in the Y_0 column will receive the $+I_M/2$ current from the Y_0 line. *Only* the X_0–Y_0 core receives $+I_M/2$ from both of these input lines for a total current of $+I_M$. Thus, only the X_0–Y_0 core will be switched to the 1 state. The other cores in the X_0 row (X_0–Y_1, X_0–X_2, X_0–Y_3) and the other cores in the Y_0 column

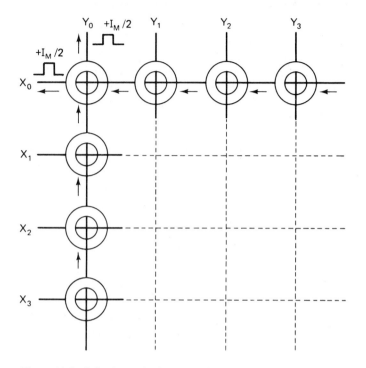

Figure 11.7 Selecting a single core to be set to the 1 state using the coincident-current technique.

$(Y_0–X_1, Y_0–X_2, Y_0–X_3)$ will *not* be set to the 1 state, because they receive only $+I_M/2$, which is not enough to switch these cores.

Using this *coincident-current* technique, any core can be selected from the plane. For instance, core $X_2–Y_3$ can be set to 1 by applying $+I_M/2$ pulses simultaneously to the X_2 and Y_3 input lines. The core that receives the full $+I_M$ current is referred to as the *full-selected core*, while those cores that receive $+I_M/2$ are called the *half-selected cores*. Only the full-selected core is affected, and each half-selected core remains in its previous state.

This same idea can be used to clear any core to the 0 state. The only difference is that *negative* current pulses are used. For instance, to clear core $X_1–Y_2$ to the 0 state would require current pulses of $-I_M/2$ simultaneously applied to the X_1 and Y_2 lines. Once again, only the full-selected core $(X_1–Y_2)$ will be switched to the 0 state while the half-selected cores are unaffected.

Sensing Any Core

We saw earlier that the state of a core is sensed by applying a $-I_M$ current pulse to clear it to 0 and then observing the sense output voltage. This same idea can be used to sense any core in the memory plane. For example, the state of the $X_2–Y_2$

core can be sensed by applying $-I_M/2$ pulses to the X_2 and Y_2 lines. The X_2–Y_2 core will receive the full $-I_M$ current, causing it to change states if it contained a 1 or remain in the same state if it contained a 0. The full-selected X_2–Y_2 core is the only one that can change states and therefore is the only core that can cause an appreciable voltage on the sense line. Thus, the output at the sense line represents the state of the X_2–Y_2 core.

Summarizing the operations that can be performed on the memory plane:

1. A 0 or 1 is written into a core by applying $-I_M/2$ or $+I_M/2$ to the correct X and Y lines.

2. The state of a core is sensed by applying $-I_M/2$ to its X and Y lines and sensing the amplitude of the output voltage on the sense line.

11.6   ADDRESSING THE CORE MEMORY

Each core in a core plane has to have a unique address that will allow it to be selected for a given operation. The total number of addresses will be equal to the number of cores in the plane. For example, the memory plane of Figure 11.6 has 16 different addresses and therefore requires a 4-bit address code. Now consider the core memory plane of Figure 11.8, which is an 8×8 array of magnetic cores.

Since there are $8 \times 8 = 64$ cores in the plane, there are 64 different addresses. Thus, a 6-bit address code is required to specify the 64 addresses. Figure 11.8 shows that the address code is stored in a 6-bit register called the *memory address register* (abbreviated MAR). The first 3 bits of the address (A_5, A_4, A_3) are used to select the correct X line; the second 3 bits (A_2, A_1, A_0) are used to select the correct Y line. Two 3-line-to-8-line decoders (Chapter 9) provide the selection logic. To illustrate, assume that the MAR holds the address 010101. The A_5, A_4, and A_3 bits (010) are decoded by the X-line decoder to produce a HIGH at the X_2 output since $X_2 = \bar{A}_5 A_4 \bar{A}_3$. The A_2, A_1, and A_0 bits (101) are decoded by the Y-line decoder to produce a HIGH at the Y_5 output. This same process is used to select any address.

The X and Y outputs from the decoders do not drive the memory core planes directly. X and Y current drivers (CD) are used to provide the high currents (usually 100–500 mA) needed to magnetize the cores.

11.7   COMPLETE CORE MEMORY

We have seen how a single core plane can be used as a memory in which any one core can be written into or read from. We can think of one plane as storing a number of 1-bit words (e.g., an 8×8 plane stores 64 1-bit words). When magnetic cores are used as a computer's internal memory, several core planes have to be

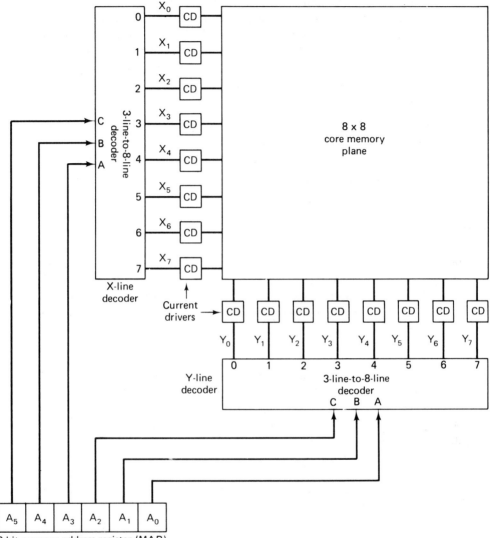

Figure 11.8   Address selection in coincident-current core memory.

combined to increase the word size. This is illustrated in Figure 11.9, where three 4 × 4 planes are combined to produce a capacity of 16 3-bit words. Note the following important points:

1.   Each plane has a separate sense line.

2.   Each plane has a separate inhibit line.

3.   The X–Y select lines are threaded through each plane so that cores in the

same relative position in each plane will be selected by the same X–Y combination. In other words, they share the same address. For example, the three cores in the X_0–Y_0 position of each plane represent a 3-bit word. Likewise, the three cores at a given X–Y coordinate position represent a 3-bit word.

Sensing a Word in Memory

The bits of a word stored at a particular address can be sensed by applying a $-I_M/2$ current to the correct X and Y lines and then detecting the sense amplifier outputs for each plane. For example, the contents of the X_0–Y_0 address can be read out by applying $-I_M/2$ current pulses to the X_0 and Y_0 lines. The X_0–Y_0 cores in *each* plane are the only ones that will be cleared to the 0 state. The sense output of a given plane will produce an output pulse if the X_0–Y_0 core in that plane contained a 1 before it was cleared to 0. Otherwise, the sense output will stay LOW. The amplified sense outputs are stored in the memory buffer register (MBR). Therefore, even though the readout process has left all the full-selected cores in the 0 state, the MBR now holds the word that was previously located at the X_0–Y_0 address. The contents of the MBR can then be transferred to other parts of the digital system which require this word.

Since the address that was just read out of is now in the all-zero state, it is necessary to put back into this address the word that was just removed. This requires a write operation, whereby the contents of the MBR (which now contains that word) are written back into this address.

Writing into Memory

We can write 0s or 1s into any address of the core memory of Figure 11.9 by applying $-I_M/2$ or $+I_M/2$ pulses to the correct X and Y lines. For instance, if the X_3 and Y_3 lines are simultaneously pulsed with $+I_M/2$, then the X_3–Y_3 cores of *each* plane will all be set to the 1 state (no other cores will be affected). If $-I_M/2$ pulses are used, then each X_3–Y_3 core will go to 0. This process has one obvious drawback: only all 0s or all 1s can be written into a given word address. This process has to be modified so that any pattern of 0s and 1s can be written into any given address. This requires the use of the inhibit line.

The single inhibit line, like the sense line, is threaded through each core in the plane and each plane has its own inhibit line. The inhibit line is used to inhibit the writing of a 1 into a core whose X and Y lines are being pulsed with $+I_M/2$. This is accomplished by simultaneously pulsing the inhibit line with a $-I_M/2$ current that opposes the X and Y currents, thereby leaving only a net current of $+I_M/2$, which is not enough to set the core. The waveforms in Figure 11.10 show how the inhibit line is used. When a 1 is to be written into a core, the inhibit line is not pulsed; when the core is to be kept at 0 (assume that it has been pre-

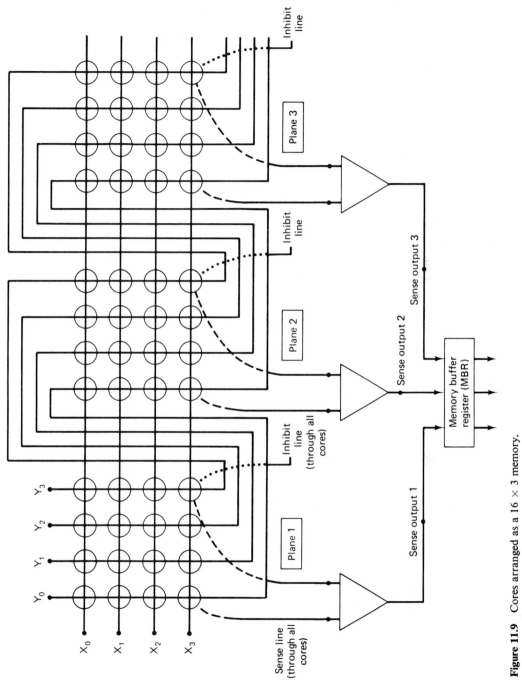

Figure 11.9 Cores arranged as a 16×3 memory.

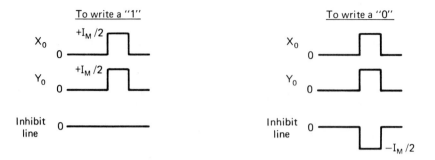

Figure 11.10   The inhibit line is used during the write operation in a core memory.

viously cleared), the inhibit line is pulsed at the same time as the X and Y lines. Note that in either case, the X and Y lines receive a $+I_M/2$ pulse.

11.8 CORE-MEMORY CHARACTERISTICS

The core-memory system just described is *nonvolatile*, which means that it will not lose the information it is storing in the event of an electrical power failure. This is a major advantage over bipolar memories, which require electrical power at all times. This undoubtedly is part of the reason that magnetic cores have been used so extensively as internal computer memories.

The core-memory system is also a random-access memory, because the time it takes to fetch a word is the same for any address in memory. The complete cycle time (for both read and write) will range from 0.3 μs to 1.0 μs for most modern computers. The access time is somewhat shorter because it includes only the time it takes for the word to be fetched from memory and placed in the MBR. Access time is typically half of the cycle time for a core memory.

We have been dealing with small, representative core memories for convenience. In practice, however, a core array can be very large. The standard size plane is 64 $\times$ 64 cores. For a large-scale memory system, up to 80 of these planes can be stacked together.

EXAMPLE 11.3   Consider a 4096 $\times$ 16 core memory system. Determine the number of: (a) cores per plane; (b) planes; (c) select lines; (d) sense lines; (e) inhibit lines; and (f) address bits.

Solution: (a) 4096; (b) 16; (c) 64 X lines and 64 Y lines; (d) 16; (e) 16; and (f) 12.

11.9 RECORDING ON MOVING MAGNETIC SURFACES

Magnetic tape, drums, and discs are memory devices that involve recording and reading magnetic spots on a moving surface of magnetic material. For each of these devices, a thin coating of magnetic material is applied to a smooth non-magnetic surface. Magnetic tapes, for example, consist of a layer of magnetic material deposited on plastic tape. Drums are thinly coated metal cylinders. Discs have the magnetic material coated on both sides of a flat disc that resembles a phonograph record.

Recording and reading of binary information on tapes, drums, and discs uses the same basic principles. Figure 11.11 illustrates the fundamental concept of recording on a moving magnetic surface. The read/write head is a core of high-permeability soft iron with a coil wound around it and a small air gap (typically 0.001 inch wide). In writing, a current is driven through the coil, establishing magnetic lines of flux in the core. These flux lines remain in the core until they encounter the air gap, which has a very high reluctance to magnetic flux. This causes the flux lines to deviate and travel through the magnetic coating on the moving surface. Thus, pulses of current in the coil result in spots of magnetism on the moving surface. These spots remain magnetized after they pass the read/write head.

The read operation is just the opposite of the write operation. During the read operation the coil is used as a sense line. As the magnetic surface is moved under the read/write head, the spots that have been magnetized produce a flux up through the air gap and into the core. This change in the core's flux induces a voltage signal in the sense winding which is then amplified and interpreted.

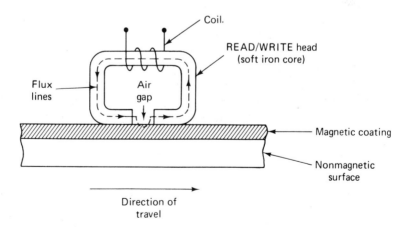

Figure 11.11 Fundamental parts for recording on moving magnetic surface.

The coil around the core usually has a center tap, so that half of the coil is used as a sense winding and the other half is heavier-gauge wire used for supplying current in the write operation.

11.10 MAGNETIC DRUMS

A magnetic drum is a metal cylinder coated with a ferromagnetic material. Drums range from 2 inches to 4 feet in diameter and up to 5 feet in length. The surface is divided into circular bands called *tracks* or *channels* (refer to Figure 11.12). There are several tracks for each inch of drum length and each track has its own READ/ WRITE head. The drums are continuously rotated at speeds of up to 10,000 revolutions per minute (rpm).

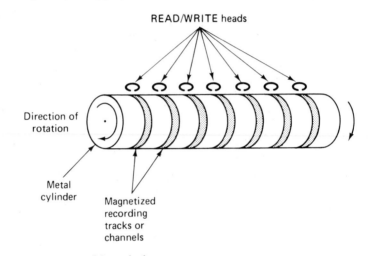

Figure 11.12   Magnetic drum.

The reading or writing at a given spot on a given track takes place as the drum rotates. Words are recorded *serially* on a track; that is, the bits of the word are recorded and read one at a time in sequence along a track. The location of a word is identified by a coded address that represents the angular displacement of the drum relative to a zero reference position. The heads are energized by address selection logic circuitry, which selects the desired track, and are synchronized by the output of a pre-recorded timing track that generates pulses at bit and word time intervals. In this manner, words can be written on or read from any word position on any track.

In a typical computer application, a magnetic drum is exchanging batches of information with an internal core memory. Thus, access time is important or there will be unacceptable delays before a batch of information can be processed by the computer. The access time could be as long as one drum revolution; on the average, it is one half of this. For example, a drum rotating at 4000 rpm has an access time of approximately 8 ms. Faster access times can be achieved by using faster drum rotation and/or by using several read/write heads at various positions on each track.

Data Transfer Rate

This is the rate at which data can be transferred to or from the drum memory once the desired location on the drum has been accessed. A typical drum memory has a data transfer rate of 2×10^6 bits per second, or about 0.5 μs per bit.

Capacity

Capacities of magnetic drums vary widely, depending on the size of the drum. Small drums might have capacities of 25,000 bits distributed over 20 tracks, while very large drums can store up to 10^8 bits on 1000 tracks. The packing density of most eurrently available drums is in the range 200–4000 bits per inch. The packing density can be increased by maintaining the heads very close to the surface.

11.11 MAGNETIC DISCS

Magnetic-disc memories provide a very large storage capacity at moderate operating speeds. The cost of the magnetic disc memory is somewhat lower than magnetic drum but somewhat higher than magnetic tape. In this method of storage, data are recorded on a layer of magnetic material that is coated on hard flat discs resembling phonograph records. A number of rotating discs are stacked with space between each disc (refer to Figure 11.13). Information is recorded *serially* on concentric bands or tracks on both surfaces of each disc by movable read/write heads. The packing density along a given track can be as high as 5000 bits per inch. Typical rotation speeds can range up to 6000 rpm. A single disc surface can have from 200 to 500 tracks and can store up to 50 million bits.

The read/write heads move radially in and out between discs in the space provided. The access time depends on rotation speed and disc diameter. For faster access time, additional heads are used to serve the same discs, some serving the inner tracks and others serving the outer tracks. The heads do not contact the disc surface, because this would cause excessive wear. Instead, most manufacturers use a *flying head*, which is shaped to ride along on a thin layer of air that adheres to the rotating disc. In most fast-acting mechanisms, a compressed-air piston is used to force the head toward the disc into or out of the air cushion.

Magnetic hard disc memories have typical access times in the 10–40 ms range and typical data transfer rates of 2.5×10^6 bits per second.

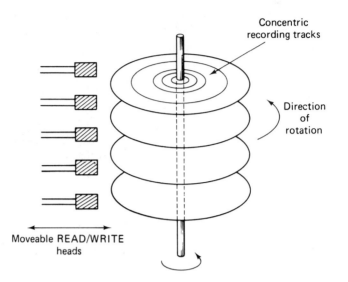

Figure 11.13   Typical disc arrangement.

Floppy Discs

A relatively recent innovation in disc storage, which was originally developed at IBM, uses a small flexible disc with a plastic base in place of the more conventional metal disc. A floppy disc (also called a *diskette*) is a flexible disc that looks like a 45-rpm phonograph record without grooves. Because they can be bent easily, floppy discs are enclosed in a protective cardboard envelope. This keeps them rigid as they are spun around by the disc drive unit in much the same way as phonograph records.

Figure 11.14 shows the standard floppy disc, which has a diameter slightly less than 8 inches. The protective envelope has cutouts for the drive spindle, the read/record head, and an index position hole. The access slot allows the read/record head to make contact with the surface of the disc as it spins inside the envelope. The index hole allows a photoelectric sensor to be used to determine a reference point for all the tracks on the disc.

Floppy discs are normally rotated at a speed of 360 rpm, which corresponds to one rotation in about 167 ms. As the disc rotates, the read/record head makes contact with it through the access slot so that it can write or sense magnetic pulses on the disc surface. The standard 8-inch disc surface is divided into 77 concentric tracks numbered 0 to 76 [see Figure 11.15(a)], with track number 0 being the outside track. These tracks are normally spaced about 0.02 inch apart, with track 0 having a total length of about 25 inches and track 77 a total length of 15 inches. This means that the longer outside tracks have space for more information than do the shorter inside tracks. For convenience in accessing a given point on the

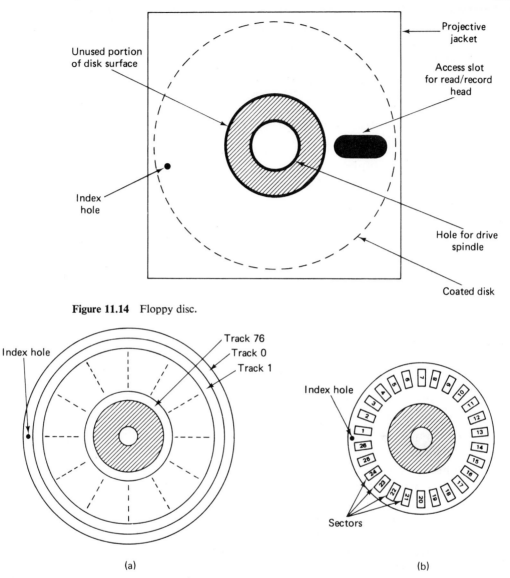

Figure 11.14 Floppy disc.

Figure 11.15 (a) Disc is divided into 77 concentric tracks; (b) each track is divided into 26 equal sectors.

disc, the disc circumference is divided into 26 equal-size *sectors* [see Figure 11.15(b)] in which information can be stored. This means that the same amount of storage space is used on each track irrespective of the length of the track; thus, more space is wasted (not used) on the longer tracks.

The index hole is placed between the last sector (26) and the first sector (1). Using a light source/photodetector combination, the disc drive circuitry will know when the disc is in its reference position with sector 1 next to the index hole. By counting each sector as it passes under the read/record head, the drive circuitry will always know which sector is passing under the read/record head.

Each sector usually stores 128 bytes of information, so that the total storage capacity of a single floppy disc surface will be

$$128 \frac{\text{bytes}}{\text{sector}} \times 26 \frac{\text{sectors}}{\text{track}} \times 77 \frac{\text{tracks}}{\text{disc}} \approx 256{,}000 \frac{\text{bytes}}{\text{disc}} = 2 \times 10^6 \frac{\text{bits}}{\text{disc}}$$

Some manufacturers use "double-density" floppy discs, which cram 256 bytes into each sector.

Access Time

Data are recorded or read from the disc surface one bit at a time at a nominal rate of 1 bit every 4 μs* or 1 byte every 32 μs. This rate, however, occurs after the read/record head has been positioned on the desired sector of the desired track. It takes approximately 6 ms for the head-positioning mechanism to move the head from one track to the next and 16 ms more to lower the head onto the track. With 77 tracks, it will take about 1/2 s in the worst case for the head to move from its present track to the track that is to be accessed for reading or writing. Once the head is on the desired track, it has to find the desired sector on that track. Since the disc makes one revolution in 167 ms, it will take 167 ms to find the right sector in the worst case. Thus, even in the worst case it takes a little over 0.6 s to find the desired record on the disc. On the average, this access time will be more like 0.3 s. This is much faster than the time required to randomly access a record on even the fastest magnetic tape system.

11.12 MAGNETIC TAPE

Currently, magnetic tape is the most popular medium for storing very large quantities of digital information. Most medium-size and large-size computers have one or more magnetic tape units. Magnetic tape represents the least expensive type of storage; the cost averages about 0.0001 cent per bit. Tapes are usually $\frac{1}{2}$–1 inch wide and have 7–14 tracks across the width of the tape. Packing density usually ranges from 200 to 1600 bits per inch along each track. A typical tape reel is about 2400 ft and can store more than 10^8 bits of information.

A major difference between tape and drums or discs is that the tape is not kept in continuous motion but is started and stopped each time it is used. The tape

---

*This is a data transfer rate of 250 kbits/sec, about 10 times slower than conventional metal disc systems.

transports have the ability to start and stop very quickly. The necessity to accelerate and decelerate the tape quickly is due to two reasons. First, since the reading and writing processes cannot begin until the tape has reached sufficient speed, a time delay occurs as the tape comes up to speed. Second, the tape which passes under the heads during the acceleration and deceleration processes is wasted, so that fast starting and stopping conserves tape space. Typical start and stop times are on the order of 1 msec.

Most modern tape units use one READ/WRITE head per track and usually these heads have *two* air gaps, one for writing and one for reading [refer to Figure 11.16(a)]. The read gap is positioned after the write gap so that the information which is written on the tape can be immediately read out and compared with the input information (usually in a register) to see that it is correct.

These are several formats for recording data on magnetic tape. Typically, one character (number, letter, symbol) is recorded on a given row along the tape.

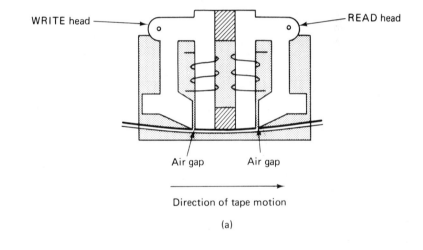

(a)

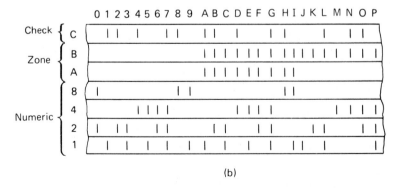

(b)

Figure 11.16   (a) Two-gap READ/WRITE head; (b) seven-track tape format.

Figure 11.16(b) shows a seven-track format where one of the tracks is for a parity bit. Data is recorded on tape in blocks called *records* with spaces between records to allow for starting and stopping of the tape.

Magnetic tape has become a very reliable storage medium, and its basic limitation continues to be its relatively long access time. However, despite this limitation, it continues to be an economical and convenient medium for bulk storage of large quantities of information.

A relatively recent entry into the bulk memory field is the *digital tape cassette*, similar to the common audio tape cassettes but usually of higher quality. Many microcomputer systems can use even inexpensive audio cassettes to store digital data in the form of different audio frequencies.

11.13   SEMICONDUCTOR MEMORIES

By far the most rapidly changing memory technology, semiconductor memories currently range from high-speed bipolar ECL memories to slower, high-density MOS memories. These memories are LSI chips (usually 14–24 pins) that have the capacity for storing thousands of bits per chip. Current state-of-the-art bipolar memory chips include a 1024-bit ECL memory with an access time of 12 ns and a 4096-bit TTL memory with an access time of 45 ns. 2K ECL and 8K TTL memory chips are expected in the near future. MOS memories are currently dominated by NMOS technology, where 8K static memory chips with access times ranging from 55 to 500 ns are readily available, and 16K chips are not far off. NMOS dynamic memory chips with capacities up to 64K are becoming commonplace. CMOS memory chips are also readily available and their very low power requirements make them ideal for battery-operated systems.

RAMs

There are two broad classifications used for semiconductor memories based on the relative frequency of read operations versus write operations which are to be performed. Memories designed so that reading and writing can take place with equal ease are called read/write memories (RWM). These semiconductor RWMs, however, are often referred to as RAMs (random access memories). Of course, RAM refers to any random access memory (such as magnetic core), but whenever we talk about semiconductor memories, RAM will always mean a semiconductor memory with equal read and write capabilities.

RAMs are used in computers for the *temporary* storage of programs and data. The contents of many of the RAM locations will be continually changing as the computer executes a program. This demands fast read and write cycle times for the RAM in order not to slow down the computer.

A major disadvantage of semiconductor RAMs is that they are *volatile*, which means that when electrical power is removed from the chip, the RAM loses all its stored information. Some RAMs, however, use such small amounts of power in the standby mode (no read or write operations taking place) that they can be powered from batteries in the event that the main power fails or is turned off.

Read-Only Memories (ROMs)

This class of semiconductor memories encompasses those types that are designed primarily for having data read from them. During *normal* operation of the computer or digital system, no new data can be written into a ROM. However, there are some types of ROMs that can be written into (programmed), but the process is slow and generally complex, and so is normally performed before the ROM is inserted into the system. These ROMs are more accurately referred to as *read-mostly memories* (RMM).

ROMs are used to store data and information that is not to change during the operation of a system. A major use for ROMs is in the storage of programs in microcomputers. Since all ROMs are *nonvolatile*, these programs are not lost when the microcomputer is turned off. When the microcomputer is turned on, it can immediately begin executing the program stored in ROM. ROMs are also used for program and data storage in microprocessor-controlled equipment such as sophisticated electronic cash registers.

ROMs can be subdivided into several types, which differ as to how information is written or *programmed* into the memory storage locations.

Mask-Programmed ROM

This type of ROM has its storage locations written into (programmed) by the manufacturer according to the customer's specifications. A photographic negative called a *mask* is used to control the electrical interconnections on the chip. A special mask is required for each different set of information to be stored in the ROM. Since these masks are expensive, this type of ROM is economical only if you need a large quantity of the same ROM. Some ROMs of this type are available as off-the-shelf devices preprogrammed with commonly used information or data such as certain mathematical tables and character generator codes for CRT displays. A major disadvantage of this type of ROM is the fact that it cannot be reprogrammed in the event of a design change requiring a modification of the stored program. The ROM would have to be replaced by a new one with the desired program written into it. Several types of user-programmable ROMs have been developed to overcome this disadvantage. Mask-programmed ROMs, however, still represent the most economical approach when a large quantity of identically programmed ROMs are needed.

Programmable ROM (PROM)

A PROM is a ROM that can be programmed by the user after he has purchased it. The most common PROM contains tiny nichrome wires which act as fuses. The user can selectively "burn out" some of the fuses with currents applied at the appropriate IC pins and thereby program the PROM according to his truth table.* Once this programming process is complete, however, the program cannot be changed, so it has to be done right the first time. The IC manufacturer or distributor can program it for you according to your truth table. This will cost extra, but it will eliminate the worry over making a mistake.

Erasable Programmable ROM (EPROM)

An EPROM can be programmed by the user and it can also be *erased* and reprogrammed as often as desired. Once programmed, the EPROM is a *nonvolatile* memory that will hold its stored data indefinitely. The process for programming an EPROM involves the application of special voltage levels (typically in the 25–50 V range) to the appropriate chip inputs for a specified amount of time (typically 50 ms per address location). The programming process is usually performed by a special programming circuit that is separate from the circuit in which the EPROM will eventually be working. The complete programming process can take up to several minutes for one EPROM chip.

The storage cells in an EPROM are field-effect transistors with a silicon gate that has no electrical connections (i.e., a floating gate). By applying a special high-voltage programming pulse to the device, high-energy electrons are injected into the floating-gate region, thus turning the transistor "on." These electrons remain trapped in this region once the pulse is terminated, since there is no discharge path. During this programming process, the chip's address and data pins are used to determine which memory cells will be affected by the programming pulse.

Once a memory cell has been programmed, it can only be erased by exposing it to ultraviolet (UV) light applied through a window on the chip. The UV light causes a flow of photocurrent from the floating gate back to the silicon substrate, thereby restoring the gate to its initial condition. Note that there is no way to expose just a single cell to the UV light without exposing all the cells. Thus, the UV will erase the entire memory. The erasure process typically requires 15–30 minutes of exposure to the UV rays.

11.14 SEMICONDUCTOR MEMORIES—ORGANIZATION

IC memory chips store binary information in groups called *words*. A word is the basic unit of information or data used in a computer. The number of bits that constitute a word will vary from computer to computer, ranging typically from 4

*"Truth table" here refers to the list of all the addresses and the words to be stored at these addresses.

bits to 36 bits. A single memory chip will store a given number of words of so many bits per word. For example, a popular RAM memory chip has a storage capacity of 1024 words of 4 bits each (1K × 4).

It is helpful to think of a memory chip as consisting of a group of registers, each register storing one word (see Figure 11.17). The width of each register is the number of bits per word. The number of registers is the number of words stored in the memory. This general diagram represents a memory that stores N words of M bits each (i.e., an $N \times M$ memory). Common values for the number of words per chip are 64, 256, 512, 1024, 2048, and 4096. Common values for the word size are 1, 4, and 8. As we shall see, both the word capacity and word size can be increased by combining memory chips.

The contents of each register is subject to two possible operations—*reading* and *writing*. Reading is the process of getting the word stored in the register and sending it to some other place where it can be used. The contents of the register are not changed by the read operation. Writing is the process of putting a new word into a particular register. Of course, this writing operation destroys the word that was previously stored in the register.

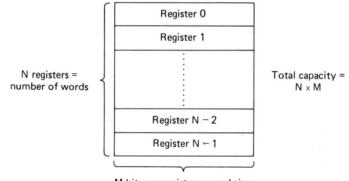

Figure 11.17 Memory arranged as group of registers.

Addressing

Each register or word is assigned a number beginning with 0 and continuing as high as needed. This number uniquely specifies the location of the register and the word it is storing, and is referred to as its *address*. For instance, address 2 refers to register 2 or word 2. Anytime that we want to refer to a particular word in the memory, we will use its address. The address of each word is an important number, because it is the means by which a device external to the memory chip can select the memory word it desires access to for a read or a write operation.

To understand how addressing is used, we must take a more detailed look at the internal organization of a typical memory chip (Figure 11.18). This particular chip stores 64 words of 4 bits each (i.e., a 64 × 4 memory). These words have addresses ranging from 0 to 63_{10}. In order to select one of the 64 address locations for reading or writing, a binary address code is applied to a decoder circuit. Since $64 = 2^6$, the decoder requires a 6-bit input code. Each address code activates one particular decoder output, which, in turn, enables its corresponding register. For example, assume an applied address code of

$$A_5 A_4 A_3 A_2 A_1 A_0 = 011010$$

Since $011010_2 = 26_{10}$, decoder output 26 will go high, selecting register 26.

Read Operation

The address code picks out one register in the memory chip for reading or writing. In order to *read* the contents of the selected register, the READ/WRITE (R/W)* input must be a 1. In addition, the CHIP SELECT (CS) input must be activated (a 1 in this case). The combination of $R/W = 1$ and $CS = 1$ enables the output buffers so that the contents of the selected register will appear at the four data outputs. $R/W = 1$ also *disables* the input buffers so that the data inputs do not affect the memory during a read operation.

Write Operation

To write a new 4-bit word into the selected register requires $R/W = 0$ and $CS = 1$. This combination *enables* the input buffers so that the 4-bit word applied to the data inputs will be loaded into the selected register. The $R/W = 0$ also *disables* the output buffers, which are tristate, so that the data outputs are in their open-circuited state during a write operation.

Chip Select

Most memory chips have one or more CS inputs which are used to enable the entire chip or disable it completely. In the disabled mode all data inputs and data outputs are disabled (Hi-Z) so that neither a read or write operation can take place. In this mode, the contents of the memory are unaffected. The reason for having CS inputs will become clear when we combine memory chips to obtain larger memories. It should be noted that many manufacturers call these inputs chip enable (CE) rather than CS.

---

*The R/W line is often labeled $R/\bar{W}$ to indicate that a 1 performs the read operation and a 0 performs a write operation.

Common Input/Output Pins

In order to conserve pins on an IC package, manufacturers often combine the data input and data output functions using common input/output pins. The R/W input controls the function of these I/O pins. During a read operation, the I/O pins act as data outputs which reproduce the contents of the selected address location. During a write operation, the I/O pins act as data inputs.

We can see why this is done by considering the chip in Figure 11.18. With separate input and output pins, a total of 18 pins is required (including ground and power supply). With four common I/O pins, only 14 pins are required.

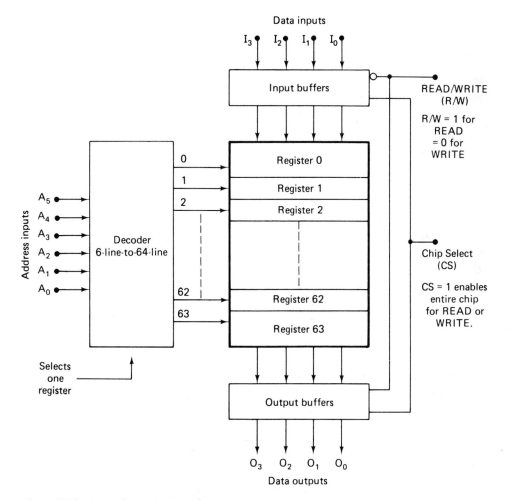

Figure 11.18 Internal organization of a 64 × 4 memory chip.

The pin savings becomes even more significant for chips with larger word size.

EXAMPLE 11.4   How many pins would be required for a memory chip that stores 256 8-bit words and has common I/O lines?

Solution: There are 256 address locations and $2^8 = 256$. Therefore, eight address inputs are needed to select any address from 00000000 to 11111111 (255_{10}). There are eight I/O lines since the word size is 8 bits. Adding one R/W line, one CS line, power, and ground gives a total of 20 pins (see Figure 11.19).

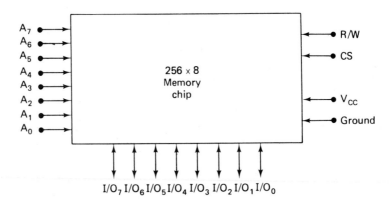

Figure 11.19   Pin function diagram for a 256 × 8 memory IC.

11.15   STATIC RAM DEVICES

The word "static" as applied to semiconductor memory devices refers to the ability of a memory cell to store data, as long as power is applied, regardless of how long it has been since the cell was written into. In other words, when a 0 or 1 is written into a memory cell, it will remain in that cell indefinitely as long as power is applied. Later we will describe "dynamic" memory devices which require that the data stored in each memory cell be periodically refreshed (rewritten) or the data will be lost. Static memory cells are essentially flip-flops that can maintain a given state indefinitely, while dynamic memory cells use capacitances that have to be periodically recharged.

There are many static RAM chips available in various capacities and pin configurations. The most common of these are listed in Table 11.1, which also indicates whether the memory has separate input/output pins or common input/output pins. Most of the devices listed are available in both NMOS and CMOS versions, although the identification numbers shown in the table are those of the NMOS devices. All of the devices, except for the 6810 and 8108, are produced by more than one IC manufacturer.

Table 11.1   Static RAMs

Device ID Number	Bit Organization	I/O Configuration	Number of Pins
6810	128 × 8	Common	24
2101	256 × 4	Separate	22
2111	256 × 4	Common	18
2112	256 × 4	Common	16
2102	1024 × 1	Separate	16
2114	1024 × 4	Common	18
2147	4096 × 1	Separate	18
8108	1024 × 8	Common	22

Static RAM Timing

There are several ways of measuring the speed of IC memories. Unfortunately, there has not been a great degree of standardization among the memory manufacturers. Thus, it sometimes happens that different manufacturers specify the speed of supposedly equivalent memory chips as being different. The three most often used specifications of memory speed are *read cycle time*, t_{RC}, *write cycle time*, t_{WC}, and *access time*, t_{ACC}. Each of these is a time measurement, so the smaller the number, the faster is the memory.

Since access time, t_{ACC}, is the most frequently used indication of memory speed, we will illustrate it in Figure 11.20, which is a generalized presentation of the read-operation timing for a typical static RAM device. The waveforms shown represent signals on the memory chip's address inputs, chip select input, R/W input, and data outputs. Note that the chip select input is labeled $\overline{CS}$, indicating that it is an active-LOW input.

Refer to the waveforms and consider the initial conditions prior to the time instant labeled 1. The R/W line is high and remains high throughout the read operation. In most semiconductor memory systems R/W is normally kept in the high state and is driven low only for a write operation. The address inputs are in some particular state, probably holding the last memory address that was accessed. The $\overline{CS}$ input is high, its inactive state, so that the memory data outputs are disabled even though $R/W = 1$ and there is an address on the address inputs. The data outputs are in a Hi-Z state, as indicated by the dashed line on the waveform.

At time 1 the address inputs are changed to the new address location from which data are to be fetched. Some address inputs will be going high and some will be going low as indicated on the waveform. As soon as the new address appears on the address inputs, the internal memory circuitry begins the process of fetching the data from the selected location and transferring it to the output buffers (see Figure 11.18). However, these data cannot appear on the data output lines until $\overline{CS}$ is driven low.

With $\overline{CS} = 0$, the data stored at the selected address location will eventually reach the data outputs. At time 2 the data outputs have become valid, and will be at either low or high logic levels. These data outputs can now be transferred to any desired destination. The delay between the time 1, when the new address was applied, and time 2, when the output data become valid, is the access time, t_{ACC}, and is a function of the memory's internal circuitry.

The $\overline{CS}$ input could have been activated prior to time 1 or could even be kept low permanently. However, it must be realized that the output data does not become valid until a time interval of t_{ACC} after the address inputs have changed to a new address.

We should point out that the chip select signal of Figure 11.20 may actually consist of more than one signal. For example, the 2101 static RAM has two chip enable inputs: $\overline{CE1}$, which is an active-LOW input, and $CE2$, which is an active-HIGH input. These inputs have to be at their respective active states before the data outputs can become valid.

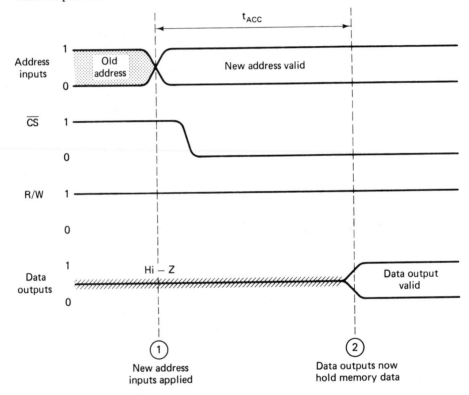

Figure 11.20   Typical static RAM timing for read operations.

11.16 DYNAMIC RAMs

Static RAMs store data in flip-flops. In dynamic RAMs the data are stored as charges on the gate-substrate capacitances of MOSFETs. The memory cell in a dynamic memory uses fewer MOSFETs than does a static memory cell, so more memory cells may be produced in a given chip area. However, the capacitances in a dynamic memory have to be periodically recharged so that the data they are storing does not leak off. The interval at which data must be refreshed varies from device to device, but it is generally about 2 ms.

If we compare the cost per bit of storage on a dynamic RAM chip versus a static RAM chip, the dynamic RAM is superior. This advantage, which is a result of the much simpler circuitry of a dynamic cell, is somewhat offset by the fact that dynamic memories require more external support circuitry than static memories, including the need for more than one supply voltage. Most static RAMs use a single supply. A general rule of thumb that currently applies is that static RAMs have the cost advantage in small memory systems where the total word capacity is 4K or less, while dynamic RAMs have the advantage in larger memory systems.

Another advantage of dynamic RAM devices is their low average power dissipation, which can be significantly lower than static devices of the same capacity. Dynamic RAMs are automatically placed in a *standby* mode of operation when they are not selected (chip select inputs inactive). In this standby mode, the device power dissipation is only a fraction (less than 5 per cent) of its dissipation when it is operating (performing a read or write). A few static RAM devices have a standby mode, but use of this mode usually requires additional circuitry.

Dynamic RAM Word Size

There is little to choose from in this regard. Most of the currently available dynamic RAM chips are $1K \times 1$, $4K \times 1$, or $16K \times 1$ devices. This means that several chips have to be connected in parallel to obtain the word size required by the system. We will see how this is done a little later.

11.17 TYPICAL 4K DYNAMIC RAM DEVICES

The operation of dynamic RAMs can be understood best by looking at several typical devices. We will describe three of the basic $4K \times 1$ chip configurations, with an emphasis on aspects of their operation that are different from those of static devices.

Internal Matrix Structure

Without going into the details of the internal cell structure of a dynamic memory, we can illustrate the basic internal organization. A $4K \times 1$ dynamic RAM has a memory cell array arranged in a 64×64 matrix (see Figure 11.21) similar to that used in a magnetic core plane. The 12 address inputs are divided into two groups of six. Address inputs A_0–A_5 are decoded to select one of the 64 rows in the matrix, while A_6–A_{11} are decoded to select one of the 64 columns. The memory

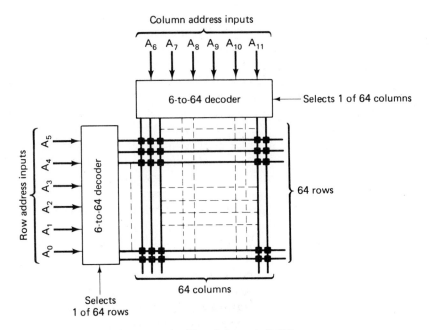

Figure 11.21   Cell arrangement in 4K × 1 dynamic RAM.

cell located at the selected row and column is the cell being accessed for a read or write operation.

22-Pin 4K × 1 Device

Figure 11.22 shows the functional diagram of a typical 22-pin 4K × 1 dynamic RAM chip. It has 12 address inputs (since 4K $= 4096 = 2^{12}$); separate data input and data output pins, D_{IN} and D_{OUT}; and three control inputs, $\overline{CS}$, $\overline{CE}$, and $\overline{WE}$. With $\overline{CS} = 0$, read and write operations are initiated by a positive-going transition on the CE (chip-enable) input; that is, CE acts like a clock input. The $\overline{WE}$ (write-enable) input determines whether a read or write operation is to be performed; $\overline{WE} = 0$ for a write and $\overline{WE} = 1$ for a read.*

The chip requires three power supplies; $+5$ V, -5 V, and $+12$ V. The need for three supplies is an important consideration when the choice between static and dynamic memories is being made.

18-Pin Device

Figure 11.22(b) shows a typical 18-pin 4K × 1 device. To reduce the package size from 22 to 18 pins, the $+5$-V power supply requirement has been eliminated, the data input and data output functions have been combined, and the CS and CE

---

*$\overline{WE}$ is just another label used for R/W or $R/\bar{W}$.

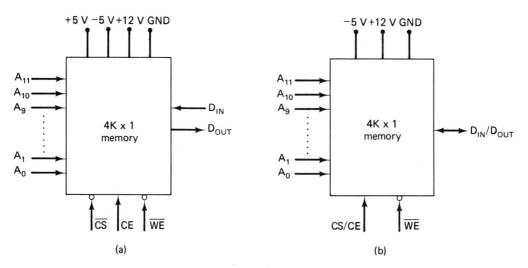

Figure 11.22 (a) 22-pin dynamic RAM; (b) 18-pin dynamic RAM.

functions have been combined into one signal. In addition, there was one pin on the 22-pin device that was not used; this unused pin is eliminated in the 18-pin device.

The changes that were made to reduce the package size by four pins will not have a limiting effect in most applications. The $\overline{WE}$ control input determines whether the single data I/O pin is acting as an input or an output. The CS/CE input performs both the function of selecting the chip and of initiating the read or write operations.

16-Pin Device

The functional representation of a typical 16-pin dynamic RAM is illustrated in Figure 11.23. Note that it retains the three supply voltages and separate data input and output pins used by the 22-pin version. The big changes are the $\overline{RAS}$ and $\overline{CAS}$ control inputs and the address input arrangement.

The device has only *six* address inputs, although a 12-bit address code is required to select one of the 4096 locations. How, then, can we use these six inputs to select which location is to be accessed? The answer is that we must apply the 12-bit address code to these inputs 6 bits at a time. In other words, the 12-bit address is divided into two halves: A_0–A_5, which is the row address (recall Figure 11.21), and A_6–A_{11}, which is the column address. The addressing operation proceeds as follows:

1. The 6-bit *row* address is applied to the chip's address inputs. The *row address strobe* $(\overline{RAS})$ input is then pulsed low. This strobe signal causes the row address to be latched into an address register in the memory chip.

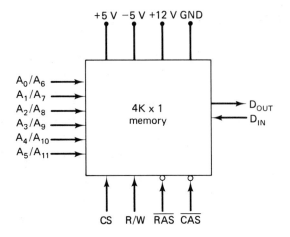

Figure 11.23 16-pin dynamic RAM chip with multiplexed address inputs.

2. The 6-bit *column* address is now applied to the address inputs and the *column address strobe* ($\overline{CAS}$) input is pulsed low. This strobe signal causes the column address to be latched into another address register in the memory chip. The complete 12-bit address is now latched into the memory chip and begins selecting the desired cell.

This process is called *address multiplexing* and requires circuitry external to the memory chip to generate $\overline{RAS}$ and $\overline{CAS}$ and to multiplex the two 6-bit addresses into the chip's six address inputs. This extra circuitry is the price that must be paid for reducing the size of the dynamic RAM chip. The address multiplexing technique is also used in 16K × 1 dynamic RAM chips to reduce the number of required address pins from 14 to 7.

11.18 DYNAMIC RAM REFRESHING

In order to maintain the information stored in a dynamic RAM, it is necessary to recharge (refresh) the memory cells at frequent intervals. Fortunately, it is not necessary to individually access each bit in order to refresh the device. This is because the design of the memory cell matrix is such that, whenever a row is selected, all cells in that row are refreshed. For most dynamic RAMs a refresh occurs automatically during a normal read or write operation. However, during normal system operation, it would be unrealistic to expect that every row in the device will be accessed within the required time interval (usually once every 2 ms). For this reason, special circuitry must be provided to perform the refresh operation by sequentially and periodically accessing each row address in the device. Some manufacturers have developed special chips, called *refresh controllers*, designed specifically for this purpose.

11.19 ROM DEVICES

A typical functional diagram for a ROM chip is shown in Figure 11.24. It has the inputs and outputs common to all ROM devices: address inputs, chip-select (or chip-enable) inputs, and data outputs. Note the absence of data inputs and a R/W control input. This is because the write operation is not part of the ROM's normal operation. PROMs, which can be programmed (written into) once by the user, and EPROMs, which can be erased and reprogrammed as often as desired, will have special inputs used for the programming operation. These special pins are not shown on the diagram since they are not used once the device has been programmed and placed in a circuit.

Once the ROM is in a circuit, its function is to permanently store the information that has previously been written into it. In order to read a word from ROM it is only necessary to apply the proper address inputs and activate the chip-select inputs. The timing is exactly the same as the static RAM timing of Figure 11.20 with the data outputs normally in a Hi-Z state until the chip-select inputs are activated.

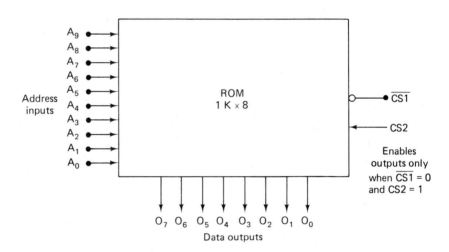

Figure 11.24   A typical functional diagram for a ROM chip.

ROM Capacities

Mask-programmed ROMs are currently produced with capacities up to $8K \times 8$, with both MOS and bipolar devices available. The bipolar devices have lower capacities but operate at higher speeds. Most PROMs are bipolar and are available in capacities up to $2K \times 8$. EPROMs are MOS devices and currently come in capacities up to $4K \times 8$.

All of these capacity limits will no doubt be exceeded as manufacturers continue to develop ways of increasing chip densities. In fact, many higher capacity devices are right now in the developmental stage and will probably hit the market before the publication of this book.

11.20 COMBINING MEMORY CHIPS

In most IC memory applications the required memory capacity or word size cannot be satisfied by one memory chip. Instead, several memory chips have to be combined to provide the desired capacity and word size. We will see how this is done through several examples that illustrate all the important concepts that will be needed when we interface memory chips to a microprocessor.

Expanding Word Size

Suppose we need a memory that can store 16 8-bit words and all we have are RAM chips which are arranged as 16×4 memories with common I/O lines. We can combine two of these 16×4 chips to produce the desired memory. The configuration for doing so is shown in Figure 11.25. Examine this diagram carefully and see what you can find out from it before reading on.

Since each chip can store 16 4-bit words and we want to store 16 8-bit words, we are using each chip to store *half* of each word. In other words, RAM-0 stores the four *higher*-order bits of each of the 16 words, and RAM-1 stores the four *lower*-order bits of each of the 16 words. A full 8-bit word is available at the RAM outputs connected to the data bus.

Any one of the 16 words is selected by applying the appropriate address code to the four-line *address bus* (AB_3, AB_2, AB_1, AB_0). For now, we will not be concerned with where these address inputs come from. Note that each address bus line is connected to the corresponding address input of each chip. This means that once an address code is placed on the address bus, this same address code is applied to both chips so that the same location in each chip is accessed at the same time.

Once the address is selected, we can read or write at this address under control of the common R/W and $\overline{CS}$ line. To read, R/W must be high and $\overline{CS}$ must be low. This causes the RAM I/O lines to act as *outputs*. RAM-0 places its selected 4-bit word on the upper four data bus lines and RAM-1 places its selected 4-bit word on the lower four data bus lines. The data bus then contains the full selected 8-bit word, which can now be transmitted to some other device (e.g., a register).

To write, $R/W = 0$ and $\overline{CS} = 0$ causes the RAM I/O lines to act as *inputs*. The 8-bit word to be written is placed on the data bus from some external device. The upper 4 bits will be written into the selected location of RAM-0 and the lower 4 bits will be written into RAM-1.

The same basic idea for expanding word size will work for many different situations. Read the following example and draw a rough diagram for what the system will look like before looking at the solution.

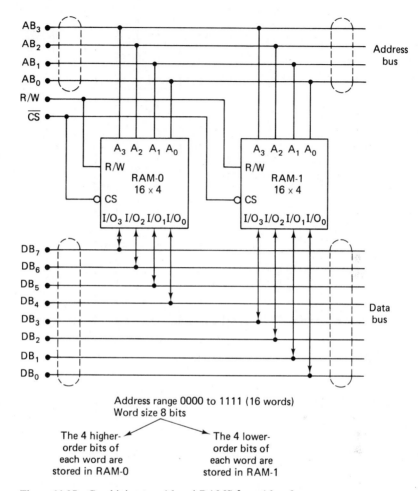

Address range 0000 to 1111 (16 words)
Word size 8 bits

The 4 higher-order bits of each word are stored in RAM-0

The 4 lower-order bits of each word are stored in RAM-1

Figure 11.25 Combining two 16 × 4 RAMS for a 16 × 8 memory.

EXAMPLE 11.5 How many 1024 × 1 RAM chips are needed to construct a 1024 × 8 memory system?

Solution: Eight chips are required, with each chip storing 1 bit of each of the 1024 8-bit words. The arrangement is shown in Figure 11.26.

Expanding Capacity

Suppose we need a memory that can store 32 4-bit words and all we have are the 16 × 4 chips. By combining two 16 × 4 chips as shown in Figure 11.27, we can produce the desired memory. Once again, examine this diagram and see what you can determine from it before reading on.

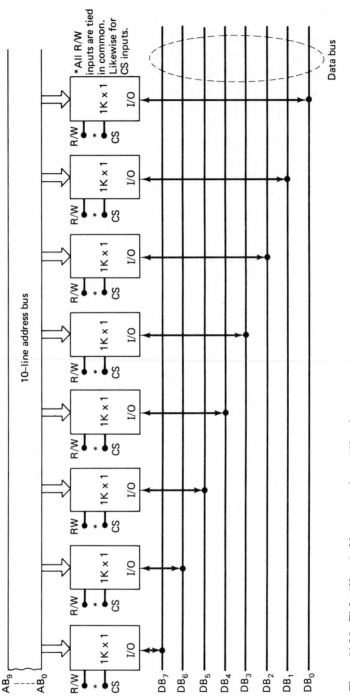

Figure 11.26 Eight 1K × 1 chips arranged as a 1K × 8 memory.

420

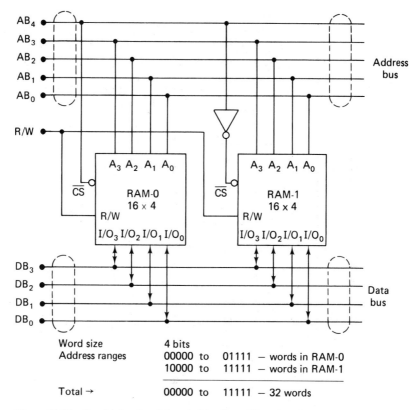

Word size   4 bits
Address ranges   00000 to   01111 — words in RAM-0
                              10000 to   11111 — words in RAM-1

Total →   00000 to   11111 — 32 words

Figure 11.27   Combining two 16 × 4 chips for a 32 × 4 memory.

To obtain a total of 32 4-bit words, we are using each RAM to store 16 words. The two RAMs share the 4-bit data bus since only one of them will be enabled at one time. The one that is enabled will place its data on the data bus during a read operation or will receive data from the data bus during a write operation.

How can we select 1 of 32 different words if each chip has only four address lines? We simply use the *CS* input as a fifth address input. There are five address bus lines that are required to access the 32 different addresses. The lower four lines, AB_3, AB_2, AB_1, and AB_0, are connected to the address inputs of each chip. These four lines select one of the 16 locations in both RAMs. The AB_4 line is used to select *which* RAM is actually going to be read from or written into at that selected location.

To illustrate, when $AB_4 = 0$, the $\overline{CS}$ of RAM-0 enables this chip for read or write. Then, any address location in RAM-0 can be accessed by AB_3–AB_0. The latter four address lines can range from 0000 to 1111 to select the desired location. Thus, the range of addresses representing locations in RAM-0 are

$$AB_4AB_3AB_2AB_1AB_0 = 00000 \text{ to } 01111$$

Note that when $AB_4 = 0$, the $\overline{CS}$ of RAM-1 is high, so that its I/O lines are disabled and cannot communicate (give or take data) with the data bus.

It should be clear that when $AB_4 = 1$, the roles of RAM-0 and RAM-1 are reversed. RAM-1 is now enabled and the AB_3–AB_0 lines select one of its locations. Thus, the range of addresses located in RAM-1 is

$$AB_4AB_3AB_2AB_1AB_0 = 10000 \text{ to } 11111$$

EXAMPLE 11.6 It is desired to combine several 256 $\times$ 8 PROMs to produce a total capacity of 1024 $\times$ 8. How many PROM chips are needed? How many address bus lines are required?

Solution: Four PROM chips are required, with each one storing 256 of the 1024 words. Since $1024 = 2^{10}$, the address bus must have 10 lines.

The configuration for the memory of Example 11.6 is similar to the 32 $\times$ 4 memory of Figure 11.27. However, it is slightly more complex, because it requires a decoder circuit for generating the CS input signals. The complete diagram for this 1024 $\times$ 8 memory is shown in Figure 11.28.

Since the total capacity is 1024 words, 10 address bus lines are required. The two highest-order lines, AB_9 and AB_8, are used to select *one* of the PROM chips; the other eight address bus lines go to each PROM to select the desired location within the selected PROM. The PROM selection is accomplished by feeding AB_9 and AB_8 into the decoder circuit. The four possible combinations of AB_9, AB_8 are decoded to generate active-LOW signals, which are applied to the CS inputs. For example, when $AB_9 = AB_8 = 0$, the 0 output of the decoder goes low (all others are high) and enables PROM-0. This causes the PROM-0 outputs to generate the data word stored at the address determined by AB_7–AB_0. All other PROMs are disabled.

Thus, all addresses in the following range are stored in PROM-0:

$$AB_9AB_8AB_7 \ldots AB_0 = 0000000000 \text{ to } 0011111111$$

These are the first 256 addresses in the memory. For convenience, these addresses can be more easily written in hexidecimal code to give a range of 000_{16} to $0FF_{16}$.

Similarly, $AB_9 = 0$, $AB_8 = 1$ selects PROM-1 to give it an address range of

$$0100000000 \text{ to } 0111111111 \quad \text{(binary)}$$

or

$$100 \text{ to } 1FF \quad \text{(hex)}$$

The reader should verify the PROM-2 and PROM-3 address ranges given in Figure 11.28.

EXAMPLE 11.7 What size of decoder would be needed to expand the memory of Figure 11.28 to 4K $\times$ 8? How many address bus lines are needed?

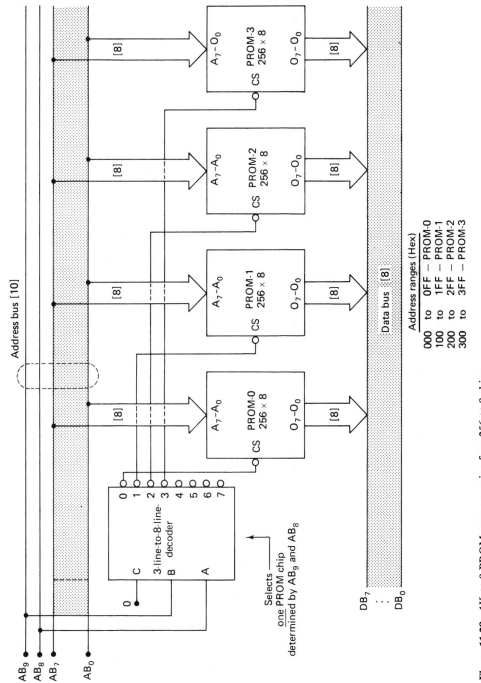

Figure 11.28 1K × 8 PROM memory using four 256 × 8 chips.

423

Solution: A 4K capacity is $4 \times 1024 = 4096$ words, which requires $4096/256 = 16$ PROM chips. To select one of 16 chips will require a 4-line-to-16-line decoder. With a capacity of 4096 words, a total of 12 address bus lines are needed—4 go to the decoder to select the PROM and the other 8 go to each PROM to select the address in the PROM.

These basic ideas for combining memory chips will be utilized further in Chapter 13, where we show how to interface (connect) semiconductor memory devices to a microprocessor to form a microcomputer. At that time we will also apply the same principles to the interfacing of input/output devices within the microcomputer structure.

11.21   NEW MEMORY TECHNOLOGIES

Two relatively new semiconductor memory technologies are beginning to make an impact in the bulk memory field. Charge-coupled devices (CCD) and magnetic bubble memories (MBM) are both somewhat slower and less expensive than semiconductor RAMs; on the other hand, they are faster than the traditional bulk memory devices (tape, disc, and drum) and require no moving parts.

CCD devices function essentially as a circulating shift register memory, but where the data are stored as charges on MOS capacitors. The capacitor charges have to be periodically refreshed through a shifting of charges from one capacitor to another. Because a CCD is a serial device, it is slower than other semiconductor RAMs (400 μs access times). Currently, CCD devices are available in capacities of 64K, and it is predicted that their simple geometry will lead to 4-Mbit chips in a few years.

MBM devices contain a thin layer of magnetic material grown on a semiconductor substrate. The magnetic material contains randomly distributed magnetic domains called "magnetic bubbles." The presence or absence of a bubble at a given position corresponds to a logic 1 or 0. Two permanent magnets are positioned on either side of the chip to provide a constant magnetic field to align all the bubbles in the same direction. Two orthogonal coils are wound around the chip to provide a rotating magnetic field to produce bubble movement. Like CCD memories, MBM devices store data in a serial shift register fashion. However, MBM devices do not require periodic refreshing of the data. The shifting of magnetic bubbles is simply a means for getting data into or out of the device. Since it is a serial or sequential access device, it has a relatively slow access time (4 ms is typical). This makes it unsuitable for computer mainframe memories, but its nonvolatility makes it ideal for bulk storage applications.

Since the permanent magnets maintain the bubble positions when power is lost, the MBM is nonvolatile and will retain its data indefinitely. In addition, it requires less power than other bulk memory devices, and it has no moving parts.

At present, commercially available bubble memory chips can be fabricated with capacities of 256,000 bits on a single chip. Within the near future, experts predict that 1-Mbit chips will be commonplace.

QUESTIONS AND PROBLEMS

11.1 Briefly define the following terms:
   (a)   Memory cell.
   (b)   Memory word.
   (c)   Byte.
   (d)   Address.
   (e)   Access time.

11.2 A certain memory has a capacity of 8K × 32. What is this capacity in words? What is the size of each word? How many memory cells does this memory contain?

11.3 How many different addresses are required by the memory of Question 11.2?

11.4 What is meant by a "volatile" memory?

11.5 Explain the difference between RAM and SAM.

11.6 Explain the difference between RWM and ROM.

11.7 Explain the difference between static and dynamic memory devices.

11.8 How many address inputs, data inputs, and data outputs are required for a 1K × 8 memory?

11.9 Explain how a magnetic core is used to store a binary value and explain how this value is sensed.

11.10 What is meant by "destructive readout" of a magnetic core?

11.11 Explain how the coincident-current technique is used to select one core from plane of cores to be set or cleared.

11.12 Describe what will happen to the cores in Figure 11.6 if lines X_1, X_2, and Y_3 are simultaneously pulsed with $+I_M/2$. What will happen to the sense output if these lines are then pulsed with $-I_M/2$?

11.13 How many X lines and Y lines are needed for a 64 × 64 core plane? How many lines would be needed if each core had it's own input select line that would be pulsed with $\pm I_M$? This comparison should make it clear why the coincident-current technique is used.

11.14 A certain digital computer has an internal core memory system capable of storing 1024 16-bit words.
   (a)   How many cores does the memory contain? How many planes?
   (b)   How many input select lines does it contain?
   (c)   How many sense lines and sense amplifiers does it require?

11.15 Answer true or false:
   (a)   In a core-memory system, each row of cores in a plane has a separate inhibit line.
   (b)   The inhibit line is not pulsed when a 1 is to be written into a core.
   (c)   A core memory storing 4096 12-bit words requires 64 address lines and 12 sense lines.

(d) When reading a given core on a plane, the operation would be unaffected by a $-I_M/2$ pulse on the inhibit line.

(e) The core memory of statement (c) would require a 12-bit address register.

11.16 How many address bits are required for a core memory that consists of 32 64×64 core planes?

11.17 A certain core memory has an 8-bit address register. What is the word capacity of this memory?

11.18 What is the basic principle of recording digital data on moving magnetic surfaces?

11.19 What are the three major *rotating* magnetic memory devices?

11.20 How do these rotating magnetic memories compare with magnetic core memories in terms of the following: access time, typical capacity, volatility, and type of access (RAM or SAM)?

11.21 What is the main disadvantage of semiconductor RAMs when compared to ROMs?

11.22 How do mask-programmed ROMs differ from PROMs and EPROMs?

11.23 Draw the pin function diagram for a 1K $\times$ 4 RAM chip that has *separate* data input and output pins, and two chip-select inputs, one active-HIGH and one active-LOW.

11.24 Which of the following statements pertain to static semiconductor memories and which pertain to dynamic memories?
(a) They are volatile.
(b) Data are stored as charges on capacitors.
(c) Data are stored in flip-flops.
(d) More than one supply voltage is required.
(e) External refresh circuitry is required.
(f) Generally require lower power.
(g) More suitable for low-capacity systems.
(h) May use multiplexed address inputs.

11.25 A 64K $\times$ 1 dynamic memory chip uses multiplexed address inputs. How many address inputs does this chip have?

11.26 Draw the pin function diagram of a 2K $\times$ 8 ROM that has three active-LOW chip-select inputs.

11.27 Show how to combine 1K $\times$ 4 RAM chips to produce a 1K $\times$ 8 memory.

11.28 Compare the 1K $\times$ 8 memory from Problem 11.27 with the memory of Figure 11.26. Aside from the obvious difference in the number of required chips, there is a significant difference in the amount of loading that signals on the address bus will experience. For each of these arrangements determine the number of loads that are being driven by each address bus line.

11.29 Show how to combine two of the ROM chips of Figure 11.24 to produce a 2K $\times$ 8 memory. Your circuit should contain no devices other than the ROM chips. This demonstrates the benefits of having two chip-select inputs.

11.30 Modify the memory of Figure 11.28 so that it has a capacity of 2K $\times$ 8. Then determine the range of addresses for each ROM.

11.31 Examine the memory circuit of Figure 11.29.
(a) Determine the total capacity and word size.

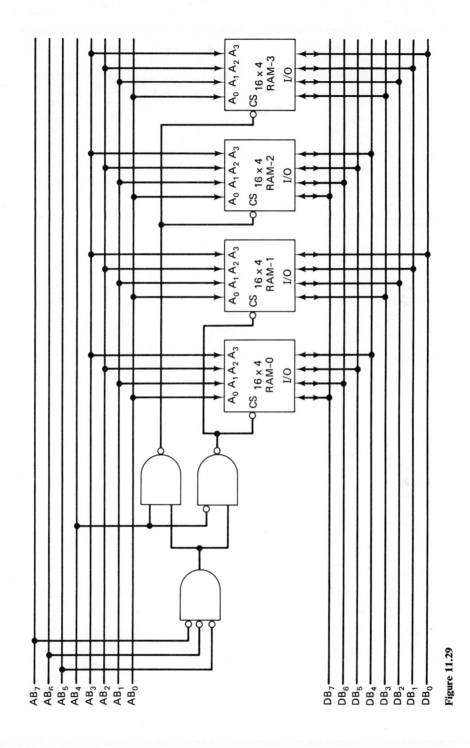

Figure 11.29

(b)   Note that there are more address bus lines than are necessary to select one of the memory locations. This is not an unusual situation, especially in small computer systems, where the actual amount of memory circuitry is much less than the maximum which the computer address bus can handle. Which RAMs will put data on the data bus when $R/W = 1$ and the address bus is at 00010110?

(c)   Determine the range of addresses stored in the RAM-0/RAM-1 combination. Repeat for the RAM-2/RAM-3 combination.

11.32   Draw the complete diagram for a 4K $\times$ 4 memory that uses static RAM chips with the following specifications: 1K $\times$ 1 capacity, common input/output line, and two active-LOW chip-select inputs. [*Hint:* The circuit can be designed using only two inverters (plus memory chips).]

12

INTRODUCTION
TO DIGITAL-COMPUTER OPERATION

By far the most significant application of digital principles is the digital computer. In the preceding chapters we studied many digital circuits and techniques and several of their applications. In this chapter we will see how many of these ideas are used in the basic operation of most digital computers. The discussion will not go into detail on any particular computer but will explain those concepts and techniques which are common to the operation of almost all digital computers. In Chapter 13 we will concentrate on microcomputers and microprocessors.*

12.1 WHAT IS A DIGITAL COMPUTER?

A digital computer is a combination of digital devices and circuits that can perform an appropriate sequence of operations with a minimum of human intervention. The list of operations that a computer is to perform in a given situation is the *program* (or set of instructions). The program is usually stored in the computer's internal mainframe memory (e.g., magnetic cores) along with all the input data (numbers) that the program requires. When the computer is started, it performs the instructions in the order in which they are stored in memory until the program is completed. It does this normally without any human intervention, and this is the major difference between a computer and a calculator. In addition, the modern computer has the capability of performing the complete program in an incredibly short period of time with virtually no errors.

*The reader might wish to review Section 1. 7 before starting this chapter.

12.2 HOW DO COMPUTERS THINK?

Computers do not think! The computer *programmer* provides a *program* of instructions and data which specifies every detail of what to do, what to do it to, and when to do it. The computer is simply a high-speed machine that can manipulate data, solve problems, and make decisions, all under the control of the program. If the programmer makes a mistake in the program or puts in the wrong data, the computer will produce wrong results. A popular saying in the computer field is "garbage in gives you garbage out."

Perhaps a better question to ask at this point is: How does a computer go about executing a program of instructions? Typically, this question is answered by showing a diagram of a computer's architecture (arrangement of its various elements) and then going through the step-by-step process which the computer follows in executing the program. We will do this—but not yet. First, we will look at a somewhat far-fetched analogy that contains many of the concepts involved in computer operation.

12.3 SECRET AGENT 89

Secret Agent 89 is trying to find out how many days before a certain world leader is to be assassinated. His contact tells him that this information is located in a series of post office boxes. To ensure that no one else gets the information, it is spread through 10 different boxes. His contact gives him the 10 keys along with the following instructions:

1. The information in each box is written in code.
2. Open box 1 first and execute the instruction located there.
3. Continue through the rest of the boxes in sequence unless instructed to do otherwise.
4. One of the boxes is wired to explode upon opening.

Agent 89 takes the 10 keys and proceeds to the post office, code book in hand.

Figure 12.1 shows the contents of the 10 post office boxes after having been decoded. Assume that you are Agent 89; begin at box 1 and go through the sequence of operations to find the number of days before the assassination attempt. Of course, it should not be as much work for you as it was for Agent 89 because you don't have to decode the messages. The answer is given in the next paragraph.

If you have proceeded correctly, you should have ended up at box 6 with an answer of 17. If you made a mistake, you might have opened box 7, in which case you are no longer with us. As you went through the sequence of operations, you essentially duplicated the types of operations and encountered many of the concepts that are part of a computer. We will now discuss these operations and

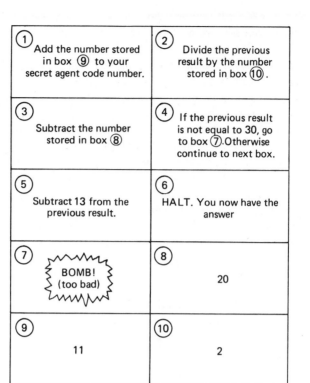

Figure 12.1   Ten post office boxes with coded message for Agent 89.

concepts in the context of the secret-agent analogy and see how they are related to actual computers.

In case you have not already guessed, the post office boxes are like the *memory* in a computer, where *instructions* and *data* are stored. Post office boxes 1–6 contain instructions to be executed by the secret agent and boxes 8–10 contain the data called for by the instructions. (The contents of box 7, to our knowledge, has no counterpart in computers.) The numbers on each box are like the *addresses* of the locations in memory.

Three different classes of instructions are present in boxes 1–6. Boxes 1, 2, 3, and 5 are instructions that call for *arithmetic operations*. Box 4 contains a *decision-making* instruction called a *conditional jump* or *conditional branch*. This instruction calls for the agent or computer) to decide whether to jump to address 7 or to continue to address 5, depending on the result of the previous arithmetic operation. Box 6 contains a simple control instruction that requires no data or refers to no other address (box number). This *halt* instruction tells the agent that the problem is finished (program is completed) and to go no further.

Each of the arithmetic and conditional jump instructions consists of two parts— an *operation* and an *address*. For example, the first part of the first instruction specifies the operation of addition. The second part gives the address (box 9) of the data to be used in the addition. These data are usually called the *operand* and its address is called the *operand address*. The instruction in box 5 is a special case in which there is no operand address specified. Instead, the operand (data) to be used in the subtraction operation is included as part of the instruction.

A computer, like the secret agent, decodes and then executes the instructions stored in memory *sequentially*, beginning with the first location. The instructions are executed in order unless some type of *branch* instruction (such as box 4) causes the operation to branch or jump to a new address location to obtain the next instruction. Once the branching occurs, instructions are executed sequentially beginning at the new address.

This is about as much information as we can extract from the secret-agent analogy. Each of the concepts we encountered will be encountered again in subsequent material. Hopefully, the analogy has furnished insights that should prove useful as we begin a more technical study of computers.

12.4 BASIC COMPUTER SYSTEM ORGANIZATION

Every computer contains five essential elements or units: the *arithmetic logic unit* (ALU), the *memory unit*, the *control unit*, the *input unit*, and the *output unit*. The basic interconnection of these units is shown in Figure 12.2. The arrows in this diagram indicate the direction in which data, information, or control signals are flowing. Two different-size arrows are used; the larger arrows represent data or information that actually consists of a relatively large number of parallel lines, and the smaller arrows represent control signals that are normally only one or a few lines. The various arrows are also numbered to allow easy reference to them in the following descriptions.

Arithmetic/Logic Unit

The ALU is the area of the computer in which arithmetic and logic operations are performed on data. The type of operation that is to be performed is determined by signals from the control unit (arrow 1). The data that are to be operated on by the ALU can come from either the memory unit (arrow 2) or the input unit (arrow 3). Results of operations performed in the ALU can be transferred to either the memory unit for storage (arrow 4) or to the output unit (arrow 5).

Memory Unit

The memory stores groups of binary digits (words) that can represent instructions (program) that the computer is to perform and the data that are to be operated on by the program. The memory also serves as storage for intermediate and final

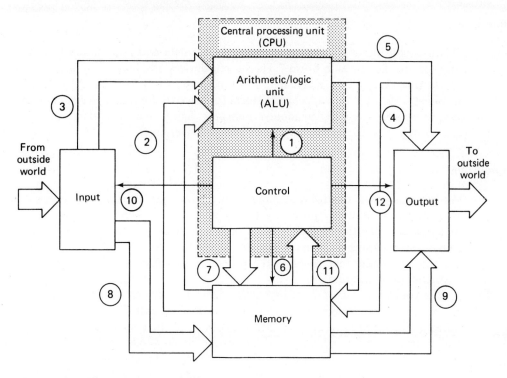

Figure 12.2   Basic computer organization.

results of arithmetic operations (arrow 4). Operation of the memory is controlled by the control unit (arrow 6), which signals for either a read or a write operation. A given location in memory is accessed by the control unit, providing the appropriate address code (arrow 7). Information can be written into the memory from the ALU or the input unit (arrow 8), again under control of the control unit. Information can be read from memory into the ALU (arrow 2) or into the output unit (arrow 9).

Input Unit

This unit consists of all of the devices used to take information and data that are external to the computer and put them into the memory unit (arrow 8) or the ALU (arrow 3). The control unit determines where the input information is sent (arrow 10). The input unit is used to enter the program and data into the memory unit prior to starting the computer. This unit is also used to enter data into the ALU from an external device during the execution of a program. Some of the common input devices are keyboards, toggle switches, teletypewriters, punched-card and punched-paper-tape readers, magnetic-tape readers, and analog-to-digital converters (ADC).

Output Unit

This unit consists of the devices used to transfer data and information from the computer to the "outside world." The output devices are directed by the control unit (arrow 12) and can receive data from memory (arrow 9) or the ALU (arrow 5), which is then put into appropriate form for external use. Examples of common output devices are LED readouts, indicator lights, teletypewriters, printers, cathode-ray-tube displays and digital-to-analog converters (DAC).

Control Unit

The function of the control unit should now be obvious. It directs the operation of all the other units by providing timing and control signals. In a sense, the control unit is like the conductor of an orchestra, who is responsible for keeping each of the orchestra members in proper synchronization. This unit contains logic and timing circuits that generate the proper signals necessary to execute each instruction in a program.

The control unit *fetches* an instruction from memory by sending an address (arrow 7) and a read command (arrow 6) to the memory unit. The instruction word stored at the memory location is then transferred to the control unit (arrow 11). This instruction word, which is in some form of binary code, is then decoded by logic circuitry in the control unit to determine which instruction is being called for. The control unit uses this information to generate the necessary signals for *executing* the instruction.

Central Processing Unit (CPU)

In Figure 12.2, the ALU and control units are shown combined into one unit called the central processing unit (CPU). This is commonly done to separate the actual "brains" of the computer from the other units. We will use the CPU designation often in our work on microcomputers because, as we shall see, in microcomputers the CPU is often contained in a single LSI chip called the *microprocessor* chip.

12.5 COMPUTER WORDS

The preceding description of how the various units in a computer interact has been, by necessity, somewhat oversimplified. To proceed in more detail, we must define the various forms of information that are continually being transferred and manipulated within the computer.

In a computer, the most elementary unit of information is the binary digit (bit). A single bit, however, can impart very little information. For this reason, the primary unit of information in a computer is a group of bits referred to as the *computer word*. Word size is so important that it is often used in describing a computer. For example, a 16-bit computer is a computer in which data and instructions are processed in 16-bit units. Of course, the word size also indicates the word size of the memory unit. Thus, a 16-bit computer has a memory unit that stores a certain number of 16-bit words.

A large variety of word sizes have been used by computer manufacturers. The larger (maxi) computers have word sizes that range from 16 to 64 bits, with 32 bits being the most common. Minicomputer word sizes run from 8 to 32, with 16 bits representing the overwhelming majority. Most microcomputers use an 8-bit word size. There are several 4-bit microcomputers which are designed for replacing digital logic circuits, and a few 16-bit microcomputers which are aimed at competing with minicomputers.

The Byte

A group of 8 bits is called a *byte* and represents a universally used unit in the computer industry. For example, a microcomputer with an 8-bit word size is said to have a word size of one byte. A 16-bit computer can be said to have a word size of two bytes. When we deal with microcomputers that have an 8-bit word size, we will use the terms "word" and "byte" interchangeably.

The 4-bit microcomputers have a word size of one-half byte. This is commonly referred to as a "nibble." Thus, each word in a 4-bit microcomputer is a nibble and two nibbles constitute a byte.

Types of Computer Words

A word stored in a computer's memory unit can contain several different types of information, depending on what the programmer intended for that particular word. We can classify computer words into three categories: (1) pure binary numerical data, (2) coded data, and (3) instructions. These will now be examined in detail.

12.6 BINARY DATA WORDS

These are words that simply represent a numerical quantity in the binary number system. For example, a certain location in the memory of an 8-bit process control microcomputer might contain the word 01110011, representing the desired process temperature in Farenheit degrees. This binary number 01110011 is equivalent to 115_{10}.

Here is an example of a 16-bit data word:

$$1010000101001001$$

which is equivalent to $41,289_{10}$.

Obviously, a wider range of numerical data can be represented with a larger word size. With an 8-bit word size, the largest data word (11111111_2) is equivalent to $2^8 - 1 = 255_{10}$. With a 16-bit word size, the largest data word is equivalent to $2^{16} - 1 = 65,535_{10}$. With 32 bits (four bytes), we can represent numbers greater than 4 billion.

Signed Data Words

A computer would not be too useful if it could only handle positive numbers. For this reason, most computers use the signed 2's-complement system. Recall that the most significant bit (MSB) is used as the *sign* bit (0 is positive and 1 is negative). Here is how the values $+9$ and -9 would be represented in an 8-bit computer:

$$+9 \longrightarrow 00001001$$
$$+ \quad \text{——binary for } 9_{10}$$

$$-9 \longrightarrow 11110111$$
$$- \quad \text{——2's complement of } 0001001$$

Here, of course, only 7 bits are reserved for the magnitude of the number. Thus, in the signed 2's-complement system, we can only represent numbers from -127_{10} to $+127_{10}$. Similarly, with 16-bit words, we can have a range from $-32,767_{10}$ to $+32,767_{10}$.

Multiword Data Units

Very often a computer needs to process data that extend beyond the range possible with a single word. For such cases, two or more memory words can be used to store the data in parts. For example, the 16-bit data word 1010101100101001_2 can be stored in two consecutive 8-bit memory locations, as shown below:

Memory Address	Contents		
0030	10101011	$\longrightarrow$	8 high-order bits of 16-bit number
0031	00101001	$\longrightarrow$	8 low-order bits of 16-bit number

Here, address location 0030_{16} stores the 8 higher-order bits of the 16-bit data word. This is also called the *high-order byte*. Similarly, address 0031_{16} stores the *low-order byte*. The two bytes combined make up the full data word.

There is no actual limit to the number of memory words that can be combined to store large numbers.

Octal and Hex Data Representation

For purposes of convenience in writing and displaying data words, they can be represented in either octal or hexadecimal codes. For example, the number $+116_{10}$ can be represented in a single byte as 01110100_2. The hex and octal representations are:

$$01110100_2 = 74_{16}$$
$$01110100_2 = 164_8$$

It is important to realize that the use of hex or octal representations is solely for convenience of the computer user; the computer memory still stores the binary numbers (0s and 1s), and these are what the computer processes.

12.7 CODED DATA WORDS

Data processed by a computer does not have to be pure binary numbers. One of the other common data forms uses the BCD code, where each group of 4 bits can represent a single decimal digit. Thus, an 8-bit word can represent two decimal digits, a 16-bit word can represent four decimal digits, and so on. Many computers can perform arithmetic operations on BCD-coded numbers as part of their normal instruction repertoire; others, especially some microcomputers, require special effort on the part of the programmer in order to do BCD arithmetic.

Data words are not restricted to representing only numbers. They are often used to represent alphabetic characters and other special characters or symbols using codes such as the 7-bit ASCII code (Chapter 5). The ASCII code is used by all minicomputer and microcomputer manufacturers. Although the basic ASCII code uses 7 bits, an extra parity bit (Chapter 5) is added to each code word, producing a one-byte ASCII code. The example below shows how a message might be stored in a sequence of memory locations using ASCII code with an *even* parity bit. The contents of each location are also given in hex code. Use Table 5.5 to determine the message. Note that the leftmost bit is the parity bit and the first character is stored in location $012A_{16}$. The decoded message is the familiar electrical Ohm's law, $I = V/R$.

Contents

Address Location	Binary	Hex
012A	11001001	C9
012B	10111101	BD
012C	01010110	56
012D	10101111	AF
012E	11010010	D2
	ASCII	

The one-byte ASCII code is particularly suited to computers with an 8-bit word size. However, computers with other word sizes still use one-byte ASCII. For example, a 16-bit computer can pack two bytes into one memory word so that each word represents two characters. This is illustrated below, where the characters I and = are stored in one 16-bit word.

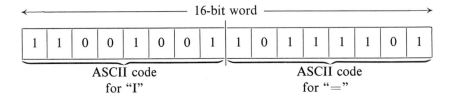

On the other hand, a 4-bit microcomputer would have to use two consecutive memory locations to represent one byte of ASCII.

Interpretation of Data Words

Suppose you are told that a particular data word in a microcomputer's memory is 01010110. This word can be interpreted in several ways. It could be the binary representation of 86_{10}; it could be the BCD representation of 56_{10}; or it could be the ASCII code for the character V. How should this data word be interpreted? It is up to the programmer, since he or she is the one who places the data in memory along with instructions that make up the program. The programmer knows what type of data word he or she is using and must make sure that the program of instructions executed by the computer interprets the data properly.

12.8 INSTRUCTION WORDS

The format used for *data* words varies only slightly among different computers, especially those with the same word size. This is not true, however, of the format for *instruction* words. These words contain the information necessary for a computer to execute its various operations, and the format and codes for these can vary widely from computer to computer. Depending on the computer, the information contained in an instruction word can be different. But, for most computers, the instruction words carry two basic units of information: the *operation* to be performed and the *address* of the *operand* (data) that is to be operated upon.

Figure 12.3 shows an example of a *single-address instruction word* for a hypothetical 20-bit computer. The 20 bits of the instruction word are divided into two parts. The first part of the word (bits 16–19) contains the *operation code* (*op code*, for short). The 4-bit op code represents the operation that the computer is being instructed to perform, such as addition or subtraction. The second part (bits 0–

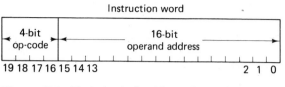

Figure 12.3 Typical single-address instruction word.

15) is the *operand address* which represents the location in memory where the operand is stored.

With 4 bits used for the op code, there are $2^4 = 16$ different possible op codes, with each one indicating a different instruction. This means that a computer using this instruction word format is limited to 16 different possible instructions which it can perform. A more versatile computer would have a greater number of instructions and would therefore require more bits in its op code. In any case, each instruction that a computer can perform has a specific op code which the computer (control unit) must interpret (decode).

The instruction word of Figure 12.3 has 16 bits reserved for the operand address code. With 16 bits, there are $2^{16} = 65,536$ different possible addresses. Thus, this instruction word can specify 16 different instructions and 65,536 operand addresses. As an example, a 20-bit instruction word might be

0 1 0 0	0 1 0 1 1 0 1 0 0 1 1 1 0 0 1 0
Op code	Address code

The op code 0100 represents one of 16 possible operations; let's assume that it is the code for *addition* (ADD). The address code is 0101101001110010 or, more conveniently, 5A72 in hexadecimal. In fact, this complete instruction word can be expressed in hexadecimal as

$$4 \quad 5 \quad A \quad 7 \quad 2$$

Op code   Address

This complete instruction word, then, tells the computer to do the following:

> Fetch the data word stored in address location 5A72, send it to the ALU and *add* it to the number in the accumulator register. The sum will then be stored in the accumulator. (Previous contents of accumulator is lost.)

We will examine this and other instructions more thoroughly later.

Multiple-Address Instructions

The single-address instruction described above is the basic type used in small computers and was once the principal type used in larger computers. The larger computers, however, have begun to use several other instruction formats which provide more information per instruction word.

Figure 12.4 shows two instruction word formats that contain more than one address. The two-address instruction has the op code plus the addresses of *both* operands which are to take part in the specified operation. The three-address instruction has the addresses of both operands plus the address in memory where the result is to be stored.

These multiple-address instruction words have the obvious advantage that they contain more information than does a single-address instruction. This means that a computer using multiple-address instructions will require fewer instructions to execute a particular program. Of course, the longer instruction words require a memory unit with a larger word size. We will not concern ourselves further with multiple-address instructions since they are not used in microcomputers.

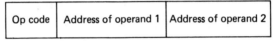

Op code	Address of operand 1	Address of operand 2

Two-address instruction

(a)

Op code	Address of operand 1	Address of operand 2	Address of where to store result

Three-address instruction

(b)

Figure 12.4 Multiple-address instruction formats.

Multibyte Instructions

We have looked at instruction word formats that contain op code and operand address information in a *single* word. In other words, a complete instruction such as those in Figure 12.3 or Figure 12.4 is stored in a *single* memory location. This is typical of computers with relatively large word sizes. For most microcomputers and many minicomputers, the smaller word size makes it impossible to provide the op code and operand address in a single word.

Since the vast majority of microcomputers use an 8-bit (one-byte) word length, we will describe the instruction formats used in 8-bit computers. With a one-byte word size there are *three* basic instruction formats: single-byte, two-byte, and three-byte instructions. These are illustrated in Figure 12.5.

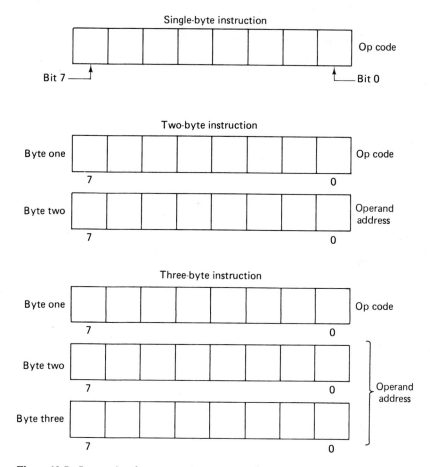

Figure 12.5 Instruction formats used in 8-bit microcomputers.

The single-byte instruction contains only an 8-bit op code, with no address portion. Clearly, this type of instruction does not specify any data from memory to be operated on. As such, single-byte instructions are used for operations that do not require memory data. An example would be the instruction *clear the accumulator register to zero* (CLA), which instructs the computer to clear all the FFs in the ALU's accumulator.

The first byte of the two-byte instruction is an op code and the second byte is an 8-bit address code specifying the memory location of the operand. In the three-byte instruction, the second and third bytes form a 16-bit operand address. For these multibyte instructions, the two or three bytes making up the complete instruction have to be stored in successive memory locations. This is illustrated below for a three-byte instruction. The left-hand column lists the address locations in memory where each byte (word) is stored. These addresses are given in hexadecimal code. The second column gives the binary word as it is actually stored in memory; the third column is the hex equivalent of this word. Examine this complete example before reading further and try to figure out what it represents.

| | Memory Word | | |
Memory Address (hex)	Binary	Hex	Description
0020	01001010	4A	Op code for ADD
0021	00110101	35	High-order address bits (HI)
0022	11110110	F6	Low-order address bits (LO)
.	.	.	
.	.	.	
.	.	.	
.	.	.	
.	.	.	
35F6	01111100	7C	Operand

The three bytes stored in locations 0020, 0021, and 0022 constitute the complete instruction for adding the data word stored in address location 35F6 to the accumulator. The second and third bytes hold the 8 high-order bits (HI) and 8 low-order bits (LO), respectively, of the operand address. Some microcomputers use the reverse order, with LO stored in the second byte and HI in the third byte of the instruction sequence. Memory location 35F6 is also shown; its contents is the data word (7C) which the control unit of the computer will fetch and send to the ALU for addition.

12.9   A SIMPLE PROGRAM EXAMPLE

Now that we have looked at some of the various types of data and instruction words, the next step is to combine data and instructions in a program. For our programming example, we will use a ficticious computer called the S-16. The "S" stands for *small* while the "16" represents the number of instructions this computer can perform and also the number of words in its memory. Although the S-16 is smaller than most practical computers, its operation is exactly the same as any larger computer. Its small size is simply a convenience for purposes of illustration.

The S-16 is an 8-bit computer. Its instruction word format is shown below.

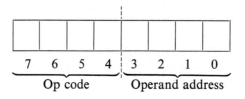

With a 4-bit op code, any of $2^4 = 16$ different instructions can be specified. The 4-bit operand address code means that there are $2^4 = 16$ different address locations in memory.

EXAMPLE 12.1   What is the total number of bits stored in the S-16 memory? How many FFs are in the S-16 accumulator register?

Solution: We know that it stores 16 words and that each word is 8 bits long. Thus, the memory capacity is 16 8-bit words, or a total of 128 bits. Since a data word is 8 bits, the accumulator register has 8 FFs.

Table 12.1 describes some of the S-16 instructions. Each instruction is accompanied by its 4-bit op code and also by a *symbolic* code made up of three or four letters. The symbolic codes are easier to remember than the op codes, since the letters are abbreviations for the operation to be performed. Read the description of each instruction carefully because each is typical of the instructions available in most computers. Remember that the *operand* referred to in these descriptions is the data word stored at the operand address location.

Table 12.1

Symbolic Code	Binary OP Code	Description of Operation
LDA	1100	*Load accumulator:* the data stored at the operand address are loaded into the accumulator register.
ADD	0100	*Add:* the operand is added to the number stored in the accumulator and the resultant sum is stored in the accumulator.
SUB	0101	*Subtract:* the operand is subtracted from the contents of the accumulator and the result is stored in the accumulator.
STA	0111	*Store accumulator:* the contents of the accumulator is stored in memory at the location specified by the operand address.
JMP	1000	*Jump (unconditionally):* the next instruction is taken from the location specified by the operand address instead of in sequence.
JPZ	1001	*Jump on zero:* the next instruction is taken from operand address *if* the accumulator contents is *zero.* Otherwise, the next instruction is taken in sequence.
HALT	0001	*Halt:* the computer operation is halted. No further instructions are executed.

Using the partial set of S-16 instructions from Table 12.1, we will write a simple program that will do the following:

1. Subtract one number (X) from another number (Y).
2. Store the result, Q, in location $1001 = 9_{16}$.
3. If the result is *zero*, halt the computer at location 0101 (5_{16}). Otherwise, halt the computer at location 0100 (4_{16}).

The complete program as it appears in the S-16 memory is shown in Table 12.2. The 16 memory locations are indicated by their hexadecimal addresses, 0–F. The words stored in locations 0–5 are the sequence of instruction words used in the program. Locations 7, 8, and 9 are used for data words. The numbers X and Y are data required by the program and are initially stored in 7 and 8, respectively. Location 9 is reserved for storing the result, Q. The locations 6 and A–F are not used in this sample program.

Table 12.2

Memory Address (hex)	Memory Word (binary)	Symbolic Code	Description
0	11001000	LDA 8	Load Y into accumulator
1	01010111	SUB 7	Subtract X from accumulator
2	01111001	STA 9	Store result Q in location 9
3	10010101	JPZ 5	If $Q = 0$, jump to 5
4	00010000	HALT	If $Q \neq 0$, halt here
5	00010000	HALT	Halt here if $Q = 0$
6	Not used		
7	$X_7 X_6 X_5 X_4 X_3 X_2 X_1 X_0$	X	Data word X
8	$Y_7 Y_6 Y_5 Y_4 Y_3 Y_2 Y_1 Y_0$	Y	Data word Y
9	????????	Q	Location where Q will be stored
A	Not used		
B			
C			
D			
E			
F			

Program Execution

Once the program is in the computer's memory, it is ready to be executed by the computer (do not be concerned for now about how the program got into memory). We will now proceed through the program and explain what the computer does at each step.

1. Operation is initiated by the computer user, usually by activating a START or RUN switch on the computer console. This causes the CONTROL unit to begin fetching instructions from memory, starting at address 0.*

*In many computers the user can set the starting address to something other than address 0.

This first instruction is LDA 8 and tells the CONTROL unit to *read* the data word stored in address 8 and load it into the ACCUMULATOR. The data word at address 8 is the value of *Y*.

2. After executing the instruction at address 0, the CONTROL unit automatically sequences to address 1 for its next instruction. This sequencing is provided by a **PROGRAM COUNTER** (PC) which starts at 0 and is automatically incremented as each instruction is executed so that it always contains the address of the next instruction.

    The instruction at address 1 is SUB 7 and causes the control unit to go to address 7 to obtain the data word (*X*). These data are then sent to the ALU to be subtracted from the contents of the ACCUMULATOR. The result of this operation (*Q*) is stored in the ACCUMULATOR.

3. The PC is incremented to address 2, so the CONTROL unit takes its next instruction from that location. This instruction, STA 9, causes the CONTROL unit to store (write) the contents of the ACCUMULATOR into address 9. The value of $Q = Y - X$ is now in memory location 9 as well as still being in the ACCUMULATOR.

4. The PC is incremented to 3 and the CONTROL unit fetches the next instruction from that location. This instruction, JPZ 5, causes the CONTROL unit to examine the contents of the ACCUMULATOR. If the value is *not* exactly equal to zero, the CONTROL unit increments PC to 4. If the ACCUMULATOR value *is* exactly zero, the CONTROL unit sets PC to 5. Thus, the next instruction may be taken from either address 4 or address 5 depending on the value of *Q*.

5. If PC is now 4, the CONTROL unit fetches the instruction at address 4. This instruction causes the computer to HALT and no further instructions will be executed. But, if PC is now 5, the CONTROL unit obtains its next instruction from address 5 (rather than 4). This instruction causes the computer to HALT so that no more instructions will be executed.

6. This completes the execution of our small program. The computer will do nothing more until the user either presses the START button to re-execute the same program or puts a new program into memory for execution.

This simple example program did not even begin to illustrate the real assets of a computer but was intended to show the sequencing action that occurs in all computers. The operation takes place one step at a time, as explained. Of course, each step is executed in a very short time interval, which depends on the speed of the computer. For example, this short program would take even the slowest microcomputer around 10 μs to execute.

12.10 COMPUTER OPERATING CYCLES

We now have the basic idea of how a computer sequences through a program. In essence, the computer is always doing one of two things: (1) fetching an instruction word from memory and interpreting that instruction, or (2) executing the operations called for by the instruction word. The computer operation, then, is comprised of two types of cycles: an *instruction* cycle and an *execution* cycle. We will examine these two cycles and describe the flow of information between the various computer units for each.

Instruction Cycle

Refer to Figure 12.6, which is a diagram of those computer elements that are involved in the instruction cycle. The arrows on the diagram are simply used to indicate the direction of information flow. The circled numbers correspond to the steps in the following description:

1. The instruction cycle begins when the address of the next instruction is transferred from the PROGRAM COUNTER (PC) to the MEMORY ADDRESS REGISTER (MAR). The PC keeps track of which instruction is being fetched from memory. It is incremented at the end of each instruction cycle. The MAR is used to hold any memory address that is being accessed for a read or write operation.

2. The CONTROL unit generates a READ pulse, which causes the MEMORY unit to read the instruction word from the memory location specified by the MAR. This instruction word is clocked into the MEMORY DATA REGISTER (MDR). The MDR functions as a buffer register for all data read from or written into memory.

3. The op-code portion of the instruction word, which is now in the MDR, is transferred to the INSTRUCTION REGISTER (IR) in the CONTROL unit. Simultaneously, the operand address portion of the instruction word is transferred to the MAR (replacing its previous contents).

4. The op code in the IR is fed to the INSTRUCTION DECODER (ID), which determines which op code is present. This information is sent to the control-signal-generating portion of the CONTROL unit to determine which control signals will be needed to execute the instruction during the execution cycle.

5. The PC is incremented to prepare for the next instruction cycle. (Note that the PC contents is not now in the MAR.)

To summarize, during the instruction cycle, the instruction word is fetched from memory, the op-code portion is decoded, the address portion is placed in the MAR, and the PC is incremented.

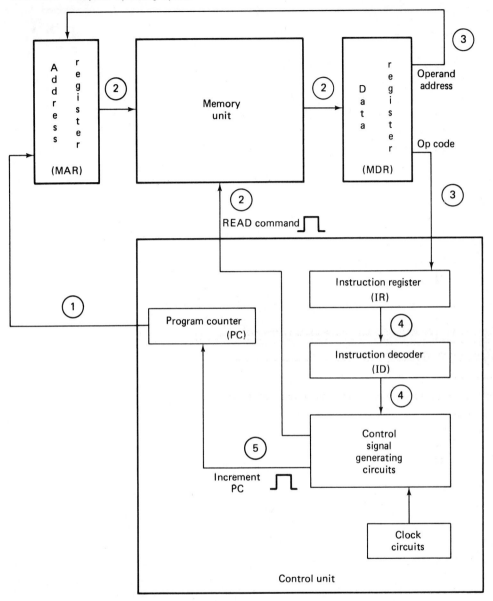

Figure 12.6 Flow of information during a computer's instruction cycle.

Execution Cycle

An instruction cycle is followed by an execution cycle during which the instruction is actually carried out. The exact sequence of operations during the execution cycle will, of course, depend on the instruction. Refer to the diagram in Figure 12.7, showing the flow of information during a typical execution cycle.

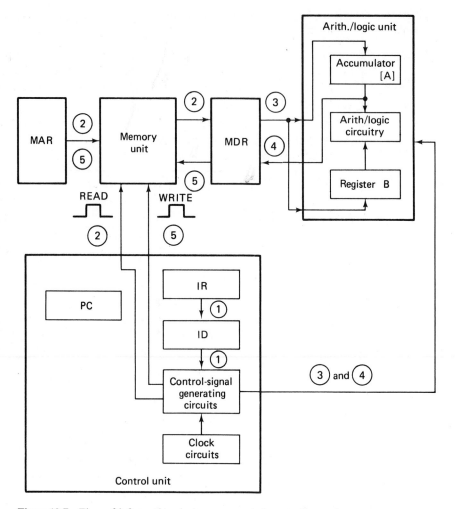

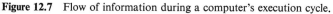

Figure 12.7   Flow of information during a computer's execution cycle.

1. The IR still holds the op code and the ID indicates which instruction is being executed; this determines which signals will be generated by the control-signal-generating circuitry during the following steps.

2. If the instruction requires fetching a data word from memory, the CONTROL unit generates a READ pulse, which takes the data word from the address specified by the MAR and places it in the MDR. Recall that the MAR is holding the operand address from the instruction cycle.

3. Once in the MDR, the data word can be transferred to the ALU, where it might be placed in the ACCUMULATOR or B-register. The CONTROL unit generates the control signals to produce this transfer and also to

determine what operation, if any, is to be performed by the ALU on the data word.

4. If the instruction requires placing data into memory, such as a STA instruction, the data are first transferred into the MDR. These data often come from the ACCUMULATOR and the transfer is initiated by a signal from the CONTROL unit.

5. Once the data are in the MDR, the CONTROL unit generates a WRITE pulse which causes the data to be written into the address location specified by the MAR (operand address).

It should be clear that an execution cycle will follow steps 1, 2, and 3 or steps 1, 4, and 5, depending on the instruction being executed. At the completion of the execution cycle, the computer immediately goes into an instruction cycle and takes its next instruction from the address specified by the PROGRAM COUNTER. Recall that the PC was incremented at the end of the instruction cycle.

EXAMPLE 12.2   A certain computer has a 16-bit word size and a 6-bit op code. How many bits does it have in each of its registers (MAR, MDR, PC, IR, A)?

Solution: With a 16-bit word size and a 6-bit op code, the operand address portion of an instruction word will be 10 bits long. Thus, the MAR and PC must be 10 bits each.
   The MDR and A both handle a complete word as it comes from memory, so these registers have to be 16 bits long.
   The IR receives only the op-code bits and is only 6 bits long.

Synchronizing and Controlling the Operations

The logic circuitry that synchronizes and controls all the computer operations is located in the CONTROL unit. This circuitry always includes one or more clock generators that generate periodic clock signals to act as the time frame for all operations. For example, one complete instruction cycle or one complete execution cycle might require four clock pulses. Figure 12.8 illustrates the timing for such a situation.

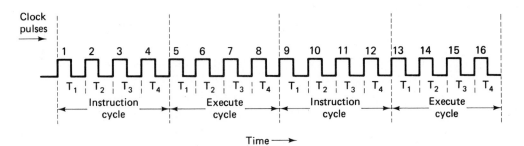

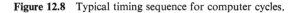

Figure 12.8   Typical timing sequence for computer cycles.

As shown in the diagram, each series of four clock pulses constitutes one operating cycle. Pulses 1–4 comprise the first instruction cycle, followed by pulses 5–8, which comprise the first execution cycle. Pulses 9–12 constitute the second instruction cycle, followed by pulses 13–16 for the second execution cycle. This sequence continues for as many cycles as are necessary to process all the instructions in a given program.

Each operating cycle is divided into four time intervals and certain operations occur at each of these times. For example, during the first interval (T_1) of each instruction cycle, the contents of the PROGRAM COUNTER is transferred to the MAR. Actually, this transfer takes place on the leading edge of the T_1 pulse in the instruction cycle. As another example, during the execution of an ADD instruction, the addition of the numbers in the ACCUMULATOR and the B register occurs during the T_4 interval of the execution cycle.

The CONTROL unit sends control signals to the other computer units during each of the clock time intervals. The sequence of control signals will be the same for each instruction cycle, but the sequence will vary during the execution cycle depending on the instruction being executed (the op code). Clearly, the ultimate speed of the computer operation depends to a great extent on the clock signal. In general, a higher-frequency clock signal indicates a faster computer. We will go into more detail on timing sequences in our study of microprocessors and microcomputers.

12.11 INPUT/OUTPUT (I/O)

We have said very little about the computer's INPUT and OUTPUT units so far, and we do not intend to say much until later. This does not mean that these units are of secondary importance. In fact, it is often the input and output devices that are used to distinguish one computer system from another.

These units represent the means through which the computer communicates with the external world. For example, the instructions and data that constitute a program must be placed in the computer's internal memory before being executed. In large computer systems, this might be done by a punched card reader which translates holes punched on "IBM" cards into the instruction and data codes required by the computer. In a microcomputer, the instructions and data might be entered from a keyboard or from sets of toggle switches. Of course, there are many other types of input devices in common usage, such as paper tape readers, magnetic tape readers, and analog-to-digital (A/D) converters.

Once the computer executes its program, it usually has results or control signals that it must present to the external world. For example, a large computer system might have a line printer as an output device. Here, the computer sends out signals to print out the results on paper. A microcomputer might display its results on indicator lights or on LED displays. Again, there are many other types of output devices, such as CRT displays, paper-tape punches, and digital-to-analog (D/A) converters.

Interfacing

The most important aspect of the I/O units involves *interfacing*, which can be defined as the joining of dissimilar devices in such a way that they are able to function in a compatible and coordinated manner. *Computer interfacing* is more specifically defined as the synchronization of digital information transmission between the computer and external input/output devices.

Many input/output devices are not directly compatible with the computer because of differences in such characteristics as operating speed, data format (e.g., hex, ASCII, binary), data transmission mode (e.g., serial, parallel), and logic signal level. Such I/O devices require special interface circuits which allow them to communicate with the CONTROL, MEMORY, and ALU portions of the computer system. A common example is the popular teletypewriter (abbreviated TTY), which can operate both as an input and an output device. The TTY transmits and receives data serially (one bit at a time) while most computers handle data in parallel form. Thus, a TTY requires interface circuitry in order to send data to or receive data from a computer.

12.12 HARDWARE AND SOFTWARE

A computer system consists of hardware and software. The hardware refers to the electronic, mechanical, and magnetic elements from which the computer is fabricated. This hardware, especially in large-scale computer systems, can be somewhat awe-inspiring. But, regardless of the complexity of the computer hardware, the computer is a useless maze of wire unless it has a function to perform and a program to tell it how to do it.

Software refers to the totality of programs and programming systems used by a computer. These programs are all initially written on paper (software) before being transferred to some storage medium (hardware), and completely control the computer's operation from startup to shutdown. Many types of programs are supplied by computer manufacturers, but the user will often have program needs that are unique and so must write his own programs. This is especially true for microcomputers, which are relatively new and are employed in such a wide variety of applications.

QUESTIONS AND PROBLEMS

12.1 List the *five* major units of a digital computer and briefly describe the function of each.

12.2 Match each of the terms in column A with its definition from column B.

A	B
1. CPU	a. Tells computer which operation to perform.
2. Instruction cycle	b. Time when computer *performs* operation called for by instruction word.
3. MAR	c. Where data to be operated on are stored.
4. Program counter	d. Keeps track of instruction addresses.
5. Microprocessor	e. Time when computer *fetches* an instruction from memory.
6. Execution cycle	f. CPU on a chip.
7. Op code	g. Holds address of memory location being accessed.
8. Operand address	h. Control and arithmetic/logic units.

12.3   A certain computer has a 22-bit instruction word consisting of an 8-bit op code and a 14-bit address code. How many different instructions can this instruction word specify? How many possible memory addresses can this instruction word specify?

12.4   Below is a listing of the memory contents of the S-16 computer. The program counter is set to 0 and the computer begins executing the program. Determine the sequence of operations that the computer performs as it executes the program. What value is stored in location F after the computer halts?

Memory Address (hex)	Memory Word (hex)
0	C9
1	5A
2	5B
3	9D
4	48
5	8D
6	00
7	00
8	02
9	09
A	05
B	03
C	00
D	7F
E	10
F	FF

12.5   Change the contents of location B to 04 and repeat Question 12.4.

12.6   List the different events that occur in a typical computer instruction cycle.

12.7   Repeat Problem 12.6 for the execution cycle of an ADD instruction.

12.8   Repeat Problem 12.6 for the execution cycle of an STA (store accumulator) instruction.

12.9   Explain the difference between hardware and software.

12.10   Define the term "interfacing."

INTRODUCTION
TO THE MICROPROCESSOR
AND MICROCOMPUTER

Because of large-scale integration (LSI), the latest major development in semi-conductor technology, it is now reasonable to predict that computers will eventually affect all aspects of human life. LSI has made it possible to integrate the control and arithmetic units of a computer on a *single* chip. In other words, the central processing unit (CPU) which we talked about in Chapter 12 is available as a single circuit component commonly referred to as a *microprocessor*. In addition, LSI has given us the capability to construct large-capacity memories from LSI memory chips, and also many types of LSI input/output interface chips.

What this means is that a relatively small number of LSI chips can be put together to operate as a computer—a *microcomputer*. Right now microcomputer prices range from $5 to $1,000, making it possible to use computers in areas where cost and size were previously prohibitive, such as in controlling appliances or automobiles or as personal computers for the home.

13.1 BASIC μC ELEMENTS

It is important that we understand the distinction between the microcomputer (μC) and the microprocessor (μP). A μC contains many elements, one of which is the μP. The μP is the central processing unit (CPU) portion of the μC. This is illustrated in Figure 13.1, where the basic elements of a μC are shown. The μP is typically a single LSI chip that contains all the control and arithmetic circuits of the μC. The μP may consist of more than one chip. This is true, for example, of bipolar μPs (TTL, Shottky TTL, ECL) which do not have the high packing densities of the MOS devices, and so require two or more chips to produce a μP with appropriate word size.

The memory unit shows both RAM and ROM devices, typical of most μCs, although one or the other might not be present in certain applications. The RAM

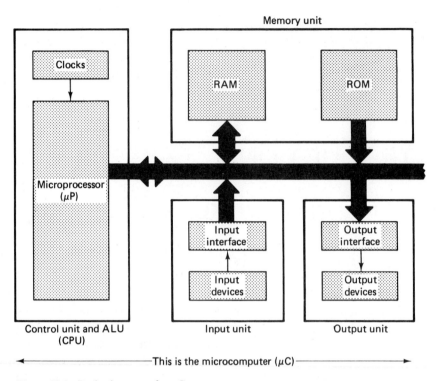

Figure 13.1   Basic elements of a μC.

section consists of one or more LSI chips arranged to provide the designed memory capacity. This section of memory is used to store programs and data, which will change often during the course of operation. It is also used as storage for intermediate and final results of operations performed during execution of a program.

The ROM section consists of one or more LSI chips to store instructions and data that do not change. For example, it might store a program that causes the μC to continually monitor a keyboard, or it might store a table of ASCII codes needed for outputting information to a teletype unit.

The input and output sections contain the interface circuits needed to allow the I/O devices to properly communicate with the rest of the computer. In some cases, these interface circuits are LSI chips designed by the μP manufacturer to interface his μP to a variety of I/O devices. In other cases, the interface circuits may be as simple as a buffer register.

13.2   WHY μPs AND μCs?

When the first single-chip μPs were introduced a few years ago, it was difficult to foresee the tremendous impact these devices would have on the creation of new products. But designers have rapidly become aware of the capabilities and versatility of these devices. Microprocessors are being utilized in new products in place of *random logic*; random logic refers to conventional logic designs using flip-flops, gates, counters, registers, and other medium-scale-integration (MSI) functions. For instance, a traffic-light controller that previously required 200 TTL chips can now be built with 12 chips using a μP system costing less than $250.

There are several fundamental reasons for the superiority of μP-based designs over random logic designs:

1. Fewer IC packages, printed-circuit boards, and connectors, thereby reducing assembly costs.

2. Greater reliability because of the decreased number of IC interconnections.

3. Lower power requirements, making power supply design easier.

4. Simpler system testing, evaluation, and redesign. Since μP-based equipment operates under the control of a program in memory (usually ROM), its operation is easily modified by simply changing the program (replacing or reprogramming the ROM). It is easier to change the *software* in such a system than to change the wiring in a random logic system.

5. Since product features can be added to μP-based equipment by adding to the software, manufacturers are increasing the capabilities and value of their products. For example, makers of μP-controlled cash registers are adding automatic tax computation by putting extra steps into the program stored in ROM. If the tax rate changes, the ROM can be replaced or reprogrammed to take care of the change.

Despite these advantages, μPs and μCs cannot compete with random logic in areas where high speed is required. Even those μPs that utilize high-speed bipolar technology are at a speed disadvantage because of the sequential nature of programmed computer control. μCs perform operations *one at a time* in anywhere from 0.1 to 20 μs per operation; in random logic systems many operations can be performed in parallel (simultaneously). Of course, μPs can be utilized in those portions of high-speed systems where speed is not critical.

13.3 TYPICAL μC STRUCTURE

We are now prepared to take a more detailed look at μC organization. The many possible μC structures are essentially the same in principle, although they vary as to the size of the data and address busses, and the types of control signals they use. In order to provide the clearest means for learning the principles of μC operation, it is necessary to choose a single type of μC structure and study it in detail. Once a solid understanding of this typical μC is obtained, it will be relatively easy to learn about any other type. The μC structure we have chosen to present here represents the most common one in use today and is shown in Figure 13.2.

This diagram shows the basic elements of an 8-bit microcomputer system and the various busses that connect them together. Although this diagram looks somewhat complex, it still does not show all the details of the μC system. For the time being, however, it will be sufficient for our purposes. We will add the pertinent details after a thorough discussion of the overall operation. Examine this diagram carefully and try to get as much information from it as possible before reading further.

The Bus System

The μC has three busses which carry all the information and signals involved in the system operation. These busses connect the microprocessor (CPU) to each of the memory and I/O elements so that data and information can flow between the μP and any of these other elements. In other words, the CPU is continually involved in sending or receiving information to or from a location in memory, an input device, or an output device.*

In the μC, all information transfers are referenced to the CPU. When the CPU is sending data to another computer element, it is called a WRITE operation and the CPU is WRITING into the selected element. When the CPU is receiving data from another element, it is called a READ operation and the CPU is READING from the selected element. It is very important to realize that the terms READ and WRITE always refer to operations performed by the CPU.

The busses involved in all the data transfers have functions that are described as follows:

Address Bus   This is a *unidirectional* bus, because information flows over it in only one direction, from the CPU to the memory or I/O elements. The CPU alone can place logic levels on the 16 lines of the address bus, thereby generating $2^{16} = 65,536$ different possible addresses. Each of these addresses corresponds to one memory location or one I/O element. For example, address $20A0_{16}$ might be a location in RAM or ROM where an 8-bit word is stored, or it might be an 8-bit buffer register that is part of the interface circuitry for a keyboard input device.

When the CPU wants to communicate with (READ or WRITE), a certain memory location or I/O device, it places the appropriate 16-bit address code on its 16 address pin outputs, A_0–A_{15}, and onto the address bus. These address bits are then *decoded* to select the desired memory location or I/O device. This decoding

---

*In some μC systems, it is possible for I/O devices to send data directly to or receive data directly from memory without the CPU being involved. This type of operation is called *direct memory access* (DMA).

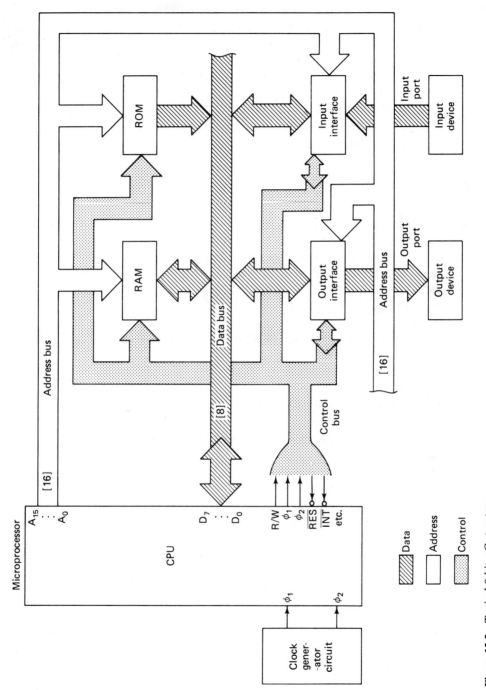

Figure 13.2 Typical 8-bit µC structure.

process usually requires decoder circuitry not shown on this diagram but which will be introduced later.

Data Bus This is a *bidirectional* bus, because data can flow to or from the CPU. The CPU's eight data pins, D_0–D_7, can be either inputs or outputs, depending on whether the CPU is performing a READ or a WRITE operation. During a READ operation they act as inputs and receive data that have been placed on the data bus by the memory or I/O element selected by the address code on the address bus. During a WRITE operation the CPU's data pins act as outputs and place data on the data bus, which are then sent to the selected memory or I/O element. In all cases, the transmitted data words are 8 bits long because the CPU handles 8-bit data words, making this an 8-bit μC.

In some microprocessors, the data pins are used to transmit other information in addition to data (e.g., address bits or CPU status information). That is, the data pins are time-shared or *multiplexed*, which means that special control signals must be generated by the CPU to tell the other elements exactly what is on the data bus at a particular time. We will not concern ourselves with this type of operation.

Control Bus This is the set of signals that is used to synchronize the activities of the separate μC elements. Some of these control signals, such as R/W, are signals the CPU sends to the other elements to tell them what type of operation is currently in progress. The I/O elements can send control signals to the CPU. An example is the reset input ($\overline{RES}$) of the CPU, which, when driven LOW, causes the CPU to reset to a particular starting state. Another example is the CPU's interrupt input ($\overline{INT}$), used by I/O devices to get the attention of the CPU when it is performing other tasks.

The control bus signals will vary widely from one μC to another. There are certain control signals that all μCs use, but there are also many control signals that are peculiar to the μP upon which the μC is based. We will include only the essential control signals in our initial discussion and then add the more specialized ones as they are needed.

Timing Signals

The most important signals on the control bus are the system clock signals that generate the time intervals during which all system operations take place. Different μCs use different kinds of clock signals, depending on the type of μP being used. Some μPs, like the 8085, the 6502, and the Z-80, do not require an external clock-generating circuit. A crystal or *RC* network connected to the appropriate μP pins sets the operating frequency for the clock signals which are generated on the μP chip. Other μPs, such as the 8080A and 6800, require an external circuit to generate the clock signals needed by the CPU and the other μC elements. These manufacturers often provide a special clock-generator chip designed to be used with their μP.

Many of the currently popular μPs (8080, 8085, 6800, 6502) use a two-phase clock system with nonoverlapping pulses such as those shown in Figure 13.3. Other widely used μPs (Z-80, RCA 1802) operate from a single clock signal. In our subsequent discussions we will use the two-phase clock system. The two clock phases, ϕ_1 and ϕ_2, are always part of the control bus. Other timing signals, derived from ϕ_1 and ϕ_2, are sometimes generated by the CPU and become part of the control bus.

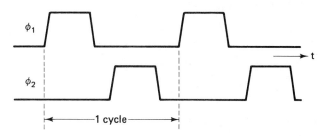

Figure 13.3   Two-phase clock system.

I/O Ports

During the execution of a program, the CPU is constantly READING or WRITING into memory. The program may also call on the CPU to READ from one of the input devices or WRITE into one of the output devices. Although the diagram of the 8-bit μC (repeated in Figure 13.4) only shows one input and one output device, there can be any number of each tied to the μC bus system. Each I/O device is normally connected to the μC bus system through some type of interface circuit. The function of the interface is to make the μC and the device compatible so that data can be easily passed between them. The interface is needed whenever the I/O device uses different signal levels, signal timing, or signal format than the μC.

For example, a typical I/O device is the standard teletype unit (abbreviated TTY), which sends ASCII-coded information to the computer in *serial* fashion (1 bit at a time over a single line). The μC, however, accepts data from the data bus as 8 *parallel* bits. Thus, an interface circuit is used to convert the TTY's serial signal to an 8-bit parallel data word, and another to convert the μC's parallel output data to a serial signal for the TTY.

It was mentioned during the discussion of the address bus that the CPU places a 16-bit address on this bus to select a certain memory location or a certain I/O device. This means that each I/O device has a specific address just like any location in memory. In many μCs, the CPU does not distinguish between memory and I/O, and it communicates with both in the same way using the same control signals. This method is called *memory-mapped I/O*. Other μCs use separate control signals and separate address decoders for I/O. This is called *isolated I/O*. We will

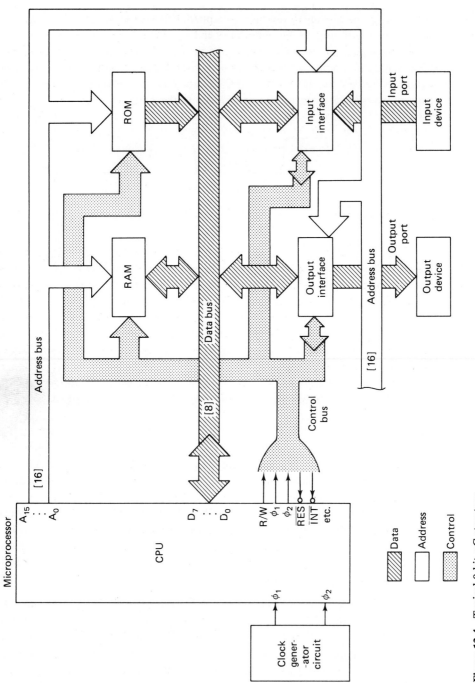

Figure 13.4 Typical 8-bit μC structure.

concentrate mainly on the memory-mapped I/O technique, since it is the most common and has several advantages over the isolated I/O technique.

Although I/O devices are treated like memory locations, they are significantly different from memory in some respects. One big difference is that I/O devices can have the capability to *interrupt* the μC while it is executing a program. What this means is that an I/O device can send a signal to the μP chip's interrupt ($\overline{INT}$) input to tell the CPU that it wishes to communicate with it. The CPU will then suspend execution of the program it is currently working on and will perform the appropriate operation with the interrupting I/O device. RAM and ROM do not normally have interrupting capability.

13.4 READ AND WRITE OPERATIONS

We are now ready to take a more detailed look at how the μP communicates with the other μC elements. Remember, the μP is the CPU and contains all the control and arithmetic/logic circuitry needed to execute a program of instructions stored in RAM or ROM. The CPU is continually performing READ and WRITE operations as it executes a program. It fetches each instruction from memory with a READ operation. After interpreting the instruction, it may have to perform a READ operation to obtain the operand from memory, or it may have to WRITE data into memory. In some cases, the instruction may call for the CPU to READ data from an input device (such as a keyboard or TTY) or to WRITE data into an output device (like an LED display or a magnetic tape cassette).

The READ Operation

The following steps take place during a READ operation:

1. The CPU generates the proper logic level on its R/W line for initiating a READ operation. Normally, $R/W = 1$ for READ. The R/W line is part of the control bus and goes to all the memory and I/O elements.

2. Simultaneously, the CPU places the 16-bit address code onto the address bus to select the particular memory location or I/O device from which the CPU wants to receive data.

3. The selected memory or I/O element places an 8-bit word on the data bus. All nonselected memory and I/O elements will not affect the data bus because their *tristate* outputs will be in the disabled (Hi-Z) condition.

4. The CPU receives the 8-bit word from the data bus on its data pins, D_0–D_7. These data pins act as inputs whenever $R/W = 1$. This 8-bit word is then latched into one of the CPU's internal registers, such as the accumulator.

This sequence can be better understood with the help of a timing diagram showing the interrelationship between the signals on the various buses (see Figure 13.5). Everything is referenced to the ϕ_1 and ϕ_2 clock signals. The complete READ operation occurs in one clock cycle. This is typically 1 μs for MOS microprocessors. The leading edge of ϕ_1 initiates the CPU's generation of the proper R/W and address signals. After a short delay, typically 100 ns for a MOS μP, the R/W line goes HIGH and the address bus holds the new address code (point 1 on timing diagram). Note that the address bus waveform shows both possible transitions (LOW to HIGH and HIGH to LOW), because some of the 16 address lines will be changing in one direction while others will be changing in the opposite direction.

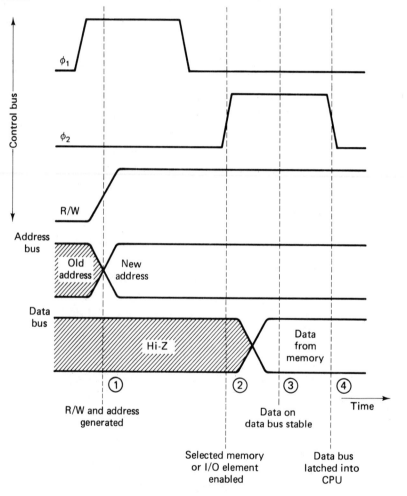

Figure 13.5 Typical μC timing for a READ operation.

During the ϕ_2 pulse, the selected memory or I/O device is enabled (point 2) and it proceeds to put its data word on the data bus. Prior to this, the data bus is in its Hi-Z state, since no device connected to it has been enabled. At some point during the ϕ_2 pulse, the data on the data bus become stable (point 3). Again, both possible data line transitions are shown on the diagram. The delay between the start of the ϕ_2 pulse and the data bus stabilizing depends on the speed of the memory and I/O elements. For memory this delay would be its *access time*. On the falling edge of ϕ_2, the data on the data bus are latched into the CPU (point 4). Clearly, then, the memory and I/O devices must be capable of putting data on the bus prior to the falling edge of ϕ_2, or proper transfer to the CPU will not occur. Thus, it is necessary to ensure that these devices have a speed compatible with the μC clock frequency.

EXAMPLE 13.1 A certain type of PROM has an access time specified as 750 ns (typical) and 1 μs (maximum). Can it be used with a μC that has a clock frequency of 1 MHz?

Solution: No, with a clock frequency of 1 MHz, the ϕ_2 pulse duration will be less than 500 ns. Thus, the PROM would have to have an access time of less than 500 ns for proper data transfer to the CPU.

The WRITE Operation

The following steps occur during a WRITE operation:

1. The CPU generates the proper logic level on the R/W line for initiating a WRITE operation. Normally, $R/W = 0$ for WRITE.

2. Simultaneously, the CPU places the 16-bit address code onto the address bus.

3. The CPU then places an 8-bit word on the data bus via its data pins D_0–D_7, which are now acting as outputs. This 8-bit word typically comes from an internal CPU register, such as the accumulator. All other devices connected to the data bus have their outputs disabled.

4. The selected memory or I/O element takes the data from the data bus. All nonselected memory and I/O elements will not have their inputs enabled.

This sequence has the timing diagram shown in Figure 13.6. Once again, the leading edge of ϕ_1 initiates the R/W and address bus signals (point 1). During the ϕ_2 pulse, the selected memory or I/O device is enabled (point 2) and the CPU places its data on the data bus. The data bus levels become stabilized a short time into the ϕ_2 pulse (typically 100 ns). These data are then written into the

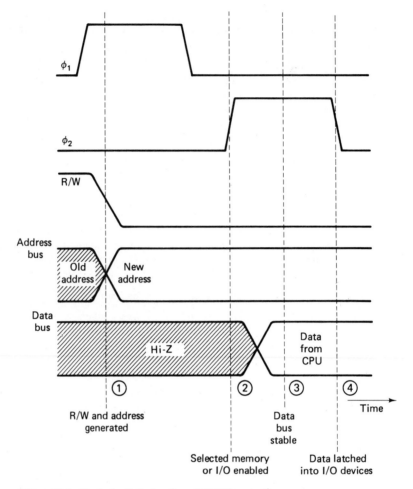

Figure 13.6   Typical μC timing for a WRITE operation.

selected memory location while ϕ_2 is high. If an I/O device has been selected, it usually latches the data from the data bus on the falling edge of ϕ_2 (point 4).

The READ and WRITE operations encompass most of the μC activity that takes place outside the CPU. The following example illustrates.

EXAMPLE 13.2   Below is a short program that is stored in memory locations 0020_{16}– 0029_{16} of an 8-bit μC. Note that since each word is one byte (8 bits), it takes two successive bytes to represent a 16-bit operand address code. Determine the total number of READ and WRITE operations that the μC will perform as it executes this program.

Memory Address (hex)	Memory Word (hex)	Symbolic Code	Description
0020	49	LDA	Load accumulator (ACC) with *X*.
0021	01⎫		⎡Address of⎤
0022	50⎭		⎣operand *X*⎦
0023	7E	ADD	Add *Y* to contents of ACC.
0024	01⎫		⎡Address of⎤
0025	51⎭		⎣operand *Y*⎦
0026	B2	STA	Store ACC contents.
0027	01⎫		⎡Address where ACC⎤
0028	52⎭		⎣will be stored ⎦
0029	EF	HLT	Halt operation.

Solution: The CPU begins executing the program by READING the contents of memory location 0020. The word stored there (49) is taken into the CPU and is interpreted as an instruction op code. (In other words, the CPU *always* interprets the first word of a program as an instruction and the programmer must *always* adhere to this format.) The CPU decodes this op code to determine the operation to be performed and to determine if an operand address follows the op code. In this case, an operand address is required and is stored in the next two successive bytes in the program. Thus, the CPU must READ locations 0021 and 0022 to obtain the address of data *X*. Once this address (0150) has been read into the CPU, the CPU proceeds to execute the LDA instruction by READING memory location 0150 and putting its contents into the accumulator. Thus, the complete execution of this first instruction requires *four* separate READ operations; one for the op code, two for the address, and one for the LDA operation.

Similarly, *four* READ operations are needed for the second instruction, which begins at 0023. This instruction also uses a two-byte operand address and requires reading the contents of this address for transfer to the CPU's arithmetic unit.

The third instruction begins at 0026 and uses a two-byte operand address. The CPU uses *three* READ operations to obtain these. However, the instruction STA calls for a WRITE operation whereby the CPU transfers the contents of the accumulator to memory location 0152.

The final instruction at 0029 is simply an op code with no operand address. The CPU READS this op code (EF), decodes it, and then halts further operations.

The total number of READ operations, then, is *twelve* and the total number of WRITE operations is *one*.

13.5 ADDRESS ALLOCATION TECHNIQUES

We mentioned earlier that some type of decoding circuitry is needed to help the CPU select the memory location or I/O device that it is trying to READ from or WRITE into. When the CPU places a 16-bit address on the address bus, we want it to activate one and only one memory location or I/O device. The actual decoding circuitry depends on the type of memory chip used for RAM and ROM. As we saw in Chapter 11, the decoders are used to drive the chip-select inputs of the various memory chips so that only the chip or group of chips that are storing the desired word will be activated. Before expanding further on the decoding circuits, we will say a few words about memory organization.

The ROM portion of an 8-bit μC memory is usually implemented in single chips which store a certain number of 8-bit words per chip. For example, a μC might have 1024 words (1K) of ROM. This could be implemented on a single 1024 $\times$ 8 ROM chip or it might be on four 256 $\times$ 8 ROM chips. On the other hand, RAMs require more logic on each chip, since they can be written into as well as read from. Thus, typically, several RAM chips have to be combined in order to handle 8-bit words. Figure 13.7 shows two common methods for implementing 1K of RAM.

In Figure 13.7(a) eight 1024 $\times$ 1 RAM chips are used, with each chip storing 1 bit of each word. For example, RAM 7 contains the MSBs and RAM 0 contains the LSBs of each of the 1024 words. This complete arrangement is called a memory "module." In this case, it is a 1024-word module. In a module, all the chips are selected simultaneously, as evidenced by the common line going to the chip-select (CS) inputs.

In Figure 13.7(b) we are using 256 $\times$ 4 RAM chips. Two of these chips are combined to form a 256 $\times$ 8 module, so that a total of four 256 $\times$ 8 modules are implemented with the eight chips shown. Each module has a separate "module select" input and only one module is selected at one time for read or write.

Memory Space

The total number of addresses that a CPU can select using a 16-bit address code is 65,536. In hexadecimal, these addresses range from 0000 to FFFF and constitute the μC's total "memory space." Rarely is this total memory space used in a typical microcomputer. A typical situation is illustrated in Figure 13.8, where we see how the total memory space might be allocated among RAM, ROM, and I/O devices.

As shown, hex addresses 0000–03FF have been allocated for RAM. This is a total of 1024_{10} address locations. Addresses F000–F7FF have been allocated for I/O devices, for a total of 2048_{10} addresses. This does not mean that 2048 different I/O devices are used with this μC, because, as we shall see, it is often convenient to use a range of addresses, rather than one address, to select a particular I/O device. Addresses F800–FFFF are used for ROM to store permanent programs and data, a total of 2048_{10} address locations. This leaves addresses 0400–EFFF, which are not being used in this example.

The actual address allocations for RAM, ROM, and I/O are chosen by the μC designer but also depend to some extent on the μP being used. For example, when the 8080A μP is reset by a low level on its $\overline{RES}$ input, its internal program counter (PC) goes to 0000_{16}, and that is where the μP takes its first instruction from. This, of course, means that the lower-numbered addresses, beginning with 0000, have to be ROM or RAM program storage rather than data storage.

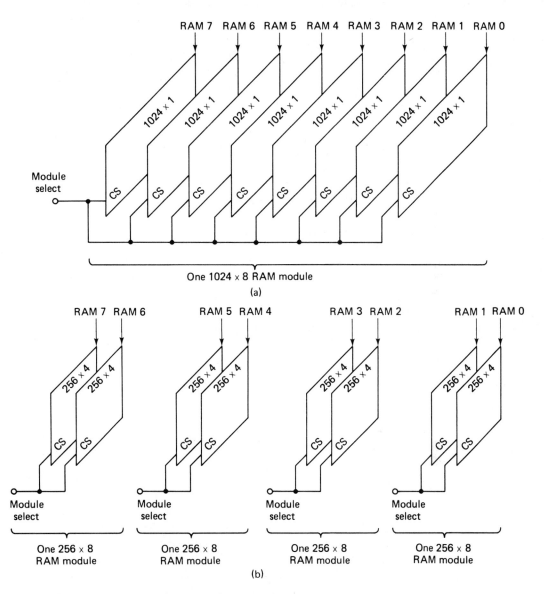

Figure 13.7   Two typical ways of implementing 1024 8-bit words of RAM.

Memory Pages

Many μCs divide their 65,536 available addresses into 256 blocks of 256 addresses. Each of these blocks is called a *page*. Figure 13.9 shows how a μC's memory space can be organized as pages. Note that each page covers a range of 256 addresses.*

*Some μCs use a different number of addresses per page.

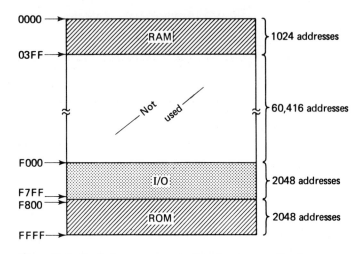

Figure 13.8 Typical memory space allocation in a small μC.

For example, page 00 includes address locations 0000–00FF and page 69 includes address locations 6900–69FF. Also note that the page number is the same as the first two hex digits of the addresses on that page. For example, all addresses starting with 69 are on page 69.

Using this page organization, we can think of each address as consisting of a page number and a word number. To illustrate, consider address 1E66.

$$\underbrace{1 \quad E}_{\substack{\text{Page} \\ \text{number}}} \quad \underbrace{6 \quad 6}_{\substack{\text{Word on} \\ \text{that page}}}$$

The first two digits specify page 1E and the last two digits specify word 66 on that page.

EXAMPLE 13.3 Refer to the typical memory space allocations given in Figure 13.8. How many pages have been allocated for RAM? Repeat for I/O and ROM.

Solution: The RAM occupies addresses 0000–03FF; this includes pages 00, 01, 02, and 03, for a total of *four* pages.

The I/O uses pages F0, F1, F2, F3, F4, F5, F6, and F7, for a total of *eight* pages.

The ROM occupies pages F8, F9, FA, FB, FC, FD, FE, and FF, for a total of *eight* pages.

EXAMPLE 13.4 Refer to Figure 13.7 and determine how many pages of memory can be stored in each module.

Solution: The 1024 × 8 module in Figure 13.7(a) can store *four* pages of memory, because one page is 256 words. The 256 × 8 modules in Figure 13.7(b) can each store *one* page of memory.

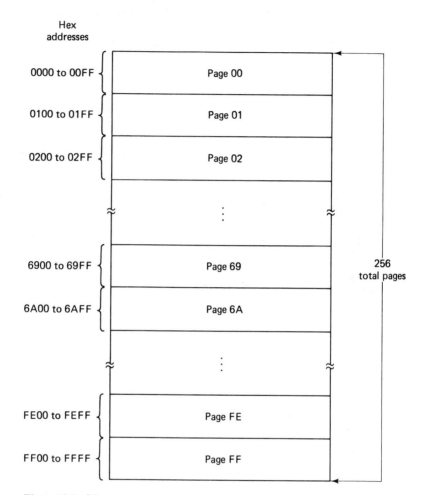

Figure 13.9 Memory space organized into pages.

13.6 ADDRESS DECODING TECHNIQUES

We are now ready to develop the address-decoding circuitry that is needed to select the single memory location or I/O device that the CPU is addressing. There are several approaches to the decoding problem, all basically the same. Rather than showing several variations of decoding logic, we have chosen to develop one method whose concepts and ideas are applicable to all decoding schemes. The approach we will use is straightforward and utilizes page-organized addressing. We will develop the complete decoding logic for an 8-bit μC that uses 16-bit addresses and whose memory-space allocation is given in Figure 13.10.

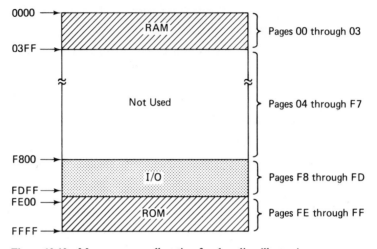

Figure 13.10 Memory-space allocation for decoding illustration.

Decoding for RAM

Before we can develop the decoding logic for the RAM addresses, we have to choose the type of chip we want to use to implement the RAM. Since we are going to use a page-organized method, we will use the RAM modules of Figure 13.7(b), each of which can store one page. With a RAM address allocation of 0000 to 03FF, ranging over four pages, we will need eight 256×4 RAM chips. This chip has eight address inputs, a R/W input, a chip-select input, and four data input/output lines.

The complete decoding circuitry for the four pages of RAM is shown in Figure 13.11. Note the following points:

1. Two RAM chips constitute a one-page module storing 256 8-bit words, with each RAM contributing 4 bits of each word.

2. The CPU R/W control output is connected to each RAM chip's R/W input. Of course, only one RAM module will actually READ or WRITE, depending on which one has its CS input activated.

3. The CPU's eight higher-order address lines $A_{15}-A_8$ are fed to the decoding logic, which includes a 3-line-to-8-line decoder. This decoder takes the 3-bit binary input code, CBA, and activates the corresponding output to select the memory page module that is being addressed.

4. The CPU's eight lower-order address lines, A_7-A_0, are connected to the address inputs of *each* RAM chip. These address lines select the memory word from the selected page module.

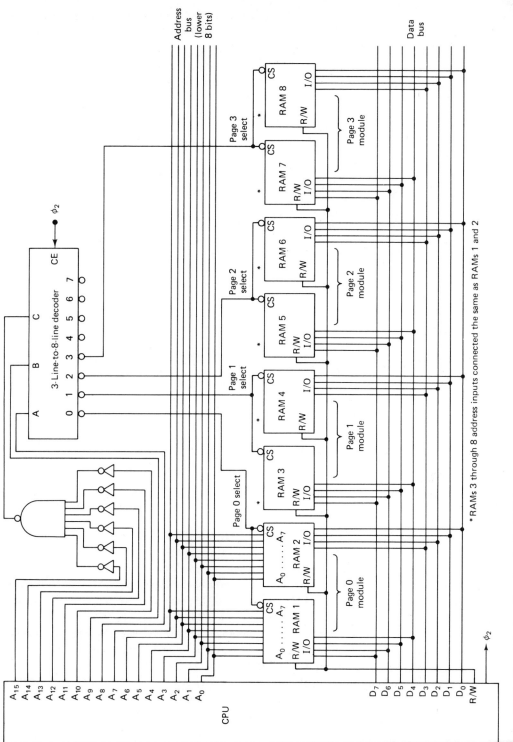

Figure 13.11 Complete decoding circuitry for RAM.

471

EXAMPLE 13.5 Assume that the CPU is performing a WRITE operation into memory location $02A5_{16}$. It places a low level on R/W and generates the address code at its address outputs. This address code in binary is

$$02A5_{16} = \underset{\underset{A_{15}}{\uparrow}}{0000} \, \underset{\underset{A_8 A_7}{\uparrow \uparrow}}{0010 \, 1010} \, \underset{\underset{A_0}{\uparrow}}{0101}$$

The six MSBs of this address code produce a LOW level at the NAND-gate output, which is applied to the decoder's C input. Address lines A_9 and A_8 place a 1 and 0 at the decoder's B and A inputs, respectively. Thus, the decoder input binary code is $CBA = 010$, which produces a LOW at decoder output 2. This LOW drives the PAGE 2 SELECT line to select RAMs 5 and 6.

The eight lower-order address lines select address 10100101 from page 2 as the location that the CPU will write into. The CPU places its data on the data bus and writes them into this location during ϕ_2. Notice that the decoder is not enabled until ϕ_2, since ϕ_2 drives its chip-enable (CE) input. This is done to ensure that the levels on the address lines have stabilized before the memory is activated. Otherwise, it is possible that one or more RAM locations might accidentally get activated as the address lines are changing from the old address code to the new address code.

This example should clearly show how the decoding circuitry selects the exact address specified by the CPU. The trick here is to notice that we are only using address locations 0000–03FF for RAM. For this address range, address lines A_{15}–A_{10} will all be LOW, so that the inverters and NAND gate will produce a LOW at decoder input C. Address lines A_9 and A_8, then, actually select the page module being addressed. It should be obvious that any combination on lines A_{15}–A_{10} other than all LOWs will produce a HIGH at decoder input C. This prevents outputs 0–3 of the decoder from being activated. Decoder outputs 4–7 are not used.

Decoding for ROM and I/O

Referring to Figure 13.10, we see that addresses F800–FDFF (pages F8–FD) are reserved for I/O and addresses FE00–FFFF (pages FE and FF) are ROM. Since the I/O and ROM allocations are adjacent, we can develop the decoding circuitry for both simultaneously. First, we must specify the type of ROM chip we want to use. A good choice would be a 256×8 ROM; two of these are needed to store two pages.

As far as I/O is concerned, let us assume that the μC is connected to *six* different peripheral devices—three *input* devices and *three* output devices. Each device will have an interface circuit that connects it to the μC. Since six pages have been allocated in memory space for I/O, we can assign one page per interface circuit. In other words, we can assign all addresses in page FD to interface 1, all addresses in page FC to interface 2, and so on. The reason for doing this is that it only requires decoding the eight higher-order address lines and thereby minimizes I/O address decoding circuitry. The table below summarizes the I/O organization we will be using.

Interface Number	Type	Addresses
1	Output	FD00–FDFF (page FD)
2	Output	FC00–FCFF (page FC)
3	Output	FB00–FBFF (page FB)
4	Input	FA00–FAFF (page FA)
5	Input	F900–F9FF (page F9)
6	Input	F800–F8FF (page F8)

The complete decoding circuitry for I/O and ROM is shown in Figure 13.12. Note the following points:

1. Two 256×8 ROM chips are used, one per page.

2. Interface circuits 1, 2, and 3 are used for output devices, while interface circuits 4, 5, and 6 are used for input devices. Typically, the six interface circuits would all be different. They could be as simple as buffer circuits or FF registers, or as complex as an LSI interface chip such as a UART (used for teletype or tape cassette).

3. The CPU's eight higher-order address lines go to the decoding logic, which includes a 3-to-8 decoder. This decoder generates eight page-select outputs, only one of which will be activated, depending on the address code.

4. The CPU's eight lower-order address lines are connected to the address inputs of each ROM to select one word from the selected page.

5. The CPU's R/W line is not connected to the ROM chips. A ROM does not have a R/W input because it cannot be written into. The R/W line is, however, connected to the interface circuits. There are two reasons for this. First, some interface circuits are capable of being written into or read from. For example, it may be an output interface through which the CPU can both send data to an output device (WRITE) and receive status information from the output device (READ).

   The second reason is to avoid accidental conflicts on the data bus. To illustrate, interface circuit 4 is selected by any address on page FA whenever the CPU wants to read data from input device 1. Suppose that through some programming error the CPU is performing a *WRITE* operation on address FA00. This address will select interface circuit 4 and will cause data from input device 1 to be placed on the data bus. However, since the CPU is performing a WRITE operation, a data word from one of its internal registers will also be placed on the data bus. These two different sets of logic levels on the data bus can cause damage to the CPU chip or the interface chip. This problem is avoided by using the R/W line from the

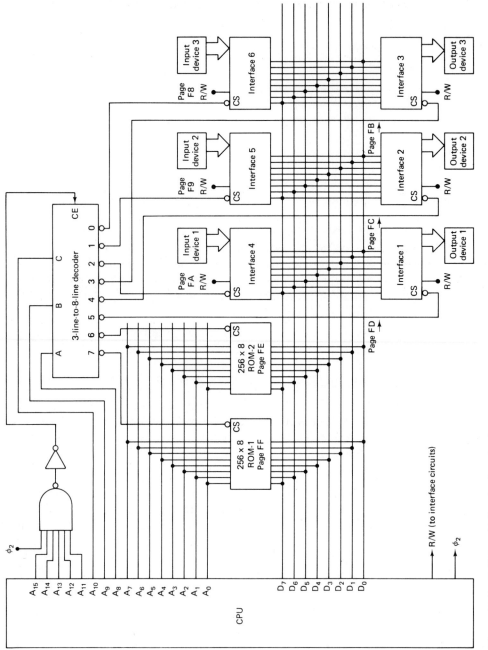

Figure 13.12 ROM and I/O selection circuitry.

CPU to disable interface circuit 4 during a CPU WRITE operation ($R/W = 0$).

EXAMPLE 13.6 Describe the process by which the decoding circuitry selects ROM address FE77.

Solution: The CPU places FE77 on the address bus. In binary, this address code is

$$FE77_{16} = 1111 \vdots 1110 \vdots 0111 \vdots 0111$$
$$\quad\quad \uparrow \quad\quad \uparrow\uparrow \quad\quad\quad \uparrow$$
$$\quad A_{15} \quad\quad A_8 A_7 \quad\quad A_0$$

The five MSBs, A_{15}–A_{11}, produce a LOW level at the NAND gate output (when $\phi_2 = 1$), which is inverted and applied to the decoder's chip-enable to activate the decoder. Address lines A_{10}, A_9, and A_8 produce a code of 110 at the decoder inputs to activate decoder output 6. The LOW at output 6 drives the PAGE FE SELECT line to select ROM-2.

The lower-order address lines (A_7–A_0) select location 01110111 in ROM-2 as the actual location whose word is placed on the data bus via ROM-2's data output pins. This word does not appear on the data bus until ϕ_2 because ϕ_2 has to be HIGH for the decoder to be activated. During ϕ_2, the CPU takes the word from the data bus and places it in one of its internal registers.

EXAMPLE 13.7 Describe the process by which the CPU writes a word into output device 2.

Solution: The chip-select input of interface circuit 2 is driven by decoder output 4, which is the PAGE FC SELECT line. This means that any address whose first two hex digits are FC will select interface circuit 2 and, hence, output device 2. Thus, to write data into output device 2, the CPU must set $R/W = 0$ and generate an address of the form FCXX, where XX can be any hex digits. Then it must place the word to be written on the data bus, from which it is transmitted by interface circuit 2 to output device 2.

These examples showed how the decoding circuitry selects the ROM location or I/O device specified by the CPU. The main point used in designing this circuitry is that for all addresses in the range F800–FFFF, address lines A_{15}–A_{11} will be HIGH. This condition produces a HIGH at the CE input of the decoder. Address lines A_{10}, A_9, and A_8, then, activate the appropriate decoder output. Notice that any combination on lines A_{15}–A_{11} other than all HIGHs will produce a LOW at CE, thereby disabling all decoder outputs (all HIGH), so that no ROM or I/O is selected.

Complete Decoding Circuitry

The complete decoding circuitry for our example μC is obtained by combining the circuits of Figures 13.11 and 13.12. The resultant circuit includes two NAND gates, seven inverters, two 3-line-to-8-line decoders, eight 256 $\times$ 4 RAMs, two

256 × 8 ROMs, and six different interface circuits. Of course, if the amount of RAM or ROM memory is to be expanded at a later time, the necessary decoder circuitry must be added. Many μC manufacturers include enough decoder circuitry on their memory boards so that future memory expansion only requires adding the necessary RAM or ROM chips.

Buffering the Address and Data Lines

Looking at Figures 13.11 and 13.12, you will notice that all of the CPU's address lines are driving two or more loads. This will usually require some buffer circuits, especially if the CPU is a MOS device and the loads are TTL. For example, CPU address line A_{15} is driving an inverter (Figure 13.11) and a NAND gate (Figure 13.12). Typically, logic gates and decoders are TTL devices because of their speed and availability. If the CPU is MOS, its outputs will normally be capable of driving only one TTL load. Thus, A_{15} cannot drive both TTL loads directly. The same is true for A_{14}–A_8. These eight address lines must, therefore, be coupled to their loads through some type of buffer circuit (such as a 74126 or 8T97) which can easily drive several TTL loads. This is illustrated in Figure 13.13(a) for line A_{15}.

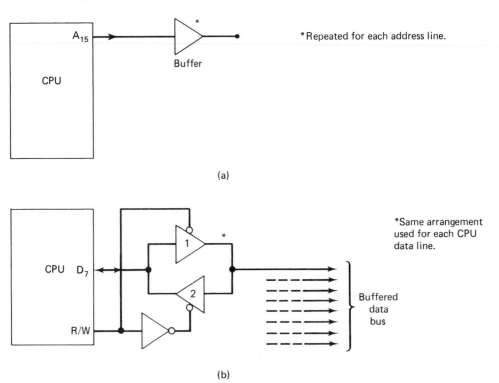

(a)

(b)

Figure 13.13   (a) Buffered address line; (b) bidirectional buffers for CPU data lines.

Address lines A_0–A_7 are each driving eight RAM chips and two ROM chips. These memory chips are normally MOS devices, so their inputs draw negligible current from the signal sources driving them. However, there is still a loading problem because of the capacitance present at each input (typically 2–5 pF). The accumulated effect of these capacitances can drastically slow down the signals on the address bus lines and cause improper operation. For this reason, buffers should be used on all the address lines. The buffer circuits will have a very low output resistance so that they can drive capacitive loads without significant signal deterioration.

The same considerations apply for the CPU's data lines, D_0–D_7. There is an important difference, however, because these lines are bidirectional. Figure 13.13(b) shows a common way to buffer these data lines using two tristate buffers. The R/W signal from the CPU is used to control which buffer is activated. For a WRITE operation R/W is LOW and buffer 1 is activated to allow data from the CPU data line onto the data bus. For a READ operation R/W is HIGH and buffer 2 is activated to allow data from the data bus into the CPU. Note that only one buffer is activated at any given time. This same arrangement is used for each CPU data line.

13.7 THE MICROPROCESSOR (μP)

All microcomputers, although they vary in their architecture, have one element in common—the microprocessor chip. As we know, the $μP$ functions as the central processing unit of the $μC$. In essence, the $μP$ is the heart of the $μC$ because its capabilities determine the capabilities of the $μC$. Its speed determines the maximum speed of the $μC$, its address and data pins determine the $μC$'s memory capacity and word size, and its control pins determine the type of I/O interfacing that must be used.

The $μP$ performs a large number of functions, including:

1. Providing timing and control signals for all elements of the $μC$.
2. Fetching instructions and data from memory.
3. Transferring data to and from I/O devices.
4. Decoding instructions.
5. Performing arithmetic and logic operations called for by instructions.
6. Responding to I/O generated control signals such as *reset* and *interrupts.*

The $μP$ contains all the logic circuitry for performing these functions, but it should be kept in mind that a great deal of the $μP$'s internal logic is not externally accessible. For example, we cannot apply an external signal to the $μP$ chip to increment the program counter (PC). Instead, the $μP$ elements are *software-*

accessible. This means that we can affect the internal μP circuitry only by the *program* we put in memory for the μP to execute. This is what makes the μP so versatile and flexible—when we want to change the μP's operation, we simply change the program (e.g., by changing the ROMs that store the program). This is generally easier than rewiring hardware.

Since we must construct programs that tell the internal μP logic what to do, we need to become familiar with the internal μP structure, its characteristics, and its capabilities. The μP logic is extremely complex, but we can divide it into three areas: the register section, the control and timing section, and the ALU (see Figure 13.14). Although there is some overlap among these three areas, for clarity we will consider them separately.

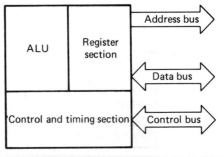

Microprocessor

Figure 13.14   Major functional areas of a μP chip.

13.8 TIMING AND CONTROL SECTION

We will not discuss this part of the μP in too much detail for two reasons. First, it is the one area on the μP chip over which we have very little control. Second, we do not have to know the detailed structure of the timing and control section to be able to develop useful programs for the μP.

The major function of this μP section is to fetch and decode instructions from program memory, and then to generate the necessary control signals required by the ALU and register section for executing these instructions. The fetching and decoding functions correspond to the *instruction cycle*, and the control signal generation takes place during the *execution cycle*. Both of these cycles were discussed in Chapter 12 and we need not elaborate on them further.

Control Bus Signals

The control section also generates *external* control signals that are sent to other μC elements as part of the system's control bus. We used some of these control bus signals in our description of the μC, namely R/W and ϕ_2. In addition to generating output signals for the control bus, the μP control section also responds to control bus signals that are sent from other μC elements to the μP chip. The RESET and INT (interrupt) mentioned earlier are examples.

Each μP has its own unique set of input and output control signals that are described in detail in the manufacturers' operation manuals. We will not attempt to define all of these here. Instead, we will describe some of the control bus signals that are common to several different microprocessors.

RESET   All μPs have this input. When this input is activated, most of the μP's internal registers are reset to 0. In many μPs, the program counter (PC) is reset to 0 so that the instruction stored at memory location 0000_{16} is the first to be executed. In some μPs, activating the RESET input does not clear the PC. Instead, the PC is loaded from two specific memory locations (such as FFFE and FFFF). In other words, upon RESET the address of the first instruction is taken from these memory locations (each one stores one byte of the 16-bit address). Usually, this starting address is stored in ROM and is often referred to as an *address vector*.

R/W   This μP output line informs the rest of the μC as to whether the μP is in a READ or WRITE operation. Some μPs use separate control lines, *RD* to indicate a READ operation and *WR* to indicate a WRITE operation.

MREQ (memory request)   This μP output indicates that a memory access is in progress.

IORQ (I/O request)   A μP output which indicates that an I/O device is being accessed. Some μPs use this signal along with MREQ to distinguish between memory and I/O operations. This allows memory and I/O to use the same address space because the IORQ and MREQ signals determine which one (I/O or memory) is enabled. This technique of treating I/O separately from memory is called *isolated I/O*.

READY   This μP input is used by slow memory or I/O devices which cannot respond to a μP access request within one μP clock cycle. When the slow device is selected by the address decoding circuitry, it immediately sends a READY signal to the μP. In response, the μP suspends all its internal operations and enters what is called a WAIT state. It remains there until the device is ready to send or receive data, which the device so indicates by removing the READY signal.

HOLD   This μP input is used for direct memory access (DMA) operations, which we will not get into. When an external device activates this input, the μP finishes executing the instruction it is currently working on and then *floats* its address and data buses. This means that it disables its tristate data and address pins so that they are effectively disconnected from the μC buses. This allows other external devices to use the address and data buses as long as HOLD is active.

HLDA (hold acknowledge)   This is an output signal which the μP generates to indicate to external logic that the μP is in a HOLD state and that the data and address buses are available. A similar μP output is BA (buses available).

INT or IRQ (interrupt request)   This is a μP input used by I/O devices to interrupt the execution of the current program and cause the μP to jump to a special program, called the *interrupt service routine*. The μP executes this special program, which normally involves servicing the interrupting device. When this execution is complete, the μP resumes execution of the program it had been working on when it was interrupted.

INTE (interrupt enable)   This is a μP output that indicates to external devices whether or not the internal μP interrupt logic is enabled or disabled. If enabled (*INTE* = 1), the μP can be interrupted as described above. If disabled (*INTE* = 0), the μP will not respond to the *INT* or *IRQ* inputs. The state of *INTE* can be software-controlled. For example, a program can contain an instruction that makes *INTE* = 0 so that the interrupt operation is disabled.

NMI (nonmaskable interrupt)   This is another μP interrupt input, but it differs from *INT* or *IRQ* in that its effect cannot be disabled. In other words, the proper signal on *NMI* will always interrupt the μP, regardless of the interrupt enable status.

13.9   THE REGISTER SECTION

The most common operation that takes place *inside* the μP chip is the transfer of binary information from one register to another. The number and types of registers that a μP contains is a key part of its architecture, and it has a major effect on the programming effort required in a given application. The register structure of different μPs varies considerably from manufacturer to manufacturer. However, the basic functions performed by the various registers is essentially the same in all μPs. They are used to store data, addresses, instruction codes, and information on the status of various μP operations. Some are used as counters which can be controlled by software (program instructions) to keep track of such things as the number of times a particular sequence of instructions has been executed or sequential memory locations from which data are to be taken.

We will briefly describe the most common types of registers and their functions.

The Instruction Register (IR)

The function of this register has already been discussed. When the CPU* fetches an instruction word from memory, it sends it to the IR. It is stored here while the instruction decoding circuitry determines which instruction is to be executed.

---

*Since the μP is the CPU of a μC, we will use the terms μP and CPU interchangeably.

The IR is automatically used by the CPU during each instruction cycle and the programmer never needs to access this register. The size of the IR will be the same as the word size. For an 8-bit μP, the IR is 8 bits long.

The Program Counter (PC)

This register has also been discussed previously. The PC always contains the address in memory of the next instruction (or portion of an instruction) that the CPU is to fetch. When the μP RESET input is activated, the PC is set to the address of the first instruction to be executed. The μP places the contents of the PC on the address bus and fetches the first byte of the instruction from that memory location. (Recall that in 8-bit μCs, one byte is used for the op code and the following one or two bytes for the operand address.) The μC automatically increments the PC after each use and in this way executes the stored program sequentially unless the program contains an instruction which alters the sequence (e.g., a JUMP instruction).

The size of the PC depends on the number of address bits the μP can handle. Most of the more common μPs use 16-bit addresses, but there are some that use 12 bits. In either case, the PC will have the same number of bits as the address bus. In many μPs, the PC is divided into two smaller registers, PCH and PCL, each of which holds one half of an address. This is illustrated in Figure 13.15.

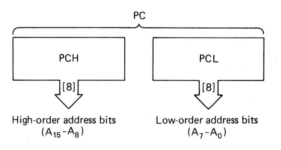

Figure 13.15 A 16-bit PC broken into two 8-bit units.

PCH holds the 8 high-order bits of the the 16-bit address and PCL holds the 8 low-order bits. The reason for using PCH and PCL is that it is often necessary to store the contents of PC in memory. Since the memory stores 8-bit words, the 16-bit PC must be broken into two halves and stored in two successive memory locations.

The programmer has no direct access to the PC. That is, there are no direct instructions to load the PC from memory or to store the PC contents in memory. However, there are many instructions that cause the PC to take on a value other than its normal sequential value. JMP (jump) and JPZ (jump on zero) are examples of instructions that change the PC so that the CPU executes instructions out of their normal sequential order.

Memory Address Register (MAR)

This is sometimes called a *storage address register* or an *address latching register*. It is used to hold the address of *data* the CPU is reading from or writing into memory. For example, when executing an ADD instruction, the CPU places the operand address portion of the ADD instruction into the MAR. The contents of the MAR are then placed on the address bus so that the CPU can fetch the data during the next clock cycle.

It should now be apparent that there are two sources of addresses for the μP's address bus, the PC and the MAR. The PC is used for *instruction* addresses and the MAR is used for *data* addresses. A multiplexer is used to switch either the PC or MAR onto the address bus, depending on whether the CPU is in an instruction cycle or an execution cycle. This is illustrated in Figure 13.16. Here the PC and MAR outputs are fed to a two-port multiplexer. The multiplexer's single output port is connected to the μP's address bus via the address bus buffers. The multiplexer's select input S is controlled by a signal from the μP's control section. One value of S selects the PC to be placed onto the address bus, and the other value of S selects the MAR.

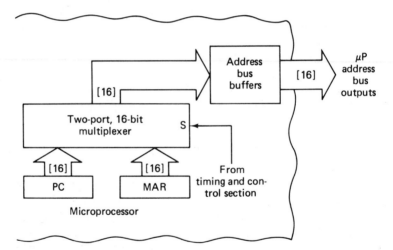

Figure 13.16 Portion of the μP showing how address source is selected for the address bus.

Accumulator

The accumulator is a register that takes part in most of the operations performed by the ALU. It is also the register in which the results are placed after most ALU operations. In many ALU instructions, the accumulator is the source of one of the operands and the destination of the result.

In addition to its use in ALU instructions, the accumulator has other uses;

for instance, it can be used as a storage register for data that are being sent to an output device or as a receiving register for data that are being read from an input device. Some μPs have more than one accumulator. In these μPs, the instruction op code specifies which accumulator is to be used.

The accumulator generally has the same number of bits as the μP's word size. Thus, an 8-bit μP will have an 8-bit accumulator. Some μPs also have an *extension register*, which is used in conjunction with the accumulator for handling binary numbers with more than 8 bits. For example, the μP has an arithmetic mode called "double-precision arithmetic," where each number is 16 bits long and is stored in two words of memory. When these 16-bit data words are sent to the CPU, the extension register stores the 8 least significant bits and the accumulator stores the 8 most significant bits. These two registers can be considered to be one 16-bit register.

General-Purpose Registers

These registers are used for many of the temporary storage functions required inside the CPU. They can be used to store data that are used frequently during a program, thereby speeding up the program execution since the CPU does not have to perform a memory READ operation each time the data are needed. These registers are often used as counters which are used to keep track of the number of times a particular instruction sequence in a program has been executed. In some μPs, the general-purpose registers can also be used as index registers (described below) or as accumulators. The number of general-purpose registers will vary from μP to μP. Some μPs have none, while others may have 12 or more.

Index Registers

An index register, like general-purpose registers, can be used for general CPU storage functions and as a counter. In addition, it has a special function which is of great use in a program where tables or arrays of data must be handled. In this function, the index register takes part in determining the addresses of data that the CPU is accessing. This operation is called *indexed addressing* and is a special form of memory addressing available to the programmer. There are several different forms of indexed addressing, but the basic idea of indexed addressing is that the actual or *effective* operand address called for by an instruction is the sum of the operand address portion of the instruction *plus* the contents of the index register.

The Status Register

Also referred to as a process status register or condition register, the status register consists of individual bits with different meanings assigned by the μP manufacturer. These bits are called *flags*, and each flag is used to indicate the status of a particular μP condition. The value of some of the flags can be examined under program control to determine what sequence of instructions to follow. Two of the most common flags are the ZERO flag, Z, and the CARRY flag, C. The value of Z will always indicate whether or not the previous data manipulation instruction produced a result of zero. Normally, a μP control signal sets Z to the HIGH state when the result of an instruction is zero and clears Z to a LOW when the result of an instruction is not zero. The value of C always indicates whether the previous instruction produced a result that exceeded the μP word-size. For example, whenever the addition of two 8-bit data words produces a sum that exceeds 8 bits (i.e., a carry out of the eighth position), the C flag will be set to 1. If the addition produces no carry, C will be 0. The C flag can be thought of as the ninth bit of any arithmetic result.

Most μP instruction sets contain several *conditional branch* instructions which determine the sequence of instructions to be executed, depending on the flag values. The jump-on-zero (JPZ) instruction is a prime example. When the CPU executes the JPZ instruction, it examines the value of the ZERO flag. If the ZERO flag is LOW, indicating that the previous instruction produced a nonzero result, the next instruction to be executed will be taken in normal sequence. If the ZERO flag is HIGH, indicating a zero result, the program will jump to the operand address for its next instruction and will continue in sequence from there. Other examples are jump-on-carry-set (JCS) and jump-on-carry-cleared (JCC), both of which examine the CARRY flag.

The Stack Pointer Register (SP)

Before defining the function of this register, we must first define the *stack*. The stack is a portion of RAM reserved for the temporary storage and retrieval of information, typically the contents of the μP's internal registers. The *stack pointer* register acts as a special memory address register used only for the stack portion of RAM. Whenever a word is to be stored on the stack, it is stored at the address contained in the stack pointer. Similarly, whenever a word is to be read from the stack, it is read from the address specified by the stack pointer. The contents of the stack pointer is initialized by the programmer at the beginning of the program. Thereafter, the SP is automatically decremented *after* a word is stored on the stack and incremented *before* a word is read from the stack. (The incrementing and decrementing is done automatically by the μP control section.)

13.10   THE ARITHMETIC/LOGIC UNIT (ALU)

Most modern μPs have ALUs capable of performing a wide variety of arithmetic and logical operations. These operations can involve two operands, such as the accumulator and a data word from memory, or the accumulator and another μP internal register. Some of the operations involve only a single operand, such as the accumulator, a register, or a word from memory.

A simplified diagram of a typical ALU is shown in Figure 13.17. The ALU block represents all the logic circuitry used to perform arithmetic, logic, and manipulation operations on the operand inputs. Two 8-bit operands are shown as inputs to the ALU, although frequently only one operand is used. Also shown as an input to the ALU is the carry flag, C, from the μP status register. The reason for this will be explained later. The functions that the ALU will perform are determined by the control signal inputs from the μP control section. The number of these control inputs varies from μP to μP.

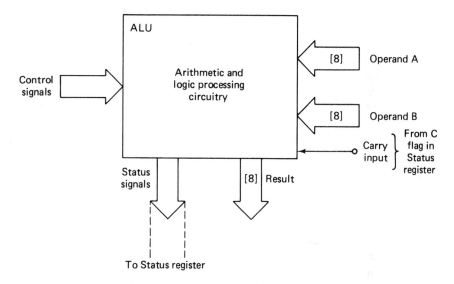

Figure 13.17 Simplified diagram of a typical μP ALU.

The ALU produces two sets of outputs. One set is an 8-bit output, representing the results of the operation performed on the operands. The other is a set of status signals which are sent to the μP status register to set or clear various flag bits. For example, if the result of an ALU operation is exactly zero, a signal is sent to the Z flag in the status register to set it to the HIGH state. If the result of the ALU operation produces a carry into the ninth bit position, a signal is sent to set the C flag in the status register. Other signals from the ALU will set or clear other flags in the status register.

The operand inputs A and B can come from several sources. When two operands are to be operated on by the ALU, one of the operands comes from the accumulator and the other operand comes from a data word fetched from memory which is stored in a data buffer register within the μP. In some μPs, the second operand can also be the contents of one of the general-purpose registers. The result of the ALU operation performed on the two operands is normally sent to the accumulator.

When only one operand is to be operated on by the ALU, the operand can be the contents of the accumulator, a general-purpose register, an index register, or a memory data word. The result of the ALU operation is then sent back to the source of the operand. When the single operand is a memory word, the result is sent to a data buffer register, from where the CPU writes it back into the memory location of the operand.

Single-Operand Operations

We will now describe some of the common ALU operations performed on a single operand.

1. Clear   All bits of the operand are cleared to 0. If the operand comes from the accumulator, for example, we can symbolically represent this operation as $0 \longrightarrow [A]$.

2. Complement (or Invert)   All bits are changed to their opposite logic level. If the operand comes from register X, for example, this operation is represented as $[\bar{X}] \longrightarrow [X]$.

3. Increment   The operand is increased by 1. For example, if the operand is 11010011, it will have 00000001 added to it in the ALU, to produce a result of 11010100. This instruction is very useful when the program is using one of the μP registers as a counter. Symbolically, this is represented as $[X + 1] \longrightarrow [X]$.

4. Decrement   The operand is decreased by 1. In other words, the number 00000001 is subtracted from the operand. This instruction is useful in programs where a register is used to count down from an initial value. Symbolically, this operation is represented as $[X - 1] \longrightarrow [X]$.

5. Shift   The bits of the operand are shifted to the left or to the right one place and the empty bit is made a 0. This process is illustrated below for a *shift-left* operation.

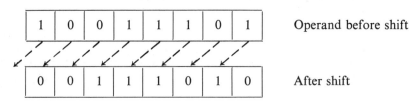

Operand before shift

After shift

Note that a 0 is shifted into the rightmost bit. Also note that the original leftmost bit is shifted out and is lost. In most μPs the bit that is shifted out of the operand is not lost; instead, it is shifted into the carry flag bit, C, of the status register. This is illustrated below for a *shift-right* operation.

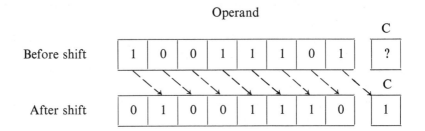

Note that a 0 is shifted into the leftmost bit and also note that the original right-most bit is shifted into C (the original value of C is lost). For the corresponding shift-left operation, the leftmost bit would be shifted into C. This type of operation is used by the programmer to test the value of a specific bit in an operand. He does this by shifting the operand the required number of times until the bit value is in the C flag. A conditional jump instruction is then used to test the C flag to determine what instruction to execute next.

6. Rotate   This is a modified shift operation in which the C flag becomes part of a circulating shift register along with the operand; that is, the value shifted out of the operand is shifted into C and the value of C is shifted into the empty bit of the operand. This is illustrated below for a "rotate-right" operation.

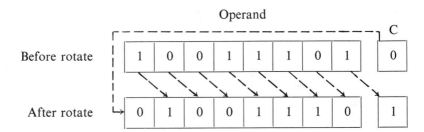

Note that the original C bit is shifted into the leftmost bit of the operand and that the rightmost bit of the operand is shifted into C. A rotate-left operates in the same manner except in the opposite direction.

Two-Operand ALU Operations

We will now describe some of the ALU operations performed on two operands.

1. Add   The ALU produces the binary sum of the two operands. Generally, one of the operands comes from the accumulator, the other from memory, and the result is sent to the accumulator. Symbolically, this operation is written [A] + [M] → [A]. If the addition of the two operands produces a carry out of the MSB position, the carry flag, C, in the status register is set to 1. Otherwise, C is cleared to 0. In other words, C serves as the ninth bit of the result.

2. Subtract   The ALU subtracts one operand (obtained from memory) from the second operand (the accumulator) and places the result in the accumulator. Symbolically, this is written $[A] - [M] \rightarrow [A]$. Most μPs use the 2's-complement method of subtraction (Chapter 1), whereby the operand from memory is 2's-complemented and then added to the operand from the accumulator. Once again, the carry bit generated by this operation is stored in the C flag.

3. Compare   This operation is the same as subtraction except that the result is not placed in the accumulator. Symbolically, this is written $[A] - [M]$. The subtraction is performed solely as a means for determining which operand is larger without affecting the contents of the accumulator. Depending on whether the result is positive, negative, or zero, various condition flags will be affected. The programmer can then use conditional jump instructions to test these flags.

4. Logical AND   The corresponding bits of the two operands are ANDed and the result is placed in the accumulator. Symbolically, this is written $[A] \cdot [M] \rightarrow [A]$. One of the operands is always the accumulator and the other comes from memory. As an example, let's assume that $[A] = 10110101$ and $[M] = 01100001$. The AND operation is performed as follows:

Bit numbers $\longrightarrow$	7	6	5	4	3	2	1	0	
	1	0	1	1	0	1	0	1	Original contents of *A*
	0	1	1	0	0	0	0	1	Operand from memory
	0	0	1	0	0	0	0	1	Result of AND that is sent to *A*

Note that each bit of the result is obtained by ANDing the corresponding bits of the operands. For example, bit 7 of the accumulator is a 1 and bit 7 of the memory word is a 0. Thus, bit 7 of the result is $1 \cdot 0 = 0$. Similarly, bit 5 of each operand is a 1, so that bit 5 of the result is $1 \cdot 1 = 1$.

5. Logical OR   The corresponding bits of the two operands are ORed and the result placed in the accumulator. Symbolically, this is shown as $[A]$ OR $[M] \rightarrow [A]$. A plus sign $(+)$ can be used in place of OR but it might cause confusion with the binary addition operation. Using the same two operands used in the illustration of the AND operation above, the result of the OR operation will be 11110101.

6. Exclusive-OR   The corresponding bits of the two operands are exclusive-ORed and the result is placed in the accumulator. Symbolically, this is written $[A] \oplus [M] \rightarrow [A]$. Using the same two operands used in the prior illustrations, the result of the EX-OR operation will be 11010100. The logic AND, OR, and EX-OR operations are very useful to a programmer.

7. Decimal (BCD) Arithmetic   Many μPs have some provision for performing addition and subtraction in the BCD system whereby an 8-bit data word is treated as two BCD-coded decimal digits. As we pointed out in Chapter 1, arithmetic operations on BCD numbers require extra steps to obtain the correct results. While in some μPs these extra steps are performed automatically whenever BCD arithmetic instructions are executed, other μPs require the programmer to insert special instructions to correct the results.

This list of arithmetic/logic operations does not include the more complex operations of multiplication, division, square roots, and so on. These operations are not explicitly performed by any currently available μP because of the extra circuitry they would require. To perform such operations, the programmer can instruct the μP to execute an appropriate series of simple arithmetic operations. For example, the multiplication operation can be obtained through a series of shifting and adding operations, and the division operation by a series of shifting and subtracting operations.

The sequences of instructions that the programmer develops for performing such complex arithmetic operations are called *subroutines*. A multiplication subroutine for obtaining the product of two 8-bit numbers might consist of as many as 20 or 30 instruction steps requiring 200 μs to execute. If this long execution time is undesirable, it is possible to use an external LSI chip containing a high-speed multiplier. Such LSI multipliers are available with execution times of only 100 ns and can be connected to the μP as I/O devices. Another alternative to slow subroutines is to use a ROM as a storage table which can store such data as multiplication tables, trigonometric tables, and logarithmic tables.

Signed Numbers

The circuitry in the ALU performs the add and subtract operations in the binary number system and always treats the two operands as 8-bit binary numbers. This is true even when the program calls for BCD arithmetic, the only difference being the extra steps needed to correct the BCD result. With an 8-bit data word, decimal numbers from 0 to 255 can be represented in binary code, assuming that all 8 bits are used for the numerical value. However, if the programmer wishes to use both positive and negative numbers, the MSB of each data word is used as the *sign bit* with 0 for $+$ and 1 for $-$. The other 7 bits of each data word represent the magnitude; for positive numbers the magnitude is in true binary form, while for negative numbers the magnitude is in 2's-complement form.

The beauty of this method for representing signed numbers is that it requires no special operations by the ALU. The circuitry in the ALU will perform addition and subtraction on the two operands in the same manner, regardless of whether the operands are unsigned 8-bit data words or 7-bit data words plus a sign bit.

As was shown in Chapter 1, the sign bit participates in the add and subtract operations just like the rest of the bits. What this means is that the μP really does not know or care whether the data it is processing are signed or unsigned. Only the programmer is concerned about the distinction, and he must at all times know the format of the data he is using in his program.

As an aid to the programmer, almost all μPs transfer the value of bit 7 of the ALU result to a flag bit in the status register. This flag bit, sometimes called the SIGN flag, S, or the NEGATIVE flag, N, will be set to 1 if the ALU result has a 1 in bit 7 and will be cleared to 0 if bit 7 is a 0. The program can then test to see if the result was positive or negative by performing a conditional jump instruction based on the SIGN flag. For example, a jump-on-negative instruction (JPN) will cause the CPU to examine the SIGN flag in the status register to determine what sequence of instructions to follow next.

13.11   FINAL COMMENTS

Our discussion has been, by necessity, only a brief introduction to the basic principles, terminology, and operations common to most microprocessors and microcomputers. We have not even begun to gain an insight into the capabilities and applications of these devices. Almost all areas of technology have started taking advantage of the inexpensive computer control which microprocessors can provide. Some typical applications include microwave ovens, traffic controllers, home computers, electronic measuring instruments, industrial process control, electronic games, automobile emission control, and a rapidly growing number of new products.

With the revolutionary impact that microprocessors have had on the electronics industry, it is not unreasonable to expect that everyone working in electronics and related areas will have to become knowledgeable in the operation of microprocessors. Hopefully, our introduction will serve as a firm foundation for further study in this important area.

QUESTIONS AND PROBLEMS

13.1   Match each of the terms in column A with its definition from column B.

A	B
1. Control bus	a. FF indicating some particular internal microprocessor condition.
2. Interrupt	b. A block of memory spaces, typically 256.
3. Page	c. Reserved area of memory.
4. R/W	d. Signal from external device to CPU.
5. Flags	e. Carries signals used to synchronize activities of microcomputer elements.
6. Stack	f. Control signal from CPU.

13.2 What are the advantages of microprocessor-based designs over random logic? What is their main drawback compared to random logic?

13.3 Describe the functions of the three buses in a typical microcomputer.

13.4 List the sequence of events that take place during a microcomputer READ operation.

13.5 Repeat Problem 13.4 for a WRITE operation.

13.6 A certain microcomputer has the following address allocations:
RAM-0600 to 0FFF; ROM-2000 to 27FF; I/O- C800-C8FF.
(a) How many pages of RAM are there? How many memory locations is this?
(b) How many pages of ROM are there?
(c) How many different I/O devices can be used?

13.7 Refer to the circuit of Figure 13.11. What is the reason for connecting ϕ_2 to the decoder *CE* input?

13.8 Determine the address allocations for each RAM module in Figure 13.11 if the inverters on the NAND inputs are removed.

13.9 Explain why a microprocessor's address and data lines might need to be buffered.

13.10 What are the major functions performed by a microprocessor in a microcomputer?

13.11 Describe the basic function of each of the following microprocessor internal registers: Accumulator; MAR; PC; Status; General purpose; Index; Stack Pointer.

13.12 Below is a list of operations which a typical microprocessor's ALU can perform. For each one indicate whether the operation requires one or two operands.
(a) Shift.
(b) Exclusive-OR.
(c) Complement.
(d) Rotate.
(e) Logical AND.
(f) Compare.
(g) Decrement.

POWERS OF 2

2^n	n	2^{-n}
1	0	1.0
2	1	0.5
4	2	0.25
8	3	0.125
16	4	0.062 5
32	5	0.031 25
64	6	0.015 625
128	7	0.070 812 5
256	8	0.003 906 25
512	9	0.001 953 125
1 024	10	0.000 976 562 5
2 048	11	0.000 488 281 25
4 096	12	0.000 244 140 625
8 192	13	0.000 122 070 312 5
16 384	14	0.000 061 035 156 25
32 768	15	0.000 030 517 578 125
65 536	16	0.000 015 258 789 062 5
131 072	17	0.000 007 629 394 531 25
262 144	18	0.000 003 814 697 265 625
524 288	19	0.000 001 907 348 632 812 5
1 048 576	20	0.000 000 953 674 316 406 25
2 097 152	21	0.000 000 476 837 158 203 125
4 194 304	22	0.000 000 238 418 579 101 562 5
8 388 608	23	0.000 000 119 209 289 550 781 25
16 777 216	24	0.000 000 059 604 644 775 390 625
33 554 432	25	0.000 000 029 802 322 387 695 312 5
67 108 864	26	0.000 000 014 901 161 193 847 656 25
134 217 728	27	0.000 000 007 450 580 596 923 828 125

Appendix II

KARNAUGH-MAP METHOD FOR SIMPLIFYING LOGIC CIRCUITS

The Karnaugh map is a graphical device used to simplify a logic equation or to convert a truth table to its corresponding logic circuit in a simple, orderly process. Although a Karnaugh map (henceforth abbreviated *K map*) can be used for problems involving any number of input variables, its practical usefulness is limited to six variables. The following discussion will be limited to problems with up to four inputs, since even five- and six-input problems are too involved and are best done by a computer program.

II.1 KARNAUGH-MAP FORMAT

The K map, like a truth table, is a means for showing the relationship between logic inputs and the desired output. Figure II.1 shows three examples of K maps for two, three, and four variables, together with the corresponding truth tables. These examples illustrate the following important points:

1. The truth table gives the value of output X for each combination of input values. The K map gives the same information in different form. Each square in the K map corresponds to one entry of the truth table. For example, the square in the upper-left-hand corner of the two-variable K map corresponds to the first entry in the truth table (the term $\bar{A}\bar{B}$). Since $X = 1$ for this case in the truth table, the corresponding K-map square contains a 1. Similarly, the square in the lower-right-hand corner corresponds to the fourth entry in the truth table (the term AB).

2. The K-map squares are labeled so that only one variable changes in going from one square to the next, either horizontally or vertically. For example,

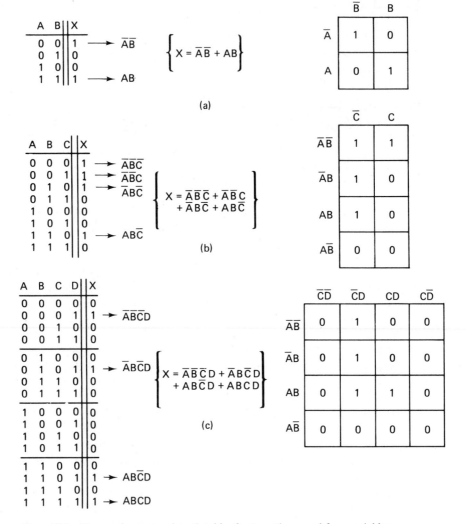

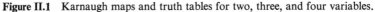

Figure II.1   Karnaugh maps and truth tables for two, three, and four variables.

the upper-left-hand square of the four-variable map is $\bar{A}\bar{B}\bar{C}\bar{D}$; the square immediately to its right is $\bar{A}\bar{B}\bar{C}D$ (only the last variable is different). Also, the square directly below it is $\bar{A}B\bar{C}\bar{D}$ (only the second variable is different).

3.   The sum-of-products expression for output X can be obtained by ORing together those squares on the K map which contain a 1. For example, in the two-variable case the $\bar{A}\bar{B}$ and AB squares contain a 1, so $X = \bar{A}\bar{B} + AB$.

II.2 LOOPING ADJACENT 1s FOR SIMPLIFICATION

The expression for output X can be simplified by properly combining those squares in the K map which contain 1s. The process for combining these 1s is called *looping*.

Looping Groups of Two

Figure II.2(a) is the K map for a particular three-variable truth table. This map contains a pair of 1s that are vertically adjacent to each other; the first one represents $\bar{A}B\bar{C}$ and the second one represents $AB\bar{C}$. Note that in these two terms only the A variable appears in both normal and complemented form (B and $\bar{C}$ remain unchanged). These two terms can be looped (combined) to give a resultant that eliminates the A variable since it appears in both uncomplemented and complemented forms. This is easily proved as follows:

$$X = \bar{A}B\bar{C} + AB\bar{C}$$
$$= B\bar{C}(\bar{A} + A)$$
$$= B\bar{C}(1) = B\bar{C}$$

This same principle holds true for any pair of vertically or horizontally adjacent 1s. Figure II.2(b) shows an example of two horizontally adjacent 1s. These two can be looped and the C variable eliminated since it appears in both its uncomplemented and complemented forms to give a resultant of $X = \bar{A}B$.

Another example is shown in Figure II.2(c). In a K map the top row and bottom row of squares are considered to be adjacent. Thus, the two 1s in this map can be looped to provide a resultant of $\bar{A}\bar{B}\bar{C} + A\bar{B}\bar{C} = \bar{B}\bar{C}$.

Figure II.2(d) shows a K map that has two pairs of 1s which can be looped. The two 1s in the top row are horizontally adjacent. The two 1s in the bottom row are also adjacent since in a K map the leftmost column and rightmost column of squares are considered to be adjacent. When the top pair of 1s is looped, the D variable is eliminated (since it appears as both D and $\bar{D}$) to give the term $\bar{A}\bar{B}C$. Looping the bottom pair eliminates the C variable to give the term $A\bar{B}\bar{D}$. These two terms are ORed to give the final result for X.

To summarize: *Looping a pair of adjacent 1s in a K map eliminates the variable that appears in complemented and uncomplemented form.*

Looping Groups of Four

A K map may contain a group of four 1s that are adjacent to each other. This group is called a *quad*. Figure II.3 shows several examples of quads. In (a) the four 1s are vertically adjacent and in (b) they are horizontally adjacent. The K map

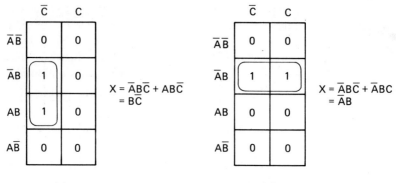

(a)

(b)

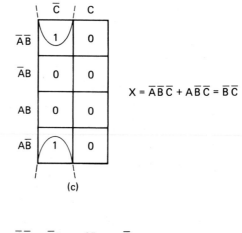

(c)

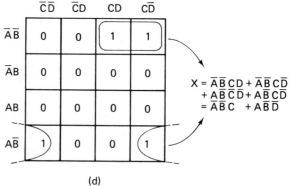

(d)

Figure II.2 Examples of looping pairs of adjacent 1s.

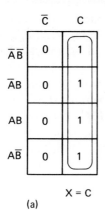

X = C

(a)

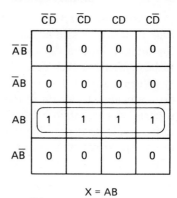

X = AB

(b)

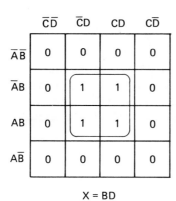

X = BD

(c)

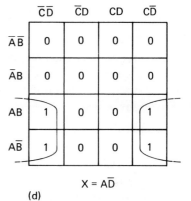

X = A$\overline{\text{D}}$

(d)

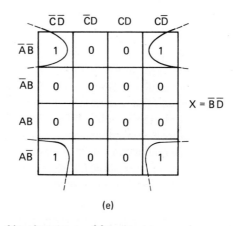

(e)

Figure II.3 Example of looping groups of four 1s.

in (c) contains four 1s in a square and are considered adjacent to each other. The four 1s in (d) are also adjacent, as are those in (e) because, as pointed out earlier, the top and bottom rows and the leftmost and rightmost columns are considered to be adjacent to each other.

When a quad is looped, the resultant term will contain only the variables that do not change form for all the squares in the quad. For example, in (a) the four squares which contain a 1 are $\bar{A}\bar{B}C$, $\bar{A}BC$, ABC, and $A\bar{B}C$. Examination of these terms reveals that only the variable C remains unchanged (both A and B appear in complemented and uncomplemented form). Thus, the resultant expression for X is simply $X = C$. This can be proved as follows:

$$X = \bar{A}\bar{B}C + \bar{A}BC + ABC + A\bar{B}C$$
$$= \bar{A}C(\bar{B} + B) + AC(B + \bar{B})$$
$$= \bar{A}C + AC$$
$$= C(\bar{A} + A) = C$$

As another example, consider Figure II.3(d), where the four squares containing 1s are $ABC\bar{D}$, $A\bar{B}C\bar{D}$, $ABC\bar{D}$, and $A\bar{B}C\bar{D}$. Examination of these terms indicates that only the variables A and $\bar{D}$ remain unchanged, so the simplified expression for X is

$$X = A\bar{D}$$

This can be proved in the same manner that was used above.

The reader should check each of the other cases in Figure II.3 to verify the indicated expressions for X. To summarize: *Looping a quad of 1s eliminates the two variables that appear in both complemented and uncomplemented form.*

Looping Groups of Eight

A group of eight 1s that are adjacent to each other is called an *octet*. Several examples of octets are shown in Figure II.4. When an octet is looped in a four-variable map, three of the four variables are eliminated because only one variable remains unchanged. For example, examination of the eight looped squares in (A) shows that only the variable B is in the same form for all eight squares; the other variables appear in complemented and uncomplemented form. Thus, for this map, $X = B$. The reader can verify the results for the other examples in the figure.

To summarize: *Looping an octet of 1s eliminates the three variables that appear in both complemented and uncomplemented form.*

II.3   COMPLETE SIMPLIFICATION PROCEDURE

We have seen how looping of pairs, quads, and octets on a K map can be used to obtain a simplified expression. We can summarize the rule for loops of *any* size: *When a variable appears in both complemented and uncomplemented form*

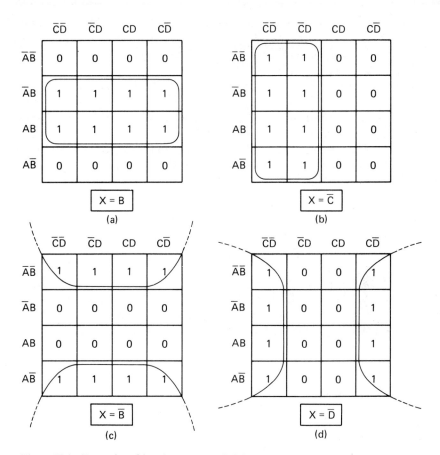

Figure II.4 Examples of looping groups of eight 1s.

within a loop, that variable is eliminated from the expression. Variables that are the same for all squares of the loop must appear in the final expression.

It should be clear that a larger loop of 1s eliminates more variables. To be exact, a loop of two eliminates one variable, a loop of four eliminates two, and a loop of eight eliminates three. This principle will now be used to obtain a simplified logic expression from a K map that contains any combination 1s and 0s.

The procedure will first be outlined and then applied to several examples. The steps below are followed in using the K-map method for simplifying a Boolean expression:

1. Construct the K map and place 1s in those squares corresponding to the 1s in the truth table. Place 0s in the other squares.

2. Examine the map for adjacent 1s and loop those 1s which are *not* adjacent to any other 1s.

3. Next, look for those 1s which are adjacent to only one other 1. Loop *any pair containing such a 1 even if the same 1 is looped more than once.*

4. Loop any octet even if some of the 1s have already been looped.

5. Loop any quad that contains one or more 1s which have not already been looped.

6. Loop any pairs necessary to include any 1s that have not yet been looped, making sure to use the minimum number of loops.

7. Form the OR sum of all the terms generated by each loop.

These steps will be followed exactly and referred to in the following examples. In each case, the resulting logic expression will be in its simplest sum-of-products form.

EXAMPLE II.1   Figure II.5(a) shows the K map for a four-variable problem. We will assume that the map was obtained from the problem truth table (step 1). The squares are numbered for convenience in identifying each loop.

Step 2:   Square 4 is the only square containing a 1 that is not adjacent to any other 1. It is looped and is referred to as loop 4.

Step 3:   Square 15 is adjacent *only* to square 11. This pair is looped and referred to as loop 11, 15.

Step 4:   There are no octets.

Step 5:   Squares 6, 7, 10, and 11 form a quad. This quad is looped (loop 6, 7, 10, 11). Note that square 11 is used again, even though it was part of loop 11, 15.

Step 6:   All 1s have already been looped.

Step 7:   Each loop generates a term in the expression for X. Loop 4 is simply $\bar{A}\bar{B}C\bar{D}$. Loop 11, 15 is ACD (the B variable is eliminated). Loop 6, 7, 10, 11 is BD (A and C are eliminated).

EXAMPLE II.2   Consider the K map in Figure II.5(b). Once again we can assume that step 1 has already been performed.

Step 2:   There are no 1s which are not adjacent to other 1s.

Step 3:   The 1 in square 3 is adjacent *only* to the 1 in square 7. Looping this pair (loop 3, 7) produces the term $\bar{A}CD$.

Step 4:   There are no octets.

Step 5:   There are two quads. Squares 5, 6, 7, and 8 form one quad. Looping this quad produces the term $\bar{A}B$. The second quad is made up of squares 5, 6, 9, and 10. This quad is looped because it contains two squares that have not been looped previously. Looping this quad produces $B\bar{C}$.

Step 6:   All 1s have already been looped.

Step 7:   The terms generated by the three loops are ORed together to obtain the expression for X.

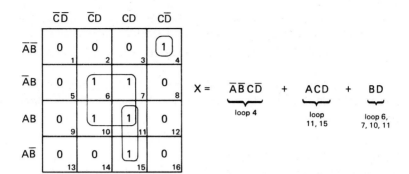

(a)

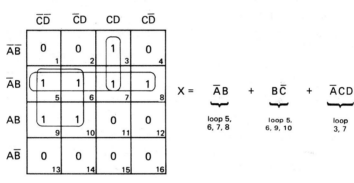

(b)

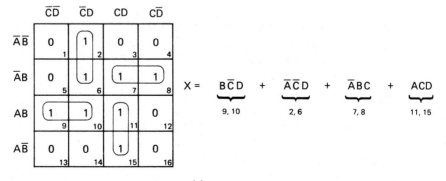

(c)

Figure II.5

EXAMPLE II.3   Consider the K map in Figure II.5(c).

Step 2:   There are no isolated 1s.

Step 3:   The 1 in square 2 is adjacent only to the 1 in square 6. This pair is looped to produce $\bar{A}\bar{C}D$. Similarly, square 9 is adjacent only to square 10. Looping this pair produces $B\bar{C}D$. Likewise, loop 7, 8 and loop 11, 15 produce the terms $\bar{A}BC$ and ACD, respectively.

Step 4:   There are no octets.

Step 5:   There is one quad formed by squares 6, 7, 10, and 11. This quad, however, is *not* looped because all the 1s in the quad have been included in other loops.

Step 6:   All 1s have already been looped.

Step 7:   The expression for X is shown in the figure.

EXAMPLE II.4   Consider the K map in Figure II.6(a).

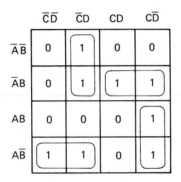

$$X = \bar{A}\bar{C}D + \bar{A}BC + A\bar{B}\bar{C} + AC\bar{D}$$

(a)

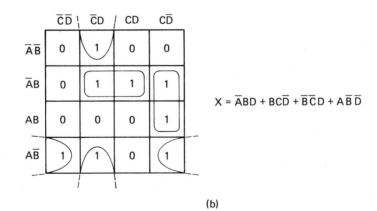

$$X = \bar{A}BD + BC\bar{D} + \bar{B}\bar{C}D + A\bar{B}\bar{D}$$

(b)

Figure II.6   The same K map with two equally good solutions.

Step 2:   There are no isolated 1s.

Step 3:   There are no 1s that are adjacent to only one other 1.

Step 4:   There are no octets.

Step 5:   There are no quads.

Steps 6 and 7:   There are many possible pairs. The looping must use the minimum number of loops to account for all the 1s. For this map there are *two* possible loopings, which require only four looped pairs. Figure II.6(a) shows one solution and its resultant expression. Figure II.6(b) shows the other. Note that both expressions are of the same complexity, so neither is better than the other.

Three more examples are shown in Figure II.7 to further illustrate the K-map procedure. The reader should study these examples and verify the results. To gain practice in using the K-map technique it would be beneficial to do any of the examples or problems from Chapters 2 and 3 which involve simplifying a Boolean expression to obtain the most efficient circuit.

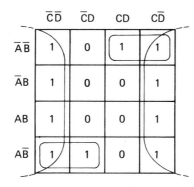

$$X = \overline{D} + \overline{A}\,\overline{B}\,C + A\,\overline{B}\,\overline{C}$$

(a)

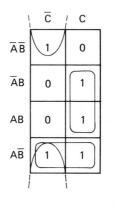

$$X = \overline{B}\,\overline{C} + BC + A\,\overline{B}$$

(b)

Figure II.7

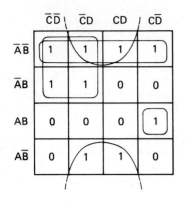

$$X = \overline{A}\,\overline{B} + \overline{A}\,\overline{C} + \overline{B}D + ABC\overline{D}$$

(c)

Figure II.7

Appendix

III

MANUFACTURERS' IC DATA SHEETS

These data sheets are presented through the courtesy of Fairchild Semi-Conductor and Solid State Scientific, Inc.

FAIRCHILD TTL/SSI • 9N00/5400, 7400

QUAD 2-INPUT NAND GATE

LOGIC AND CONNECTION DIAGRAM

DIP (TOP VIEW)

FLATPAK (TOP VIEW)

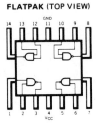

Positive logic: Y = $\overline{AB}$

SCHEMATIC DIAGRAM
(EACH GATE)

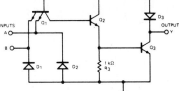

Component values shown are typical

RECOMMENDED OPERATING CONDITIONS

PARAMETER	9N00XM/5400XM			9N00XC/7400XC			UNITS
	MIN.	TYP.	MAX.	MIN.	TYP.	MAX.	
Supply Voltage V$_{CC}$	4.5	5.0	5.5	4.75	5.0	5.25	Volts
Operating Free-Air Temperature Range	−55	25	125	0	25	70	°C
Normalized Fan-Out from Each Output, N			10			10	U.L.

X = package type; F for Flatpak, D for Ceramic Dip, P for Plastic Dip. See Packaging Information Section for packages available on this product.

ELECTRICAL CHARACTERISTICS OVER OPERATING TEMPERATURE RANGE (Unless Otherwise Noted)

SYMBOL	PARAMETER	LIMITS			UNITS	TEST CONDITIONS (Note 1)		TEST FIGURE
		MIN.	TYP. (Note 2)	MAX.				
V$_{IH}$	Input HIGH Voltage	2.0			Volts	Guaranteed Input HIGH Voltage		1
V$_{IL}$	Input LOW Voltage			0.8	Volts	Guaranteed Input LOW Voltage		2
V$_{OH}$	Output HIGH Voltage	2.4	3.3		Volts	V$_{CC}$ = MIN., I$_{OH}$ = 0.4 mA, V$_{IN}$ = 0.8 V		2
V$_{OL}$	Output LOW Voltage		0.22	0.4	Volts	V$_{CC}$ = MIN., I$_{OL}$ = 16 mA, V$_{IN}$ = 2.0 V		1
I$_{IH}$	Input HIGH Current			40	µA	V$_{CC}$ = MAX., V$_{IN}$ = 2.4 V	Each Input	4
				1.0	mA	V$_{CC}$ = MAX., V$_{IN}$ = 5.5 V		
I$_{IL}$	Input LOW Current			−1.6	mA	V$_{CC}$ = MAX., V$_{IN}$ = 0.4 V, Each Input		3
I$_{OS}$	Output Short Circuit Current	−20		−55	mA	9N00/5400	V$_{CC}$ = MAX.	5
	(Note 3)	−18		−55	mA	9N00/7400		
I$_{CCH}$	Supply Current HIGH		4.0	8.0	mA	V$_{CC}$ = MAX., V$_{IN}$ = 0V		6
I$_{CCL}$	Supply Current LOW		12	22	mA	V$_{CC}$ = MAX., V$_{IN}$ = 5.0 V		6

SWITCHING CHARACTERISTICS (T$_A$ = 25°C)

SYMBOL	PARAMETER	LIMITS			UNITS	TEST CONDITIONS	TEST FIGURE
		MIN.	TYP.	MAX.			
t$_{PLH}$	Turn Off Delay Input to Output		11	22	ns	V$_{CC}$ = 5.0 V	A
t$_{PHL}$	Turn On Delay Input to Output		7.0	15	ns	C$_L$ = 15 pF R$_L$ = 400Ω	

NOTES:
(1) For conditions shown as MIN. or MAX., use the appropriate value specified under recommended operating conditions for the applicable device type.
(2) Typical limits are at V$_{CC}$ = 5.0 V, 25°C.
(3) Note more than one output should be shorted at a time.

FAIRCHILD SUPER HIGH SPEED TTL/SSI • 9S00/54S00, 74S00

QUAD 2-INPUT NAND GATE

LOGIC AND CONNECTION DIAGRAM

DIP (TOP VIEW)

Positive logic: Y = $\overline{AB}$

SCHEMATIC DIAGRAM

(EACH GATE)

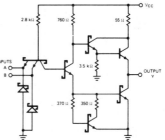

Component values shown are typical.

RECOMMENDED OPERATING CONDITIONS

PARAMETER	9S00XM/54S00XM			9S00XC/74S00XC			UNITS
	MIN.	TYP.	MAX.	MIN.	TYP.	MAX.	
Supply Voltage V_{CC}	4.5	5.0	5.5	4.75	5.0	5.25	Volts
Operating Free-Air Temperature Range	–55	25	125	0	25	75	°C
Input Loading for Each Input			1.25			1.25	U.L.

X = package type; F for Flatpak, D for Ceramic Dip, P for Plastic Dip. See Packaging Information Section for packages available on this product.

ELECTRICAL CHARACTERISTICS OVER OPERATING TEMPERATURE RANGE (Unless Otherwise Noted)

SYMBOL	PARAMETER		LIMITS			UNITS	TEST CONDITIONS (Note 1)	
			MIN.	TYP. (Note 2)	MAX.			
V_{IH}	Input HIGH Voltage		2.0			Volts	Guaranteed Input HIGH Voltage	
V_{IL}	Input LOW Voltage				0.8	Volts	Guaranteed Input LOW Voltage	
V_{CD}	Input Clamp Diode Voltage			–0.65	–1.2	Volts	V_{CC} = MIN., I_{IN} = -18mA	
V_{OH}	Output HIGH Voltage	XM	2.5	3.4		Volts	V_{CC} = MIN., I_{OH} = -1.0mA, V_{IN} = 0.8V	
		XC	2.7	3.4				
V_{OL}	Output LOW Voltage			0.35	0.5	Volts	V_{CC} = MIN., I_{OL} = 20mA, V_{IN} = 2.0V	
I_{IH}	Input HIGH Current			1.0 •	50	µA	V_{CC} = MAX., V_{IN} = 2.7V	Each Input
					1.0	mA	V_{CC} = MAX., V_{IN} = 5.5V	
I_{IL}	Input LOW Current			–1.4	–2.0	mA	V_{CC} = MAX., V_{IN} = 0.5 V	Each Input
I_{OS}	Output Short Circuit Current (Note 3)		–40	–65	–100	mA	V_{CC} = MAX., V_{OUT} = 0V	
I_{CCH}	Supply Current HIGH			10.8	16.0	mA	V_{CC} = MAX., V_{IN} = 0V	
I_{CCL}	Supply Current LOW			25.2	36.0	mA	V_{CC} = MAX., Inputs Open	

SWITCHING CHARACTERISTICS (T_A = 25°C)

SYMBOL	PARAMETER	LIMITS			UNITS	TEST CONDITIONS	TEST FIGURES
		MIN.	TYP.	MAX.			
t_{PLH}	Turn Off Delay Input to Output	2.0	3.0	4.5	ns	V_{CC} = 5.0V	DD
t_{PHL}	Turn On Delay Input to Output	2.0	3.0	5.0	ns	C_L = 15pF	

NOTES:
(1) For conditions shown as MIN. or MAX., use the appropriate value specified under recommended operating conditions for the applicable device type.
(2) Typical limits are at V_{CC} = 5.0V, 25°C.
(3) Not more than one output should be shorted at a time.

FAIRCHILD • 9LS00 (54LS/74LS00)

QUAD 2-INPUT NAND GATE

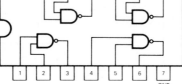

GUARANTEED OPERATING RANGES

PART NUMBERS	SUPPLY VOLTAGE			TEMPERATURE
	MIN	TYP	MAX	
9LS00XM/54LS00XM	4.5 V	5.0 V	5.5 V	−55°C to 125°C
9LS00XC/74LS00XC	4.75 V	5.0 V	5.25 V	0°C to 75°C

X = package type; F for Flatpak, D for Ceramic Dip, P for Plastic Dip. See Packaging Information Section for packages available on this product.

DC CHARACTERISTICS OVER OPERATING TEMPERATURE RANGE (unless otherwise specified)

SYMBOL	PARAMETER		LIMITS			UNITS	TEST CONDITIONS (Note 1)
			MIN	TYP	MAX		
V_{IH}	Input HIGH Voltage		2.0			V	Guaranteed Input HIGH Voltage
V_{IL}	Input LOW Voltage	XM			0.7	V	Guaranteed Input LOW Voltage
		XC			0.8		
V_{CD}	Input Clamp Diode Voltage			−0.65	−1.5	V	V_{CC} = MIN, I_{IN} = −18 mA
V_{OH}	Output HIGH Voltage	XM	2.5	3.4		V	V_{CC} = MIN, I_{OH} = −400 µA, V_{IN} = V_{IL}
		XC	2.7	3.4			
V_{OL}	Output LOW Voltage	XM,XC		0.25	0.4	V	V_{CC} = MIN, I_{OL} = 4.0 mA, V_{IN} = 2.0 V
		XC		0.35	0.5	V	V_{CC} = MIN, I_{OL} = 8.0 mA, V_{IN} = 2.0 V
I_{IH}	Input HIGH Current			1.0	20	µA	V_{CC} = MAX, V_{IN} = 2.7 V
					0.1	mA	V_{CC} = MAX, V_{IN} = 10 V
I_{IL}	Input LOW Current				−0.36	mA	V_{CC} = MAX, V_{IN} = 0.4 V
I_{OS}	Output Short Circuit Current (Note 3)		−15		−100	mA	V_{CC} = MAX, V_{OUT} = 0 V
I_{CCH}	Supply Current HIGH			0.8	1.6	mA	V_{CC} = MAX, V_{IN} = 0 V
I_{CCL}	Supply Current LOW			2.4	4.4	mA	V_{CC} = MAX, Inputs Open

AC CHARACTERISTICS: T_A = 25°C (See Page 4-50 for Waveforms)

SYMBOL	PARAMETER	LIMITS			UNITS	TEST CONDITIONS
		MIN	TYP	MAX		
t_{PLH}	Turn Off Delay, Input to Output	3.0	5.0	10	ns	V_{CC} = 5.0 V
t_{PHL}	Turn On Delay, Input to Output	3.0	5.0	10	ns	C_L = 15 pF

NOTES:
1. For conditions shown as MIN or MAX, use the appropriate value specified under recommended operating conditions for the applicable device type.
2. Typical limits are at V_{CC} = 5.0 V, T_A = 25°C.
3. Not more than one output should be shorted at a time.

FAIRCHILD TTL/SSI • 9N04/5404, 7404

HEX INVERTER

LOGIC AND CONNECTION DIAGRAM

DIP (TOP VIEW)

Positive logic: Y = $\overline{A}$

FLATPAK (TOP VIEW)

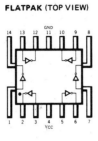

SCHEMATIC DIAGRAM
(EACH INVERTER)

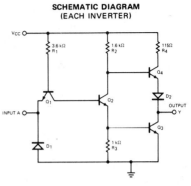

Component values shown are typical.

RECOMMENDED OPERATING CONDITIONS

PARAMETER	9N04XM/5404XM			9N04XC/7404XC			UNITS
	MIN.	TYP.	MAX.	MIN.	TYP.	MAX.	
Supply Voltage V$_{CC}$	4.5	5.0	5.5	4.75	5.0	5.25	Volts
Operating Free-Air Temperature Range	−55	25	125	0	25	70	°C
Normalized Fan-Out from Each Output, N		10				10	U.L.

X = package type; F for Flatpak, D for Ceramic Dip, P for Plastic Dip. See Packaging Information Section for packages available on this product.

ELECTRICAL CHARACTERISTICS OVER OPERATING TEMPERATURE RANGE (Unless Otherwise Noted)

SYMBOL	PARAMETER	LIMITS			UNITS	TEST CONDITIONS (Note 1)	TEST FIGURE
		MIN.	TYP. (Note 2)	MAX.			
V$_{IH}$	Input HIGH Voltage	2.0			Volts	Guaranteed Input HIGH Voltage	15
V$_{IL}$	Input LOW Voltage			0.8	Volts	Guaranteed Input LOW Voltage	16
V$_{OH}$	Output HIGH Voltage	2.4	3.3		Volts	V$_{CC}$ = MIN., I$_{OH}$ = −0.4 mA, V$_{IN}$ = 0.8 V	16
V$_{OL}$	Output LOW Voltage		0.22	0.4	Volts	V$_{CC}$ = MIN., I$_{OL}$ = 16 mA, V$_{IN}$ = 2.0 V	15
I$_{IH}$	Input HIGH Current			40	µA	V$_{CC}$ = MAX., V$_{IN}$ = 2.4 V	18
				1.0	mA	V$_{CC}$ = MAX., V$_{IN}$ = 5.5 V	
I$_{IL}$	Input LOW Current			−1.6	mA	V$_{CC}$ = MAX., V$_{IN}$ = 0.4 V	18
I$_{OS}$	Output Short Circuit Current (Note 3)	−20		−55	mA	9N04/5404	V$_{CC}$ = MAX.	19
		−18		−55	mA	9N04/7404		
I$_{CCH}$	Supply Current HIGH		6.0	12	mA	V$_{CC}$ = MAX., V$_{IN}$ = 0 V	20
I$_{CCL}$	Supply Current LOW		18	33	mA	V$_{CC}$ = MAX., V$_{IN}$ = 5.0 V	20

SWITCHING CHARACTERISTICS (T$_A$ = 25°C)

SYMBOL	PARAMETER	LIMITS			UNITS	TEST CONDITIONS	TEST FIGURE
		MIN.	TYP.	MAX.			
t$_{PLH}$	Turn Off Delay Input to Output		12	22	ns	V$_{CC}$ = 5.0 V C$_L$ = 15 pF R$_L$ = 400Ω	A
t$_{PHL}$	Turn On Delay Input to Output		8.0	15	ns		

NOTES:
(1) For conditions shown as MIN. or MAX., use the appropriate value specified under recommended operating conditions for the applicable device type.
(2) Typical limits are at V$_{CC}$ = 5.0 V, 25°C.
(3) Not more than one output should be shorted at a time.

FAIRCHILD TTL/SSI • 9N01/5401, 7401

QUAD 2-INPUT NAND GATE
(WITH OPEN-COLLECTOR OUTPUT)

LOGIC AND CONNECTION DIAGRAM

SCHEMATIC DIAGRAM
(EACH GATE)

DIP (TOP VIEW)   **FLATPAK** (TOP VIEW)

*OPEN COLLECTOR

Positive logic: Y = $\overline{AB}$

*OPEN COLLECTOR

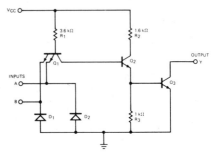

Component values shown are typical.

RECOMMENDED OPERATING CONDITIONS

PARAMETER	9N01XM/5401XM			9N01XC/7401XC			UNITS
	MIN.	TYP.	MAX.	MIN.	TYP.	MAX.	
Supply Voltage V_{CC}	4.5	5.0	5.5	4.75	5.0	5.25	Volts
Operating Free-Air Temperature Range	−55	25	125	0	25	70	°C
Normalized Fan-Out from Each Output, N			10			10	U.L.

X = package type; F for Flatpak, D for Ceramic Dip, P for Plastic Dip. See Packaging Information Section for packages available on this product.

ELECTRICAL CHARACTERISTICS OVER OPERATING TEMPERATURE RANGE (Unless Otherwise Noted)

SYMBOL	PARAMETER	LIMITS			UNITS	TEST CONDITIONS (Note 1)	TEST FIGURE
		MIN.	TYP. (Note 2)	MAX.			
V_{IH}	Input HIGH Voltage	2.0			Volts	Guaranteed Input HIGH Voltage	1
V_{IL}	Input LOW Voltage			0.8	Volts	Guaranteed Input LOW Voltage	7
I_{OH}	Output HIGH Current			0.25	mA	V_{CC} = MIN., V_{OH} = 5.5 V, V_{IL} = 0.8 V	7
V_{OL}	Output LOW Voltage			0.4	Volts	V_{CC} = MIN., I_{OL} = 16 mA, V_{IN} = 2.0 V (On Level)	1
I_{IH}	Input HIGH Current			40	μA	V_{CC} = MAX., V_{IN} = 2.4 V Each Input	4
				1.0	mA	V_{CC} = MAX., V_{IN} = 5.5 V	
I_{IL}	Input LOW Current			−1.6	mA	V_{CC} = MAX., V_{IN} = 0.4 V, Each Input	3
I_{CCH}	Supply Current HIGH		4.0	8.0	mA	V_{CC} = MAX., V_{IN} = 0 V	6
I_{CCL}	Supply Current LOW		12	22	mA	V_{CC} = MAX., V_{IN} = 5.0 V	6

SWITCHING CHARACTERISTICS (T_A = 25°C)

SYMBOL	PARAMETER	LIMITS			UNITS	TEST CONDITIONS		TEST FIGURE
		MIN.	TYP.	MAX.				
t_{PLH}	Turn Off Delay Input to Output		35	45	ns	R_L = 4.0 kΩ	V_{CC} = 5.0 V	A
t_{PHL}	Turn On Delay Input to Output		8.0	15	ns	R_L = 400Ω	C_L = 15 pF	

NOTES:
(1) For conditions shown as MIN. or MAX., use the appropriate value specified under recommended operating conditions for the applicable device type.
(2) Typical limits are at V_{CC} = 5.0 V, 25°C.

FAIRCHILD HIGH SPEED TTL/SSI • 9H04/54H04, 74H04

HEX INVERTER

LOGIC AND CONNECTION DIAGRAM

DIP (TOP VIEW)

SCHEMATIC DIAGRAM
(EACH INVERTER)

FLATPAK (TOP VIEW)

Positive logic: Y = $\overline{A}$

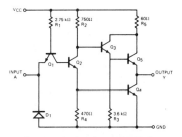

Component values shown are typical.

RECOMMENDED OPERATING CONDITIONS

PARAMETER	9H04XM/54H04XM			9H04XC/74H04XC			UNITS
	MIN.	TYP.	MAX.	MIN.	TYP.	MAX.	
Supply Voltage V$_{CC}$	4.5	5.0	5.5	4.75	5.0	5.25	Volts
Operating Free-Air Temperature Range	−55	25	125	0	25	70	°C
Input Loading for Each Input			1.25			1.25	U.L.

X = package type; F for Flatpak, D for Ceramic Dip, P for Plastic Dip. See Packaging Information Section for packages available on this product.

ELECTRICAL CHARACTERISTICS OVER OPERATING TEMPERATURE RANGE (Unless Otherwise Noted)

SYMBOL	PARAMETER	LIMITS			UNITS	TEST CONDITIONS (Note 1)	TEST FIGURE
		MIN.	TYP. (Note 2)	MAX.			
V$_{IH}$	Input HIGH Voltage	2.0			Volts	Guaranteed Input HIGH Voltage	15
V$_{IL}$	Input LOW Voltage			0.8	Volts	Guaranteed Input LOW Voltage	16
V$_{OH}$	Output HIGH Voltage	2.4			Volts	V$_{CC}$ = MIN., I$_{OH}$ = −0.5 mA, V$_{IN}$ = 0.8 V	16
V$_{OL}$	Output LOW Voltage			0.4	Volts	V$_{CC}$ = MIN., I$_{OL}$ = 20 mA, V$_{IN}$ = 2.0V	15
I$_{IH}$	Input HIGH Current			50	µA	V$_{CC}$ = MAX., V$_{IN}$ = 2.4 V	18
				1.0	mA	V$_{CC}$ = MAX., V$_{IN}$ = 5.5 V	
I$_{IL}$	Input LOW Current			−2.0	mA	V$_{CC}$ = MAX., V$_{IN}$ = 0.4 V,	18
I$_{OS}$	Output Short Circuit Current (Note 3)	−40		−100	mA	V$_{CC}$ = MAX.	19
I$_{CCH}$	Supply Current HIGH		16	26	mA	V$_{CC}$ = MAX., V$_{IN}$ = 0 V	20
I$_{CCL}$	Supply Current LOW		40	58	mA	V$_{CC}$ = MAX., V$_{IN}$ = 4.5 V	20

SWITCHING CHARACTERISTICS (T$_A$ = 25°C)

SYMBOL	PARAMETER	LIMITS			UNITS	TEST CONDITIONS	TEST FIGURE
		MIN.	TYP.	MAX.			
t$_{PLH}$	Turn Off Delay Input to Output		6.5	10	ns	V$_{CC}$ = 5.0V C$_L$ = 25 pF R$_L$ = 280Ω	T
t$_{PHL}$	Turn On Delay Input to Output		9.0	13	ns		

NOTES:
(1) For conditions shown as MIN. or MAX., use the appropriate value specified under recommended operating conditions for the applicable device type.
(2) Typical limits are at V$_{CC}$ = 5.0 V, 25°C.
(3) Not more than one output should be shorted at a time, and duration of short-circuit test should not exceed 1 second.

FAIRCHILD SUPER HIGH SPEED TTL/SSI • 9S04/54S04, 74S04 • 9S04A

HEX INVERTER

LOGIC AND CONNECTION DIAGRAM
DIP (TOP VIEW)

Positive logic: Y = $\overline{A}$

SCHEMATIC DIAGRAM
(EACH INVERTER)

Component values
shown are typical.

RECOMMENDED OPERATING CONDITIONS

PARAMETER	9S04XM/54S04XM 9S04AXM			9S04XC/74S04XC 9S04AXC			UNITS
	MIN.	TYP.	MAX.	MIN.	TYP.	MAX.	
Supply Voltage V_{CC}	4.5	5.0	5.5	4.75	5.0	5.25	Volts
Operating Free-Air Temperature Range	−55	25	125	0	25	75	°C
Input Loading for Each Input			1.25			1.25	U.L.

X = package type; F for Flatpak, D for Ceramic Dip, P for Plastic Dip. See Packaging Information Section for packages available on this product.

ELECTRICAL CHARACTERISTICS OVER OPERATING TEMPERATURE RANGE (Unless Otherwise Noted)

SYMBOL	PARAMETER		LIMITS			UNITS	TEST CONDITIONS (Note 1)
			MIN.	TYP. (Note 2)	MAX.		
V_{IH}	Input HIGH Voltage		2.0			Volts	Guaranteed Input HIGH Voltage
V_{IL}	Input LOW Voltage				0.8	Volts	Guaranteed Input LOW Voltage
V_{CD}	Input Clamp Diode Voltage			−0.65	−1.2	Volts	V_{CC} = MIN., I_{IN} = −18 mA
V_{OH}	Output HIGH Voltage	XM	2.5	3.4		Volts	V_{CC} = MIN., I_{OH} = −1.0 mA, V_{IN} = 0.8 V
		XC	2.7	3.4			
V_{OL}	Output LOW Voltage			0.35	0.5	Volts	V_{CC} = MIN., I_{OL} = 20 mA, V_{IN} = 2.0 V
I_{IH}	Input HIGH Current			1.0	50	μA	V_{CC} = MAX., V_{IN} = 2.7 V
					1.0	mA	V_{CC} = MAX., V_{IN} = 5.5 V
I_{IL}	Input LOW Current			−1.4	−2.0	mA	V_{CC} = MAX., V_{IN} = 0.5 V
I_{OS}	Output Short Circuit Current (Note 3)		−40	−65	−100	mA	V_{CC} = MAX., V_{OUT} = 0 V
I_{CCH}	Supply Current HIGH			16.2	24.0	mA	V_{CC} = MAX., V_{IN} = 0 V
I_{CCL}	Supply Current LOW			37.8	54.0	mA	V_{CC} = MAX., Inputs Open

SWITCHING CHARACTERISTICS (T_A = 25°C)

SYMBOL	PARAMETER		LIMITS			UNITS	TEST CONDITIONS	TEST FIGURES
			MIN.	TYP.	MAX.			
t_{PLH}	Turn Off Delay Input	9S04	2.0	3.0	4.5	ns	V_{CC} = 5.0 V	DD
	to Output	9S04A	1.0	2.5	3.5			
t_{PHL}	Turn On Delay Input	9S04	2.0	3.0	5.0	ns	C_L = 15 pF	
	to Output	9S04A	1.0	2.5	4.0			

NOTES:
(1) For conditions shown as MIN. or MAX., use the appropriate value specified under recommended operating conditions for the applicable device type.
(2) Typical limits are at V_{CC} = 5.0 V, 25°C.
(3) Not more than one output should be shorted at a time.

FAIRCHILD TTL/SSI • 9N05/5405, 7405

HEX INVERTER
(WITH OPEN-COLLECTOR OUTPUT)

LOGIC AND CONNECTION DIAGRAM

DIP (TOP VIEW)

*OPEN COLLECTOR

Positive logic: Y = $\overline{A}$

FLATPAK (TOP VIEW)

*OPEN COLLECTOR

SCHEMATIC DIAGRAM
(EACH INVERTER)

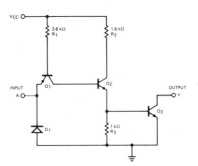

Component values shown are typical.

RECOMMENDED OPERATING CONDITIONS

PARAMETER	9N05XM/5405XM			9N05XC/7405XC			UNITS
	MIN.	TYP.	MAX.	MIN.	TYP.	MAX.	
Supply Voltage V_{CC}	4.5	5.0	5.5	4.75	5.0	5.25	Volts
Operating Free-Air Temperature Range	–55	25	125	0	25	70	°C
Normalized Fan-Out from Each Output, N			10			10	U.L.

X = package type; F for Flatpak, D for Ceramic Dip, P for Plastic Dip. See Packaging Information Section for packages available on this product.

ELECTRICAL CHARACTERISTICS OVER OPERATING TEMPERATURE RANGE (Unless Otherwise Noted)

SYMBOL	PARAMETER	LIMITS			UNITS	TEST CONDITIONS (Note 1)	TEST FIGURE
		MIN.	TYP. (Note 2)	MAX.			
V_{IH}	Input HIGH Voltage	2.0			Volts	Guaranteed Input HIGH Voltage	15
V_{IL}	Input LOW Voltage			0.8	Volts	Guaranteed Input LOW Voltage	17
I_{OH}	Output HIGH Current			0.25	mA	V_{CC} = MIN., V_{OH}= 5.5 V, V_{IN} = 0.8 V	17
V_{OL}	Output LOW Voltage			0.4	Volts	V_{CC} = MIN., I_{OL} = 16 mA, V_{IN} = 2.0 V (On Level)	15
I_{IH}	Input HIGH Current			40	µA	V_{CC} = MAX., V_{IN} = 2.4 V	18
				1.0	mA	V_{CC} = MAX., V_{IN} = 5.5 V	
I_{IL}	Input LOW Current			–1.6	mA	V_{CC} = MAX., V_{IN} = 0.4 V	18
I_{CCH}	Supply Current HIGH		6	12	mA	V_{IN} = 0 V V_{CC} = 5.0 V, T_A = 25°C	20
I_{CCL}	Supply Current LOW		18	33	mA	V_{IN} = 5 V	20

SWITCHING CHARACTERISTICS (T_A = 25°)

SYMBOL	PARAMETER	LIMITS			UNITS	TEST CONDITIONS	TEST FIGURE
		MIN.	TYP.	MAX.			
t_{PLH}	Turn Off Delay Input to Output		40	55	ns	R_L = 4kΩ V_{CC} = 5.0 V	A
t_{PHL}	Turn On Delay Input to Output		8.0	15	ns	R_L = 400Ω C_L = 15 pF	

NOTES:
(1) For conditions shown as MIN. or MAX., use the appropriate value specified under recommended operating conditions for the applicable device type.
(2) Typical limits are at V_{CC} = 5.0 V, 25°C.

FAIRCHILD TTL/SSI • 9N20/5420, 7420

DUAL 4-INPUT NAND GATE

LOGIC AND CONNECTION DIAGRAM

DIP (TOP VIEW)   **FLATPAK** (TOP VIEW)

SCHEMATIC DIAGRAM
(EACH GATE)

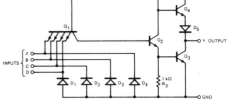

Positive logic: Y = $\overline{ABCD}$

NC — No internal connection.

Component values shown are typical.

RECOMMENDED OPERATING CONDITIONS

PARAMETER	9N20XM/5420XM			9N20XC/7420XC			UNITS
	MIN.	TYP.	MAX.	MIN.	TYP.	MAX.	
Supply Voltage V_{CC}	4.5	5.0	5.5	4.75	5.0	5.25	Volts
Operating Free-Air Temperature Range	−55	25	125	0	25	70	°C
Normalized Fan Out From Each Output, N			10			10	U.L.

X = package type; F for Flatpak, D for Ceramic Dip, P for Plastic Dip. See Packaging Information Section for packages available on this product.

ELECTRICAL CHARACTERISTICS OVER OPERATING TEMPERATURE RANGE (Unless Otherwise Noted)

SYMBOL	PARAMETER	LIMITS			UNITS	TEST CONDITIONS (Note 1)		TEST FIGURE
		MIN.	TYP. (Note 2)	MAX.				
V_{IH}	Input HIGH Voltage	2.0			Volts	Guaranteed Input HIGH Voltage		1
V_{IL}	Input LOW Voltage.			0.8	Volts	Guaranteed Input LOW Voltage		2
V_{OH}	Output HIGH Voltage	2.4	3.3		Volts	V_{CC} = MIN., I_{OH} = −0.4 mA, V_{IN} = 0.8 V		2
V_{OL}	Output LOW Voltage		0.22	0.4	Volts	V_{CC} = MIN., I_{OL} = 16 mA, V_{IN} = 2.0 V		1
I_{IH}	Input HIGH Current			40	μA	V_{CC} = MAX., V_{IN} = 2.4 V	Each Input	4
				1.0	mA	V_{CC} = MAX., V_{IN} = 5.5 V		
I_{IL}	Input LOW Current			−1.6	mA	V_{CC} = MAX., V_{IN} = 0.4 V, Each Input		3
I_{OS}	Output Short Circuit Current	−20		−55	mA	9N20/5420	V_{CC} = MAX.	5
	(Note 3)	−18		−55	mA	9N20/7420		
I_{CCH}	Supply Current HIGH		2.0	4.0	mA	V_{CC} = MAX., V_{IN} = 0 V		6
I_{CCL}	Supply Current LOW		6.0	11	mA	V_{CC} = MAX., V_{IN} = 5.0 V		6

SWITCHING CHARACTERISTICS (T_A = 25°C)

SYMBOL	PARAMETER	LIMITS			UNITS	TEST CONDITIONS	TEST FIGURE
		MIN.	.TYP.	MAX.			
t_{PLH}	Turn Off Delay Input to Output		12	22	ns	V_{CC} = 5.0 V	A
t_{PHL}	Turn On Delay Input to Output		8.0	15	ns	C_L = 15 pF R_L = 400Ω	

NOTES:
(1) For conditions shown as MIN. or MAX., use the appropriate value specified under recommended operating conditions for the applicable
device type.
(2) Typical limits are at V_{CC} = 5.0 V, 25°C.
(3) Not more than one output should be shorted at a time.

FAIRCHILD TTL/SSI • 9N37/5437, 7437 • 9N38/5438, 7438

QUAD 2-INPUT NAND BUFFER

LOGIC AND CONNECTION DIAGRAM

DIP (TOP VIEW)

Positive logic: Y = $\overline{AB}$

FLATPAK (TOP VIEW)

SCHEMATIC DIAGRAM
(EACH BUFFER)

9N37/5437, 7437
(TOTEM-POLE OUTPUT)

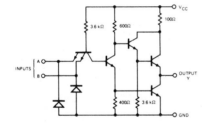

9N38/5438, 7438
(OPEN-COLLECTOR OUTPUT)

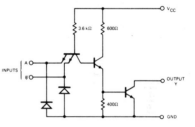

RECOMMENDED OPERATING CONDITIONS

PARAMETER	9N37XM/5437XM 9N38XM/5438XM			9N37XC/7437XC 9N38XC/7438XC			UNITS
	MIN.	TYP.	MAX.	MIN.	TYP.	MAX.	
Supply Voltage V$_{CC}$	4.5	5.0	5.5	4.75	5.0	5.25	Volts
Operating Free-Air Temperature Range	−55	25	125	0	25	70	°C
Normalized Fan Out from Each Output, N			30			30	U.L.

X = package type; F for Flatpak, D for Ceramic Dip, P for Plastic Dip. See Packaging Information Section for packages available on this product.

FAIRCHILD TTL/SSI • 9N37/5437, 7437 • 9N38/5438, 7438

9N37/5437, 7437

ELECTRICAL CHARACTERISTICS OVER OPERATING TEMPERATURE RANGE (Unless Otherwise Noted)

SYMBOL	PARAMETER	LIMITS			UNITS	TEST CONDITIONS (Note 1)
		MIN.	TYP. (Note 2)	MAX.		
V_{IH}	Input HIGH Voltage	2.0			Volts	Guaranteed Input HIGH Voltage
V_{IL}	Input LOW Voltage			0.8	Volts	Guaranteed Input LOW Voltage
V_{CD}	Input Clamp Diode Voltage			−1.5	Volts	V_{CC} = MIN., I_I = −12 mA
V_{OH}	Output HIGH Voltage	2.4			Volts	V_{CC} = MIN., I_{OH} = −1.2 mA, V_{IL} = 0.8 V
V_{OL}	Output LOW Voltage		0.22	0.4	Volts	V_{CC} = MIN., I_{OL} = 48 mA, V_{IH} = 2.0 V
I_I	Input Current at Maximum Input Voltage			1.0	mA	V_{CC} = MAX., V_{IN} = 5.5 V
I_{IH}	Input HIGH Current			40	μA	V_{CC} = MAX., V_{IN} = 2.4 V
I_{IL}	Input LOW Current			−1.6	mA	V_{CC} = MAX:, V_{IN} = 0.4 V
I_{OS}	Output Short Circuit Current (Note 3)	−20		−70	mA	V_{CC} = MAX., V_{IN} = 0 V
I_{CCH}	Supply Current HIGH		9.0	15.5	mA	V_{CC} = MAX., All Inputs at 0 V
I_{CCL}	Supply Current LOW		34	54	mA	V_{CC} = MAX., All Inputs at 5.0 V

SWITCHING CHARACTERISTICS (T_A = 25°C)

SYMBOL	PARAMETER	LIMITS			UNITS	TEST CONDITIONS
		MIN.	TYP.	MAX.		
t_{PLH}	Turn Off Delay Input to Output		13	22	ns	V_{CC} = 5.0 V
t_{PHL}	Turn On Delay Input to Output		8.0	15	ns	C_L = 45 pF R_L = 133Ω

9N38/5438, 7438

ELECTRICAL CHARACTERISTICS OVER OPERATING TEMPERATURE RANGE (Unless Otherwise Noted)

SYMBOL	PARAMETER	LIMITS			UNITS	TEST CONDITIONS (Note 1)
		MIN.	TYP. (Note 2)	MAX.		
V_{IH}	Input HIGH Voltage	2.0			Volts	Guaranteed Input HIGH Voltage
V_{IL}	Input LOW Voltage			0.8	Volts	Guaranteed Input LOW Voltage
V_{CD}	Input Clamp Diode Voltage			−1.5	Volts	V_{CC} = MIN., I_I = −12 mA
I_{OH}	Output HIGH Current			250	μA	V_{CC} = MIN., V_{OH} = 5.5 V, V_{IL} = 0.8 V
V_{OL}	Output LOW Voltage		0.22	0.4	Volts	V_{CC} = MIN., I_{OL} = 48 mA, V_{IH} = 2.0 V
I_I	Input Current at Maximum Input Voltage			1.0	mA	V_{CC} = MAX., V_{IN} = 5.5 V
I_{IH}	Input HIGH Current			40	μA	V_{CC} = MAX., V_{IN} = 2.4 V
I_{IL}	Input LOW Current			−1.6	mA	V_{CC} = MAX., V_{IN} = 0.4 V
I_{CCH}	Supply Current HIGH		5.0	8.5	mA	V_{CC} = MAX., All Inputs at 0 V
I_{CCL}	Supply Current LOW		34	54	mA	V_{CC} = MAX., All Inputs at 5.0 V

SWITCHING CHARACTERISTICS (T_A = 25°C)

SYMBOL	PARAMETER	LIMITS			UNITS	TEST CONDITIONS
		MIN.	TYP.	MAX.		
t_{PLH}	Turn Off Delay Input to Output		14	22	ns	V_{CC} = 5.0 V
t_{PHL}	Turn On Delay Input to Output		11	18	ns	C_L = 45 pF R_L = 133Ω

NOTES:
(1) For conditions shown as MIN. or MAX., use the appropriate value specified under recommended operating conditions for the applicable device type.
(2) Typical limits are at V_{CC} = 5.0 V, 25°C.
(3) Not more than one output should be shorted at a time, and duration of the short-circuit test should not exceed one second.

FAIRCHILD TTL/SSI • 9N73/5473, 7473 • 9N107/54107, 74107

DUAL JK MASTER/SLAVE FLIP-FLOP WITH SEPARATE CLEARS AND CLOCKS

DESCRIPTION — The TTL/SSI 9N73/5473, 7473 and 9N107/54107, 74107 are Dual JK Master/Slave flip-flops with a separate clear and a separate clock for each flip-flop. Inputs to the master section are controlled by the clock pulse. The clock pulse also regulates the state of the coupling transistors which connect the master and slave sections. The sequence of operation is as follows: 1) Isolate slave from master. 2) Enter information from J and K inputs to master. 3) Disable J and K inputs. 4) Transfer information from master to slave.

LOGIC AND CONNECTION DIAGRAM

9N73/5473, 7473

DIP (TOP VIEW)

FLATPAK (TOP VIEW)

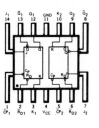

9N107/54107, 74107

DIP (TOP VIEW)

Positive logic:
    LOW input to clear sets Q to LOW level
    Clear is independent of clock

TRUTH TABLE

t_n		t_{n+1}
J	K	Q
L	L	Q_n
L	H	L
H	L	H
H	H	$\bar{Q}_n$

NOTES:
t_n = Bit time before clock pulse.
t_{n+1} = Bit time after clock pulse.

CLOCK WAVEFORM

LOGIC DIAGRAM
(EACH FLIP-FLOP)

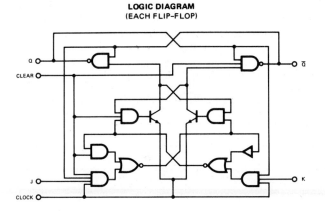

FAIRCHILD TTL/SSI • 9N73/5473, 7473 • 9N107/54107, 74107

RECOMMENDED OPERATING CONDITIONS

PARAMETER	9N73XM/5473XM 9N107XM/54107XM			9N73XC/7473XC 9N107XC/74107XC			UNITS
	MIN.	TYP.	MAX.	MIN.	TYP.	MAX.	
Supply Voltage V_{CC}	4.5	5.0	5.5	4.75	5.0	5.25	Volts
Operating Free-Air Temperature Range	−55	25	125	0	25	70	°C
Normalized Fan Out from Each Output, N			10			10	U.L.
Width of Clock Pulse, $t_{p(clock)}$ (See Fig. E)	20			20			ns
Width of Clear Pulse, $t_{p(clear)}$ (See Fig. F)	25			25			ns
Input Setup Time, t_{setup} (See Fig. E)	$\geqslant t_{p(clock)}$			$\geqslant t_{p(clock)}$			
Input Hold Time, t_{hold}	0			0			

X = package type; F for Flatpak, D for Ceramic Dip, P for Plastic Dip. See Packaging Information Section for packages available on this product.

ELECTRICAL CHARACTERISTICS OVER OPERATING TEMPERATURE RANGE (Unless Otherwise Noted)

SYMBOL	PARAMETER	LIMITS			UNITS	TEST CONDITIONS (Note 1)		TEST FIGURE
		MIN.	TYP. (Note 2)	MAX.				
V_{IH}	Input HIGH Voltage	2.0			Volts	Guaranteed Input HIGH		46 & 47
V_{IL}	Input LOW Voltage			0.8	Volts	Guaranteed Input LOW		46 & 47
V_{OH}	Output HIGH Voltage	2.4	3.5		Volts	V_{CC} = MIN., I_{OH} = −0.4 mA		46
V_{OL}	Output LOW Voltage		0.22	0.4	Volts	V_{CC} = MIN., I_{OL} = 16 mA		47
I_{IH}	Input HIGH Current at J or K			40	µA	V_{CC} = MAX., V_{IN} = 2.4 V		49
				1.0	mA	V_{CC} = MAX., V_{IN} = 5.5 V		
	Input HIGH Current at Clock or Clear			80	µA	V_{CC} = MAX., V_{IN} = 2.4 V		49
				1.0	mA	V_{CC} = MAX., V_{IN} = 5.5 V		
I_{IL}	Input LOW Current at J or K			−1.6	mA	V_{CC} = MAX., V_{IN} = 0.4 V		48
	Input LOW Current at Clear or Clock			−3.2	mA	V_{CC} = MAX., V_{IN} = 0.4 V		48
I_{OS}	Output Short Circuit Current (Note 3)	−20		−57	mA	9N73/5473; 9N107/54107	V_{CC} = MAX.	50
		−18		−57	mA	9N73/7473; 9N107/74107	V_{IN} = 0 V	
I_{CC}	Supply Current		20	40	mA	V_{CC} = MAX.		49

SWITCHING CHARACTERISTICS (T_A = 25°C)

SYMBOL	PARAMETER	LIMITS			UNITS	TEST CONDITIONS	TEST FIGURE
		MIN.	TYP.	MAX.			
f_{max}	Maximum Clock Frequency	15	20		MHz		E
t_{PLH}	Turn Off Delay Clear to Output		16	25	ns	V_{CC} = 5.0 V	F
t_{PHL}	Turn On Delay Clear to Output		25	40	ns	C_L = 15 pF	F
t_{PLH}	Turn Off Delay Clock to Output	10	16	25	ns	R_L = 400Ω	E
t_{PHL}	Turn On Delay Clock to Output	10	25	40	ns		E

NOTES:
(1) For conditions shown as MIN. or MAX., use the appropriate value specified under recommended operating conditions for the applicable device type.
(2) Typical limits are at V_{CC} = 5.0 V, 25°C.
(3) Not more than one output should be shorted at a time.

FAIRCHILD TTL/SSI • 9N86/5486, 7486

QUAD 2-INPUT EXCLUSIVE OR GATE

DESCRIPTION — The TTL/SSI 9N86/5486, 7486 is a Quad 2-input Exclusive OR gate designed to perform the function: $Y = A\overline{B} + \overline{A}B$. When the input states are complementary, the output goes to the HIGH level.

Input clamping diodes are provided to minimize transmission line effects. On chip input buffers are also provided to lower the fan in requirement to only 1 U.L. (unit load). The 9N86/5486, 7486 is fully compatible with all members of the Fairchild TTL family.

LOGIC AND CONNECTION DIAGRAM

DIP (TOP VIEW)

FLATPAK (TOP VIEW)

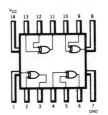

TRUTH TABLE

INPUTS		OUTPUT
A	B	Y
L	L	L
L	H	H
H	L	H
H	H	L

Positive logic: $Y = A \oplus B$

SCHEMATIC DIAGRAM

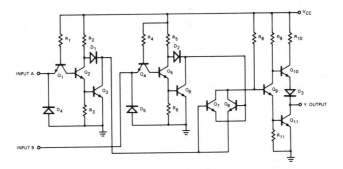

1/4 of Circuit shown.

FAIRCHILD TTL/SSI • 9N86/5486, 7486

RECOMMENDED OPERATING CONDITIONS

PARAMETER		9N86XM/5486XM			9N86XC/7486XC			UNITS
		MIN.	TYP.	MAX.	MIN.	TYP.	MAX.	
Supply Voltage V_{CC}		4.5	5.0	5.5	4.75	5.0	5.25	Volts
Operating Free-Air Temperature Range		–55	25	125	0	25	70	°C
Normalized Fan Out	LOW Level			10			10	U.L.
from Each Output N	HIGH Level			20			20	U.L.

X = package type; F for Flatpak, D for Ceramic Dip, P for Plastic Dip. See Packaging Information Section for packages available on this product.

ELECTRICAL CHARACTERISTICS OVER OPERATING TEMPERATURE RANGE (Unless Otherwise Noted)

SYMBOL	PARAMETER	LIMITS			UNITS	TEST CONDITIONS (Note 1)		TEST FIGURE
		MIN.	TYP. (Note 2)	MAX.				
V_{IH}	Input HIGH Voltage	2.0			Volts	Guaranteed Input HIGH Voltage		98
V_{IL}	Input LOW Voltage			0.8	Volts	Guaranteed Input LOW Voltage		98
V_{OH}	Output HIGH Voltage	2.4			Volts	V_{CC} = MIN., I_{OH} = –800 μA, V_{IH} = 2.0 V, V_{IL} = 0.8 V		98
V_{OL}	Output LOW Voltage			0.4	Volts	V_{CC} = MIN., I_{OL} = 16 mA, V_{IH} = 2.0 V, V_{IL} = 0.8 V		99
I_{IH}	Input HIGH Current			40	μA	V_{CC} = MAX., V_{IN} = 2.4 V	(Each Input)	100
				1.0	mA	V_{CC} = MAX., V_{IN} = 5.5 V		
I_{IL}	Input LOW Current			–1.6	mA	V_{CC} = MAX., V_{IN} = 0.4 V (Each Input)		101
I_{OS}	Output Short Circuit Current (Note 3)	–20		–55	mA	9N86/5486	V_{CC} = MAX., V_{IH} = 4.5 V	102
		–18		–55	mA	9N86/7486	V_{IL} = 0 V	
I_{CC}	Supply Current		30	43	mA	9N86/5486	V_{CC} = MAX., V_{IN} = 4.5 V	103
			30	50	mA	9N86/7486		

SWITCHING CHARACTERISTICS (T_A = 25°C)

SYMBOL	PARAMETER	LIMITS			UNITS	TEST CONDITIONS		TEST FIGURE
		MIN.	TYP.	MAX.				
t_{PLH}	Turn Off Delay Input to Output		15	23	ns	Other Input Low	V_{CC} = 5.0 V	S
			18	30		Other Input High	C_L = 15 pF	
t_{PHL}	Turn On Delay Input to Output		11	17		Other Input Low	R_L = 400Ω	
			13	22		Other Input High		

NOTES:
(1) For conditions shown as MIN. or MAX., use the appropriate value specified under recommended operating conditions for the applicable device type.
(2) Typical limits are at V_{CC} = 5.0 V, 25°C.
(3) Not more than one output should be shorted at a time.

TTL/MSI 9390/5490, 7490
DECADE COUNTER

DESCRIPTION — The TTL/MSI 9390/5490, 7490 is a Decade Counter which consists of four dual rank, master slave flip-flops internally interconnected to provide a divide-by-two counter and a divide-by-five counter. Count inputs are inhibited, and all outputs are returned to logical zero or a binary coded decimal (BCD) count of 9 through gated direct reset lines. The output from flip-flop A is not internally connected to the succeeding stages, therefore the count may be separated into these independent count modes:

A. If used as a binary coded decimal decade counter, the $\overline{CP}_{BD}$ input must be externally connected to the Q_A output. The $\overline{CP}_A$ input receives the incoming count, and a count sequence is obtained in accordance with the BCD count for nine's complement decimal application.

B. If a symmetrical divide-by-ten count is desired for frequency synthesizers or other applications requiring division of a binary count by a power of ten, the Q_D output must be externally connected to the $\overline{CP}_A$ input. The input count is then applied at the $\overline{CP}_{BD}$ input and a divide-by-ten square wave is obtained at output Q_A.

C. For operation as a divide-by-two counter and a divide-by-five counter, no external interconnections are required. Flip-flop A is used as a binary element for the divide-by-two function. The $\overline{CP}_{BD}$ input is used to obtain binary divide-by-five operation at the Q_B, Q_C, and Q_D outputs. In this mode, the two counters operate independently; however, all four flip-flops are reset simultaneously.

PIN NAMES   **LOADING**

R_0	Reset-Zero Inputs	1 U.L.
R_9	Reset-Nine Inputs	1 U.L.
$\overline{CP}_A$	Clock Input	2 U.L.
$\overline{CP}_{BD}$	Clock Input	4 U.L.
Q_A, Q_B, Q_C, Q_D	Outputs	10 U.L.

1 Unit Load (U.L.) = 40μA HIGH/1.6mA LOW.

TRUTH TABLES

BCD COUNT SEQUENCE (Note 1)

COUNT	OUTPUT			
	Q_D	Q_C	Q_B	Q_A
0	L	L	L	L
1	L	L	L	H
2	L	L	H	L
3	L	L	H	H
4	L	H	L	L
5	L	H	L	H
6	L	H	H	L
7	L	H	H	H
8	H	L	L	L
9	H	L	L	H

RESET/COUNT (see Note 2)

RESET INPUTS				OUTPUT			
$R_{0(1)}$	$R_{0(2)}$	$R_{9(1)}$	$R_{9(2)}$	Q_D	Q_C	Q_B	Q_A
H	H	L	X	L	L	L	L
H	H	X	L	L	L	L	L
X	X	H	H	H	L	L	H
X	L	X	L	COUNT			
L	X	L	X	COUNT			
L	X	X	L	COUNT			
X	L	L	X	COUNT			

NOTES:
1. Output Q_A connected to input CP_{BD} for BCD count.
2. X indicates that either a HIGH level or a LOW level may be present.

LOGIC DIAGRAM

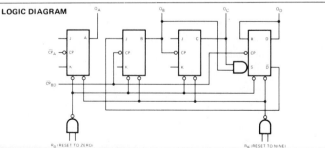

R_0 (RESET TO ZERO) R_9 (RESET TO NINE)

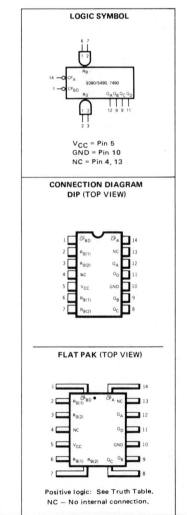

LOGIC SYMBOL

V_{CC} = Pin 5
GND = Pin 10
NC = Pin 4, 13

CONNECTION DIAGRAM
DIP (TOP VIEW)

1	$\overline{CP}_{BD}$	$\overline{CP}_A$	14
2	$R_{0(1)}$	NC	13
3	$R_{0(2)}$	Q_A	12
4	NC	Q_D	11
5	V_{CC}	GND	10
6	$R_{9(1)}$	Q_B	9
7	$R_{9(2)}$	Q_C	8

FLAT PAK (TOP VIEW)

Positive logic: See Truth Table.
NC — No internal connection.

TTL/MONOSTABLE 9603/54121, 74121
MONOSTABLE MULTIVIBRATOR

DESCRIPTION — The 9603/54121, 74121 is a TTL Monostable Multivibrator with dc triggering from positive or gated negative going inputs and with inhibit facility. Both positive and negative going output pulses are provided with full fan out to 10 normalized loads.

Pulse triggering occurs at a particular voltage level and is not directly related to the transition time of the input pulse. Schmitt-trigger input circuitry for the B input allows jitter-free triggering from inputs with transition times as slow as 1.0 V/S, providing the circuit with an excellent noise immunity of typically 1.2 V. A high immunity to V_{CC} noise of typically 1.5 V is also provided by internal latching circuitry.

Once fired, the outputs are independent of further transitions on the inputs and are a function only of the timing components. Input pulses may be of any duration relative to the output pulse. Output pulse lengths may be varied from 40 ns to 40 s by choosing appropriate timing components. With no external timing components (i.e., pin 9 connected to pin 14, pins 10, 11 open) an output pulse of typically 30 ns is achieved which may be used as a dc triggered reset signal. Output rise and fall times are TTL compatible and independent of pulse length.

Pulse width is achieved through internal compensation and is virtually independent of V_{CC} and temperature. In most applications, pulse stability will only be limited by the accuracy of external timing components.

Jitter-free operation is maintained over the full temperature and V_{CC} range for more than six decades of timing capacitance (10 pF to 10 μF) and more than one decade of timing resistance (2 kΩ to 40 kΩ). Throughout these ranges, pulse width is defined by the relationship $t_{p(out)} = C_T \cdot R_T \log_e 2$.

Circuit performance is achieved with a nominal power dissipation of 90 mW at 5.0 V (50% duty cycle) and a quiescent dissipation of typically 65 mW.

Duty cycles as high as 90% are achieved when using R_T = 40 kΩ. Higher duty cycles are achievable if a certain amount of pulse-width jitter is allowed.

$$\approx .7\ C_T R_T$$

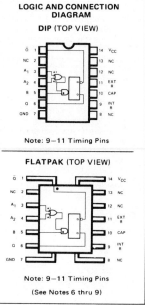

LOGIC AND CONNECTION DIAGRAM

DIP (TOP VIEW)

Note: 9–11 Timing Pins

FLATPAK (TOP VIEW)

Note: 9–11 Timing Pins

(See Notes 6 thru 9)

TRUTH TABLE (See Notes 1 thru 3)

t_n INPUT			t_{n+1} INPUT			OUTPUT
A_1	A_2	B	A_1	A_2	B	
H	H	L	H	H	H	Inhibit
L	X	H	L	X	L	Inhibit
X	L	H	X	L	L	Inhibit
L	X	L	L	X	H	One Shot
X	L	L	X	L	H	One Shot
H	H	H	H	X	H	One Shot
H	H	H	L	X	H	One Shot
X	L	L	X	H	L	Inhibit
L	X	L	H	X	L	Inhibit
X	L	H	H	H	H	Inhibit
L	X	H	H	H	H	Inhibit
H	H	L	X	L	L	Inhibit
H	H	L	L	X	L	Inhibit

Positive logic: See truth table and notes 5 and 6

$H = V_{IH} \geq 2\ V$
$L = V_{IL} \leq 0.8\ V$

NOTES:
1. t_n = time before input transition.
2. t_{n+1} = time after input transition.
3. X indicates that either a HIGH or LOW, may be present.
4. NC = No Internal Connection.
5. A_1 and A_2 are negative edge triggered-logic inputs, and will trigger the one shot when either or both go to LOW level with B at HIGH level.
6. B is a positive Schmitt-trigger input for slow edges or level detection and will trigger the one shot when B goes to HIGH level with either A_1 or A_2 at LOW level. (See Truth Table.)
7. External timing capacitor may be connected between pin 10 (positive) and pin 11. With no external capacitance, an output pulse width of typically 30 ns is obtained.
8. To use the internal timing resistor (2 kΩ nominal), connect pin 9 to pin 14.
9. To obtain variable pulse width connect external variable resistance between pin 9 and pin 14. No external current limiting is needed.
10. For accurate repeatable pulse widths connect an external resistor between pin 11 and pin 14 with pin 9 open-circuit.

TTL/MONOSTABLE • 9603/54121, 74121

ABSOLUTE MAXIMUM RATINGS (above which the useful life may be impaired)

Storage Temperature	-65°C to $+150^\circ$C
Temperature (Ambient) Under Bias	-55°C to $+125^\circ$C
V_{CC} Pin Potential to Ground Pin	-0.5 V to $+7.0$ V
*Input Voltage (dc)	-0.5 V to $+5.5$ V
*Input Current (dc)	-30 mA to $+5.0$ mA
Voltage Applied to Outputs (Output HIGH)	-0.5 V to $+V_{CC}$ value
Output Current (dc) (Output LOW)	$+30$ mA

 *Either Input Voltage limit or Input Current limit is sufficient to protect the inputs.

RECOMMENDED OPERATING CONDITIONS

PARAMETER	9603XM/54121XM			9603XC/74121XC			UNITS
	MIN.	TYP.	MAX.	MIN.	TYP.	MAX.	
Supply Voltage V_{CC}	4.5	5.0	5.5	4.75	5.0	5.25	Volts
Operating Free-Air Temperature Range	-55	25	125	0	25	70	$^\circ$C
Normalized Fan Out from Each Output, N			10			10	U.L.
Input Pulse Rise/Fall Time: Schmitt Input (B)			1.0			1.0	V/s
Logic Inputs (A_1, A_2)			1.0			1.0	V/μs
Input Pulse Width	50			50			ns
External Timing Resistance Between Pins 11 and 14							
(Pin 9 open)	1.4			1.4			kΩ
External Timing Resistance			30			40	kΩ
Timing Capacitance	0		1000	0		1000	μF
Output Pulse Width			40			40	s
Duty Cycle: R_T = 2 kΩ			67%			67%	
R_T = 30 kΩ			90%				
R_T = 40 kΩ						90%	

X= package type; F for Flatpak, D for Ceramic Dip, P for Plastic Dip. See Packaging Information Section for packages available on this product.

ELECTRICAL CHARACTERISTICS OVER OPERATING TEMPERATURE RANGE (Unless Otherwise Noted)

SYMBOL	PARAMETER		LIMITS			UNITS	TEST CONDITIONS (Note 1)	TEST* FIGURE
			MIN.	TYP. (Note 2)	MAX.			
V_{T+}	Positive-Going Threshold Voltage at A Input			1.4	2.0	Volts	V_{CC} = MIN.	57
V_{T-}	Negative-Going Threshold Voltage at A Input		0.8	1.4		Volts	V_{CC} = MIN.	57
V_{T+}	Positive-Going Threshold Voltage at B Input			1.55	2.0	Volts	V_{CC} = MIN.	57
V_{T-}	Negative-Going Threshold Voltage at B Input		0.8	1.35		Volts	V_{CC} = MIN.	57
V_{OH}	Output HIGH Voltage		2.4	3.3		Volts	V_{CC} = MIN., I_{OH} = -0.4 mA	57
V_{OL}	Output LOW Voltage			0.22	0.4	Volts	V_{CC} = MIN., I_{OL} = 16 mA	57
I_{IH}	Input HIGH Current	at A_1 or A_2		2.0	40	μA	V_{CC} = MAX., V_{IN} = 2.4 V	60
				0.05	1.0	mA	V_{CC} = MAX., V_{IN} = 5.5 V	
		at B		4.0	80	μA	V_{CC} = MAX., V_{IN} = 2.4 V	61
				0.05	1.0	mA	V_{CC} = MAX., V_{IN} = 5.5 V	
I_{IL}	Input LOW Current	at A_1 or A_2		-1.0	-1.6	mA	V_{CC} = MAX., V_{IN} = 0.4 V	58
		at B		-2.0	-3.2	mA	V_{CC} = MAX., V_{IN} = 0.4 V	59
I_{OS}	Output Short Circuit Current at Q or $\overline{Q}$ (Note 3)		-20	-25	-55	mA	9603/54121 V_{CC} = MAX.	62 & 63
			-18	-25	-55	mA	9603/74121	
I_{CC}	Supply Current	in Quiescent (Unfired) State		13	25	mA	V_{CC} = MAX.	64
		in Fired State		23	40	mA	V_{CC} = MAX.	64

NOTES:
(1) For conditions shown as MIN. or MAX., use the appropriate value specified under recommended operating conditions for the applicable device type.
(2) Typical limits are at V_{CC} = 5.0 V, 25°C.
(3) Not more than one output should be shorted at a time.

*See parameter measurement information in series 9N/54, 74TTL section.

<table>
<tr><td></td><td></td></tr>
</table>

LOW POWER CMOS HEX BUFFER/LOGIC LEVEL CONVERTERS

SCL4009A
SCL4010A

SCL4009 HEX BUFFER, INVERTING TYPE
SCL4010 HEX BUFFER, NON-INVERTING TYPE

DESCRIPTION

SCL4009 and SCL4010 are single chip monolithic silicon integrated circuits containing eighteen N-channel and twelve P-channel enhancement type MOS transistors connected to form six independent buffer/converter configurations.

SCL4009A and SCL4010A are designed for use as hex CMOS to DTL or TTL logic level converters or hex CMOS current drivers. Conversion ranges are from CMOS logic operating to +15V supply levels to DTL or TTL logic operating at +3.8V to +6V supply levels. Conversion to logic output levels greater than +6V is permitted providing V_{CC} (DTL/TTL) $\leqslant V_{DD}$ (CMOS)

ADD THE FOLLOWING SUFFIX LETTER TO INDICATE THE PACKAGE DESIRED:
D 16 LEAD DUAL IN-LINE CERAMIC
E 16 LEAD DUAL IN-LINE PLASTIC
F 16 LEAD FLAT PACKAGE
H CHIP

FEATURES

- Low static power dissipation 100nW (typ)/pkg with 10 volt supply.
- High current sinking capability, 6mA min at V_{out} "0" = 0.5V and V_{DD} = +10V.
- Medium Speed Operation, tpd 0 = 25ns (typ) and tpd1 = 40ns (typ) at C_L = 50 pf and V_{CC} = V_{DD} = 10V.
- Operation from 3 to 15 volts (1 or 2 power supplies)

LOGIC DIAGRAM

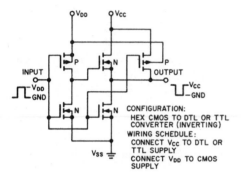

CONFIGURATION:
HEX CMOS TO DTL OR TTL CONVERTER (INVERTING)
WIRING SCHEDULE:
CONNECT V_{CC} TO DTL OR TTL SUPPLY
CONNECT V_{DD} TO CMOS SUPPLY

SCL4009A Circuit Diagram for 1 of 6 identical stages

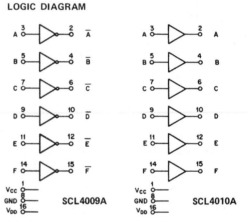

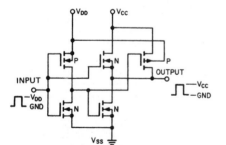

CONFIGURATION:
HEX CMOS TO DTL OR TTL CONVERTER (NON-INVERTING)
WIRING SCHEDULE:
CONNECT V_{CC} TO DTL OR TTL SUPPLY
CONNECT V_{DD} TO CMOS SUPPLY

SCL4010A Circuit Diagram for 1 of 6 identical stages

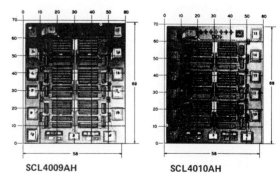

SCL4009AH   SCL4010AH

ABSOLUTE MAXIMUM LIMITS

Storage Temp. Range	$-65°C$ to $+150°C$	DC Supply Voltage ($V_{DD} -V_{SS}$)	+15 volts, -0.5 volts
Operating Temp. Range F & D	$-55°C$ to $+125°C$	Power Dissipation /pkg.	200 mW
Operating Temp. Range E	$-40°C$ to $+85°C$	All Inputs $V_{SS} \leqslant V_{IN} \leqslant V_{DD}$	
		$V_{CC} \leqslant V_{DD}$ must be maintained	

STATIC ELECTRICAL CHARACTERISTICS — SCL4009A — SCL4010A

Characteristics	Symbol	Test Condition	V_{DD} Volts	Limits									Units
				$-55°C$ F & D $-40°C$ E			$+25°C$			$+125°C$ F & D $+85°C$ E			
				Min.	Typ.	Max.	Min.	Typ.	Max.	Min.	Typ.	Max.	
Quiescent Device Current	I_L		5	—	0.005	0.3	—	0.005	0.3	—	0.3	18	µA
			10	—	0.010	0.5	—	0.010	0.5	—	0.6	30	
Quiescent Device Dissipation per Pkg.	P_D		5	—	0.025	1.5	—	0.025	1.5	—	1.5	90	µW
			10	—	0.10	5.0	—	0.10	5.0	—	6.0	300	
Output Voltage Low Level	V_{OL}		5	—	—	0.01	—	—	0.01	—	—	0.05	Volts
			10	—	—	0.01	—	—	0.01	—	—	0.05	
Output Voltage High Level	V_{OH}		5	4.99	—	—	4.99	5.0	—	4.95	—	—	Volts
			10	9.99	—	—	9.99	10.0	—	9.95	—	—	
Noise Immunity (all inputs)	V_{NL}	SCL4009A Only	5	1.0	—	—	1.0	2.25	—	0.9	—	—	Volts
			10	2.0	—	—	2.0	4.5	—	1.2	—	—	
	V_{NL}	SCL4010A Only	5	1.6	—	—	1.6	2.25	—	1.4	—	—	Volts
			10	3.2	—	—	3.2	4.5	—	2.9	—	—	
	V_{NH}		5	1.4	—	—	1.5	2.25	—	1.5	—	—	Volts
			10	2.9	—	—	3.0	4.5	—	3.0	—	—	
Output Drive Current N-Channel	I_{DN}	V_O=0.4 V_O=0.5	5	3.75	—	—	3.0	4.0	—	2.1	—	—	mA
			10	10.0	—	—	8.0	12.0	—	5.6	—	—	
Output Drive Current P-Channel	I_{DP}	V_O=2.5 V_O=9.5	5	-1.85	—	—	-1.25	-2.0	—	-0.9	—	—	mA
			10	-0.9	—	—	-0.6	-1.2	—	-0.4	—	—	
Input Current	I_I		—	—	—	—	—	10	—	—	—	—	pA

DYNAMIC ELECTRICAL CHARACTERISTICS at T_A = 25°C and C_L = 15 pF
Typical Temperature Coefficient for all values of V_{DD} = 0.3%/°C

Characteristics	Symbol		V_{DD} Volts	Limits			Units
				Min.	Typ.	Max.	
CLOCKED OPERATION							
Propagation Delay Time High to Low	t_{PHL}	$V_{CC} = V_{DD}$	5	—	40	70	ns
			10	—	15	30	
	t_{PHL}	V_{DD} = 10V V_{CC} = 5V	—	—	15	25	ns
Propagation Delay Time Low to High	t_{PLH}	$V_{CC} = V_{DD}$	5	—	50	85	ns
			10	—	20	55	
	t_{PLH}	V_{DD} = 10V V_{CC} = 5V	—	—	15	30	ns
Transition Time High to Low	t_{THL}	$V_{CC} = V_{DD}$	5	—	20	45	ns
			10	—	10	40	
Transition Time Low to High	t_{TLH}	$V_{CC} = V_{DD}$	5	—	60	125	ns
			10	—	30	60	
Input Capacitance	C_I	Any Input	—	—	5	—	pF

CMOS QUAD BILATERAL SWITCH   SCL4016A

DESCRIPTION

The SCL4016A is a single chip monolithic silicon integrated circuit containing eight N-channels and eight P-channel enhancement mode MOS transistors connected to form four independent bilateral signal switches. Each switch consists of both "P" and an "N" channel device with common source and drain connections. A single control signal is required per switch. Direct control of the "N" unit is provided while an internal inverter provides simultaneous control of the "P" channel device. Both "P" and "N" devices in a given switch are biased "ON" or "OFF" together by the control signal. The CMOS switch permits peak input-signal voltage swings equal to the full supply voltage, a considerable advantage over the single-channel types.

ADD THE FOLLOWING SUFFIX LETTER TO INDICATE THE PACKAGE DESIRED:
D 14 LEAD DUAL IN-LINE CERAMIC
E 14 LEAD DUAL IN-LINE PLASTIC
F 14 LEAD FLAT PACKAGE
H CHIP

FEATURES

- Wide range of digital and analog signal levels — Digital or analog signal to 15 V peak Analog signal ±7.5 V peak
- Low "ON" resistance — 200 Ω typ. over 15 V_{p-p} signal input range, for V_{DD}-V_{SS} = 15 V
- Matched switch characteristics — 20 Ω typ. difference between R_{ON} values at a fixed bias point over 15 V_{p-p} signal input range V_{DD}-V_{SS} = 15 V
- High "ON/OFF" output voltage ratio—65 dB typ. @ f_{is} = 10 kHz, R_L = 10K Ω
- High degree of linearity — < 0.5% distortion typ. @ f_{is} = 1 kHz, V_{is} = 5 V_{p-p}, V_{DD}-$V_{SS} \geqslant$ 10 V, R_L = 10 k Ω
- Extremely low "OFF" switch leakage resulting in very low off current and high effective "OFF" resistance — 10 pA typ. = V_{DD}-V_{SS} = 10 V, T_A = 25°C
- Extremely high control input impedance (control circuit isolated from signal circuit) — 10^{12} Ω typ.
- Low crosstalk between switch –50 dB typ. @ f_{is} = 0.9 MHz, R_L = 1 k Ω
- Matched control-input to signal-output capacitances — Reduces output signal transients
- Transmits frequencies up to 10 MHz

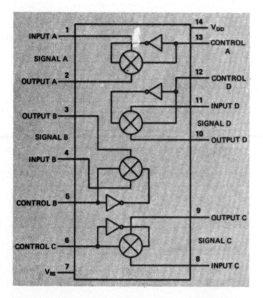

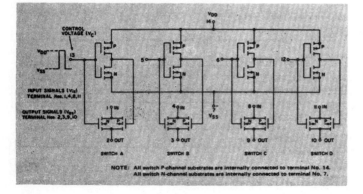

NOTE: All switch P-channel substrates are internally connected to terminal No. 14. All switch N-channel substrates are internally connected to terminal No. 7.

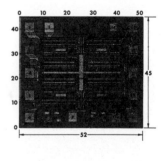

SCL4016AH

LOW POWER CMOS NOR GATES

♦ SCL4000 Dual 3–Input positive NOR
  plus Inverter
♦ SCL4001 Quad 2–Input positive NOR
♦ SCL4002 Dual 4–Input positive NOR
♦ SCL4025 Triple 3–Input positive NOR

ADD THE FOLLOWING SUFFIX LETTER TO INDICATE
THE PACKAGE DESIRED:
  D 14 LEAD DUAL IN-LINE CERAMIC
  E 14 LEAD DUAL IN-LINE PLASTIC
  F 14 LEAD FLAT PACKAGE
  H CHIP

DESCRIPTION

The SCL4000A, SCL4001A, SCL4002A, and SCL4025A
are single chip monolithic integrated circuits containing
N-Channel and P-Channel enhancement mode MOS Transistors. The output of each gate contains two tapered buffers
to provide symmetrical outputs with additional drive
capability.

FEATURES

- Operation from a single power supply from 3 to 15 volts
- Ultra-low static power consumption 10nW(typ)/pkg with
  10 volt supply
- High Fan-out capability
- Symmetrical output swing independent of system and
  fanout, V_{DD} to V_{SS}
- Medium speed operation, tpd = 25 ns(typ) at C_L = 15 pf
- Low output impedance, 400Ω(typ) independent of out-
  put logic level

LOGIC DIAGRAMS

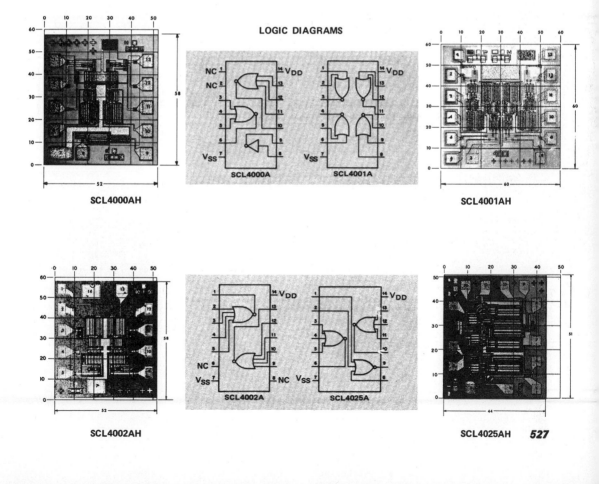

SCL4000AH   SCL4001AH

SCL4002AH   SCL4025AH   **527**

ABSOLUTE MAXIMUM LIMITS

Storage Temp. Range	$-65°C$ to $+150°C$	DC Supply Voltage ($V_{DD} - V_{SS}$)	$+15$ Volts, -0.5 Volts
Operating Temp. Range F & D	$-55°C$ to $+125°C$	Power Dissipation/pkg.	200 mW
Operating Temp. Range E	$-40°C$ to $+85°C$	All Inputs $V_{SS} \leqslant V_{IN} \leqslant V_{DD}$	

STATIC ELECTRICAL CHARACTERISTICS — SCL4000A — SCL4001A — SCL4002A — SCL4025A

Characteristics	Symbol	Test Condition	V_{DD} Volts	$-55°C$ F & D $-40°C$ E Min.	Typ.	Max.	$+25°C$ Min.	Typ.	Max.	$+125°C$ F & D $+85°C$ E Min.	Typ.	Max.	Units
Quiescent Device Current	I_L		5	—	—	0.05	—	0.001	0.05	—	—	3	μA
			10	—	—	0.10	—	0.001	0.10	—	—	6	
Quiescent Device Dissipation/Package	P_D		5	—	—	0.25	—	0.005	0.25	—	—	15	μW
			10	—	—	1.00	—	0.010	1.00	—	—	60	
Output Voltage Low Level	V_{OL}	$I_0 = 0$	5	—	—	0.01	—	—	0.01	—	—	0.05	V
			10	—	—	0.01	—	—	0.01	—	—	0.05	
Output Voltage High Level	V_{OH}	$I_0 = 0$	5	4.99	—	—	4.99	5	—	4.95	—	—	V
			10	9.99	—	—	9.99	10	—	9.95	—	—	
Noise Immunity (All Inputs)	V_{NL}		5	1.5	—	—	1.5	2.25	—	1.4	—	—	V
			10	3.0	—	—	3.0	4.50	—	2.9	—	—	
Noise Immunity (All Inputs)	V_{NH}		5	1.4	—	—	1.5	2.25	—	1.5	—	—	V
			10	2.9	—	—	3.0	4.50	—	3.0	—	—	
Output Drive Current N-Channel	I_{DN}	$V_0 = 0.4$	5	0.5	—	—	0.4	1.5	—	0.28	—	—	mA
		$V_0 = 0.5$	10	1.1	—	—	0.9	3.5	—	0.65	—	—	
Output Drive Current P-Channel	I_{DP}	$V_0 = 2.5$	5	-0.62	—	—	-0.5	-3.0	—	-0.35	—	—	mA
		$V_0 = 9.5$	10	-0.62	—	—	-0.5	-2.3	—	-0.35	—	—	
Input Current	I_I		—	—	—	—	—	10	—	—	—	—	pA

DYNAMIC ELECTRICAL CHARACTERISTICS at $T_A = 25°C$ and $C_L = 15$ pF
Typical Temperature Coefficient for all values of $V_{DD} = 0.3\%/°C$

Characteristics	Symbol		V_{DD} Volts	Limits Min.	Typ.	Max.	Units
Propagation Delay Time High to Low	t_{PHL}		5	—	55	120	ns
			10	—	25	40	
Low to High	t_{PLH}		5	—	55	120	ns
			10	—	25	40	
Transition Time High to Low	t_{THL}		5	—	40	125	ns
			10	—	20	70	
Low to High	t_{TLH}		5	—	50	175	ns
			10	—	25	75	
Input Capacitance	C_I	Any Input		—	5	—	pF

TTL/MSI 9318
EIGHT-INPUT PRIORITY ENCODER

DESCRIPTION — The TTL/MSI 9318 is a Multipurpose Encoder designed to accept eight inputs and produce a binary weighted code of the highest order input. The circuit uses TTL for high speed and high fanout capability, and is compatible with all members of the Fairchild TTL family.

- **MULTI-FUNCTION CAPABILITY**
    - **CODE CONVERSIONS**
    - **MULTI-CHANNEL D/A CONVERTER**
    - **DECIMAL TO BCD CONVERTER**
    - **CASCADING FOR PRIORITY ENCODING OF N BITS**
- **INPUT ENABLE CAPABILITY**
- **PRIORITY ENCODING** — AUTOMATIC SELECTION OF HIGHEST PRIORITY INPUT LINE
- **OUTPUT ENABLE** — ACTIVE LOW WHEN ALL INPUTS HIGH
- **GROUP SIGNAL OUTPUT** — ACTIVE WHEN ANY INPUT IS LOW
- **TYPICAL POWER DISSIPATION OF 250 mW**
- **INPUT/OUTPUT CHARACTERISTICS PROVIDE EASY INTERFACING WITH FAIRCHILD DTL, LPDTL, TTL, AND MSI FAMILIES**
- **ALL CERAMIC HERMETIC 16-LEAD DUAL IN-LINE PACKAGE**
- **INPUT CLAMP DIODES LIMIT HIGH SPEED TERMINATION EFFECTS**

LOGIC SYMBOL

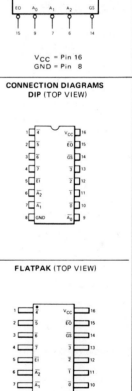

V_{CC} = Pin 16
GND = Pin 8

CONNECTION DIAGRAMS
DIP (TOP VIEW)

FLATPAK (TOP VIEW)

PIN NAMES

		LOADING (Note a)
$\overline{0}$	Priority (Active LOW) Input	1 U.L.
$\overline{1}$ to $\overline{7}$	Priority (Active LOW) Inputs	2 U.L.
$\overline{EI}$	Enable (Active LOW) Input	2 U.L.
$\overline{EO}$	Enable (Active LOW) Output	10 U.L.*
$\overline{GS}$	Group Select (Active LOW) Output	10 U.L.*
$\overline{A_0}, \overline{A_1}, \overline{A_2}$	Address (Active LOW) Outputs	10 U.L.*

NOTES:
a. 1 Unit Load (U.L.) = 40 μA HIGH/1.6 mA LOW
b. 10 U.L. is the output LOW drive factor and 20 U.L. is the output HIGH drive factor.

LOGIC DIAGRAM

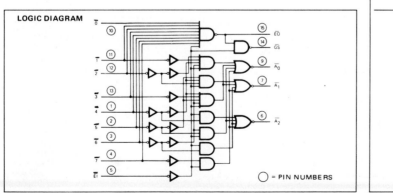

○ = PIN NUMBERS

FAIRCHILD LPTTL/MSI • 93L18

FUNCTIONAL DESCRIPTION – The LPTTL/MSI 93L18 8-input priority encoder accepts data from eight active LOW inputs and provides a binary representation on the three active LOW outputs. A priority is assigned to each input so that when two or more inputs are simultaneously active, the input with the highest priority is represented on the output, with input line 7 having the highest priority.

A HIGH on the input enable ($\overline{EI}$) will force all outputs to the inactive (HIGH) state and allow new data to settle without producing erroneous information at the outputs.

Provided with the three data outputs are a group signal output ($\overline{GS}$) and an enable output ($\overline{EO}$). The $\overline{GS}$ is active level LOW when any input is LOW; this indicates when any input is active. The $\overline{EO}$ is active level LOW when all inputs are HIGH. Using the output enable along with the input enable allows priority encoding of N input signals. Both $\overline{EO}$ and $\overline{GS}$ are inactive when the input enable is HIGH.

TRUTH TABLE

$\overline{EI}$	$\overline{0}$	$\overline{1}$	$\overline{2}$	$\overline{3}$	$\overline{4}$	$\overline{5}$	$\overline{6}$	$\overline{7}$	$\overline{GS}$	$\overline{A}_0$	$\overline{A}_1$	$\overline{A}_2$	$\overline{EO}$
H	X	X	X	X	X	X	X	X	H	H	H	H	H
L	H	H	H	H	H	H	H	H	H	H	H	H	L
L	X	X	X	X	X	X	X	L	L	L	L	L	H
L	X	X	X	X	X	X	L	H	L	H	L	L	H
L	X	X	X	X	X	L	H	H	L	L	H	L	H
L	X	X	X	X	L	H	H	H	L	H	H	L	H
L	X	X	X	L	H	H	H	H	L	L	L	H	H
L	X	X	L	H	H	H	H	H	L	H	L	H	H
L	X	L	H	H	H	H	H	H	L	L	H	H	H
L	L	H	H	H	H	H	H	H	L	H	H	H	H

H = HIGH Voltage Level
L = LOW Voltage Level
X = Don't Care

TYPICAL INPUT AND OUTPUT CIRCUITS

INPUTS **EQUIVALENT CIRCUIT**	**OUTPUTS** **EQUIVALENT CIRCUIT**

OUTPUT LOW   **OUTPUT HIGH**

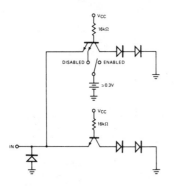
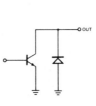
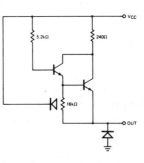

ABSOLUTE MAXIMUM RATINGS (above which the useful life may be impaired)

Storage Temperature	$-65°$C to $+150°$C
Temperature (Ambient) Under Bias	$-55°$C to $+125°$C
V_{CC} Pin Potential to Ground Pin	-0.5 V to $+7.0$ V
*Input Voltage (dc)	-0.5 V to $+5.5$ V
*Input Current (dc)	-30 mA to $+5.0$ mA
Voltage Applied to Outputs (Output HIGH)	-0.5 V to $+V_{CC}$ value
Output Current (dc) (Output LOW)	$+30$ mA

*Either Input Voltage limit or Input Current limit is sufficient to protect the inputs.

ANSWERS TO
SELECTED PROBLEMS

CHAPTER ONE

1.2 (a) 25; (b) 9.5625

1.5 time

1.4 1023

1.7 $t_r \approx 1.6 \ \mu s, t_f \approx 1.1 \ \mu s, t_w \approx 6.1 \ \mu s$

CHAPTER TWO

2.5 $x = A + B + C + D + E + F$

2.10 No

2.12 $\bar{A}\bar{B}\bar{C} + A\bar{B}\bar{C} + \bar{A}\bar{B}C$

2.8 $x = \overline{ABCD}$

2.11 $AB(\overline{\bar{A} + \bar{B}})$

2.21 (a) NOR
  (b) AND, NOR
  (c) OR, AND
  (d) NAND
  (e) NOR, NAND

2.24 (a) $A + \bar{B} + C$
  (b) $\bar{A}(\bar{B} + C)$

2.26 (a) RS
  (c) $P(M + N)$

2.25 (a) AB
  (b) $\bar{B}(\bar{A} + \bar{C})$

CHAPTER THREE

3.1 $x = A\bar{B}C$

3.5 $x = \bar{A}B + \bar{A}C + \bar{B}C$

3.8 $x = (A + B + C)(\bar{A} + \bar{B} + \bar{C})$

3.17 $Z_3 = y_1 y_0 x_1 x_0; \ Z_2 = y_1 x_1 (\overline{y_0 x_0})$
  $Z_1 = y_0 x_1 (\overline{y_1 x_0}) + y_1 x_0 (\overline{y_0 x_1})$
  $Z_0 = y_0 x_0$

3.3 $x = A\bar{B}C + \bar{A}BC$

3.7 $x = (\bar{B} + \bar{C})(\bar{A} + \bar{C})(\bar{A} + \bar{B})$

3.9 $A_3 + A_2 A_1 A_0$

3.19 $x = AB(\overline{C \oplus D})$

3.20 $x = (A + B) \cdot (C + D)$

3.21 $N - S = \bar{C}\bar{D}(A + B) + AB(\bar{C} + \bar{D})$
$E - W = \overline{N - S}$

CHAPTER FOUR

4.7 500 Hz squarewave

4.8 (a) minimum = 30 ns; no maximum limit
(b) 5 ns

4.11 5 kHz squarewave

4.17 (a) F   (f) T
(b) T   (g) T
(c) F   (h) F
(d) F   (i) F
(e) F   (j) T

4.19 1010, 0101, 1010, 0101, 1010, 0101, 1010, 0101, 1010, etc.

4.21 111, 110, 101, 100, 011, 010, 001, 000, 111, 110, 101, 100, 011, 010, 001, 000, etc.

4.22 seven

4.23 (a) 5   (b) 20 kHz

4.25 one

4.28 Circuit operation ceases (all outputs stop changing) after 14 input pulses.

CHAPTER FIVE

5.1 (a) 22; (b) 141; (c) 23.6875

5.2 (a) 100101   (d) 1001000.01110
(g) 110100   (h) 11001101

5.3 (a) 10111;   (c) 1111.0101

5.4 (a) 010000;   (b) 11110

5.5 (a) 010000;   (b) 10001

5.6 (a) 010000;   (b) 10010

5.7 (a) −6;   (b) +6.5625

5.8 (a) −9;   (b) +6.5625

5.9 (a) −10;   (b) +6.5625

5.10 eight bits

5.11 (a) 01111;   (b) 1101
(e) 00001.1;   (f) 100001

5.13 (a) 100011;   (c) 100011.00101

5.14 (a) 11;   (c) 101.11

5.15 (a) 483;   (b) 30.5

5.16 (a) 111011;
(c) 0.100101000111

5.17 (b) 011110.100

5.18 (a) 5463; B33   (b) 135.54; 5D.5

5.19 (a) 001010101100

5.20 (a) 01000111
(b) 100100100100.01100010

5.21 (a) 96.51

5.24 (a) 10110101

5.25 111011011

5.27 1001000110

5.31 (a) 1;   (b) 0

5.32 (a) 011101001

5.33 (a) no error;   (b) single error
(c) double error;   (d) no error

5.35 Add inverter to output.

5.38 DC + DB = D(C + B)

5.39 165, 166, 167, 170, 171, 172, 173, 174, 175, 176, 177, 200, 201, 202, 203, 204, 205

5.40 2E8, 2E9, 2EA, 2EB, 2EC, 2ED, 2EE, 2EF, 2FO, 2F1, 2F2, 2F3, 2F4, 2F5, 2F6, 2F7, 2F8, 2F9, 2FA, 2FB, 2FC, 2FD, 2FE, 2FF, 300

CHAPTER SIX

6.6 Overflow $= C_3 \bar{B}_3 \bar{A}_3 + \bar{C}_3 B_3 A_3$

6.7 $C_1 = A_0 B_0 + A_0 C_0 + B_0 C_0$
$C_2 = A_1 B_1 + A_1 C_1 + B_1 C_1$
$C_3 = A_2 B_2 + A_2 C_2 + B_2 C_2$
Final expression for C_3 contains 15 terms in S-of-P form. Two levels of gates compared to six.

6.8 320 ns

6.9 To form the 2's complement of subtrahend.

6.11 (a) 0111;   (b) 1010

6.12 14.3 MHz; 280 ns

6.13 $\Sigma_{11}\Sigma_{10}\Sigma_9\Sigma_8 = 1000$
$\Sigma_7\Sigma_6\Sigma_5\Sigma_4 \ \ = 0100$
$\Sigma_3\Sigma_2\Sigma_1\Sigma_0 \ \ = 0101$

CHAPTER SEVEN

7.2 seven

7.3 63_{10}

7.4 1024

7.5 250 kHz; 50%

7.7 10000

7.10 (a) 100 (four)

7.12 100 Hz

7.17 1000 and 0000 will not occur.

7.18 12.5 MHz; 8.33 MHz

7.19 (b) 33 MHz   (c) 14.3 MHz

7.20 Counter sequence: 0000, 0001, 0010, 0011, 0100, 0101, 0110, 0111, 1000, 1001, 0000, 0001, etc.

7.24 000, 001, 010, 011, 100, 101, 110, 000, 001, 010, 011, 100, 101, 110, 000, etc.

7.29 Going from 1111 to 0000.

7.30 Going from 0101 to 0110 and from 0111 to 1000.

7.33 12; 0111, 1001, 0101

7.34 16; 13

7.41 000, 100, 110, 111, 011, 000, 100, 110, 111, 011, etc.

7.43 16 kHz; 1 kHz; 40 Hz; 5 Hz

7.44 (a) F   (h) T
    (b) T   (i) T
    (c) T   (j) F
    (d) F   (k) T
    (e) F   (l) F
    (f) T   (m) T
    (g) T

CHAPTER EIGHT

8.1 (a) A, B;   (b) A;
    (c) A;   (d) B

8.6 0.4 V

8.8 (a) 74L   (b) 74L
    (c) 74S   (d) 74S, 74LS

8.9 74SO4 has lowest t_{pd}
7404 has lowest P_D

8.10 800 μA; 32 mA

8.11 (a) 1 U.L.
    (b) 2 U.L.
    (c) 5

8.12 (a) 30   (b) 48 mA

8.19 NO; YES

8.22 .092 μF

8.28 1.4 K

8.34 b, c, e, f

8.39 b, c, d, g

8.41 b

8.46 I^{2}L

8.18 125 ohms

8.20 18 mA

8.26 $X = AB + CD + FG$

8.29 (a) $X = 0$;   (c) $X = 0$

8.38 a, c, d, e, g, h

8.40 a, c, e, f, g, h

8.44 (a) $\approx$ two 7400 loads;
     (b) $\approx$ eight 74LS00 loads

8.47 AND gate

CHAPTER NINE

9.1 Five inputs, 32 outputs

9.13 $\bar{A}_0 = \bar{A}_1 = \bar{A}_2 = 0$

9.28 at t_3: A $= 1011$; B $= 1011$; and C $= 1011$

9.7 $R = 250\ \Omega$ for each LED segment

9.9 111

9.21 $Z = \bar{A}\bar{B}C + \bar{A}B\bar{C} + ABC$

CHAPTER TEN

10.1 3.58 V

10.3 20 mV, 0.4%

10.5 Eight

10.9 2 mV/step; 0.1%; 1.39 V

10.15 a, c, d, f, g

10.17 (a) 2 V, 3.6 V, 5.2 V, 0.4 V
     (b) 3 msec
     (c) .03 μF
     (d) 1.33 kHz

10.20 (a) Seven

10.23 (a) 10010111   (b) 10010111
     (c) max $= 102\ \mu$s; 51 μs

10.2 MSB $= 2.56$ V;
     LSB $= 20$ mV

10.4 5 mV

10.6 14.3%; 0.286 V

10.11 800 Ω; NO

10.16 a, b, e

10.19 20 μA; YES

10.22 6.3744 V

10.26 (a) 1.2 mV   (b) 2.7 mV

CHAPTER ELEVEN

11.2 8192; 32 bits; 262144

11.8 10; 8; 8

11.13 64 of each; 4096

11.15 (a) F   (d) F
     (b) T   (e) T
     (c) F

11.17 256 words

11.22 programmed by the manufacturer in place of the user

11.3 8192

11.10 010001

11.14 (a) 16,384 cores; 16 planes
     (b) 64;   (c) 16

11.16 Twelve

11.21 Volatility

11.25 eight

11.31 (a) 32×8
(b) RAM-2 and RAM-3
(c) 00000000 to 00001111;
00010000 to 00011111

CHAPTER TWELVE

12.2 1–h, 2–e, 3–g, 4–d, 5–f, 6–b, 7–a, 8–c

12.3 256; 16,384

12.4 Address F holds the value 00000011

12.5 Address F holds the value 00000000

CHAPTER THIRTEEN

13.1 1–e, 2–d, 3–b, 4–f, 5–a, 6–c

13.6 (a) ten pages; 2560 locations
(b) eight pages
(c) 256

13.8 RAMs 1 and 2—page FC
RAMs 3 and 4—page FD
RAMs 5 and 6—page FE
RAMs 7 and 8—page FF

13.12 (a) 1   (e) 2
(b) 2   (f) 2
(c) 1   (g) 1
(d) 1

INDEX